T0310212

OPERATOR-BASED NONLINEAR CONTROL SYSTEMS

IEEE Press
445 Hoes Lane
Piscataway, NJ 08854

OPERATOR-BASED NONLINEAR CONTROL SYSTEMS

Design and Applications

Mingcong Deng
Tokyo University of Agriculture and Technology

WILEY

Published by John Wiley & Sons, Inc., Hoboken, New Jersey.
Published simultaneously in Canada.

Library of Congress Cataloging-in-Publication Data:

Deng, Mingcong.
Operator-based nonlinear control systems design and applications / Mingcong Deng. – First edition.
pages cm. – (IEEE Press series on systems science and engineering)
ISBN 978-1-118-13122-0 (hardback)
1. Automatic control. 2. Nonlinear control theory. I. Title.
TJ213.D435 2013
629.8′36–dc23

2013026695

Printed in the United States of America

10 9 8 7 6 5 4 3 2

CONTENTS

CHAPTER 1

Introduction

1.1 DEFINITION OF NONLINEAR SYSTEMS

In general, a nonlinear system is one that does not satisfy the superposition principle or whose output is not directly proportional to its input. That is, a nonlinear system is any problem where the variable(s) to be solved for cannot be written as a linear combination of independent components. Since most economic, social, and many industrial systems are inherently nonlinear in nature, where mathematical analysis is unable to provide general solutions, nonlinear system problems, especially nonlinear systems dynamics analysis and control problems for industrial systems, are of interest to mathematicians, physicists, and engineers.

1.2 NONLINEAR SYSTEM DYNAMICS ANALYSIS AND CONTROL

Nonlinear systems control is the discipline that applies control theory to design systems with desired behaviors. It can be broadly defined or classified as nonlinear control theory and application. It seeks to understand nonlinear systems dynamics, using mathematical modeling, in terms of inputs, outputs, and various components with different behaviors and to use nonlinear control systems design schemes to develop controllers for those systems in one or many time, frequency, and complex domains, depending on the nature of the design problem. As a result, control of nonlinear systems is a multidisciplinary research field involving the synergistic integration of mechanical and electrical engineering, computer science, and even biological engineering. Control of nonlinear systems will become mainstream consumer products within the next decade, providing a significant growth opportunity for the above-mentioned engineering systems. So far, there are several significant techniques for analyzing nonlinear systems, for example, describing the function method, the phase plane method, Lyapunov-based analysis, the singular perturbation method, the Popov criterion, the center manifold theorem, the small-gain theorem, and passivity analysis. Based on the above techniques, significant results were introduced extending to nonlinear feedback systems design and control. Some cornerstone control

Operator-Based Nonlinear Control Systems: Design and Applications, First Edition. Mingcong Deng.

methods, for example, Lyapunov function method, sliding-mode control method, and nonlinear damping method, are proposed. In view of the input–output nature of the nonlinear system concept itself, it seems useful to establish computer-oriented approaches to nonlinear control systems analysis and design. Addressing this problem, the robust right coprime factorization technique of nonlinear operators, in addition to the above significant techniques, which is based on real and complex variable theory, has been a promising technique, where the operators can be either linear or nonlinear, continuous time or discrete time, finite dimensional or infinite dimensional, and in the frequency domain or time domain [1].

1.3 WHY OPERATOR-BASED NONLINEAR CONTROL SYSTEM?

As a basis for the possible next generation of control of nonlinear systems, the theoretical concept of operator-based nonlinear control has been introduced in recent years. In the operator-based nonlinear control system research approach, since the 1990s, some researchers started with the operator-theoretic nonlinear control approach, and mathematical background was provided. As for the development of the design principle, it is forecasted that the operator-theoretic nonlinear control approach will be applied significantly. As a result, research on operator-based nonlinear system control has great potential to the application for industry and daily life. However, the nonlinear control system analysis design might be difficult and impossible because of the complex uncertain nonlinearities. There was lack of a quantitative stability result, which may guarantee stability and performance of the control system with the uncertain nonlinearities. Addressing the above problem, this book aims to develop a systematic methodology using operator-based design of nonlinear control systems.

1.4 OVERVIEW OF THE BOOK

This book concerns uncertain nonlinearity in this important research field. Starting with major goals and reviews, the book gives a perspective as to how plants can be modeled as operator-based plants. The primary objectives of this book are to guide modeled plants to obtain robust right coprime factorization, provide state-of-the art research on robust stability conditions, and discuss system output tracking and fault detection issues for researchers working in this field. Considering the broad set of the readers whom I would like to reach, I some applications are included for a good understanding. The intent is to help beginning graduate students learn several developments of operator-based nonlinear control system design. This book also summarizes our understanding of the current trend and the likely future of the operator-theoretic approach reported in latest research results on several frontier problems. Motivated by the above consideration, a detailed analysis of nonlinear feedback control systems based on an operator-theoretic approach is considered in this book. Based on the operator theory, nonlinear feedback control systems can be designed and applied, that is, operator-based nonlinear feedback control using robust right coprime factorization [1–2, 8]. For instance, application of the proposed designs to networked control processes is considered and vibration control using piezoelectric actuators, ionic polymer metal composite actuators, and shape memory alloy actuators has been successfully conducted. Meanwhile, a fault detection technique based on an operator-theoretic approach is also developed. In describing these aspects of the operator-based nonlinear control system, it is assumed that the reader is familiar with Banach spaces, linear operator theory, and right coprime factorization and has some elementary knowledge of nonlinear control, found in the excellent text by de Figueiredo and Chen [1]. Some of the work described in this book is based upon a series of recent publications by the author.

ACKNOWLEDGMENTS

The author would like to acknowledge his colleagues in Japan, especially Emeritus Professor Akira Inoue for valuable comments, suggestions, and criticisms during this research. In particular, the author would like to thank his former Ph.D. students, Dr. Changan Jiang, Dr. Lihua Jiang, Dr. Shuhui Bi, Dr. Osunleke A. Saheeb, Dr. Ni Bu, Dr. Shengjun Wen, Dr. Seiji Saito, and Dr. Aihui Wang, for their work in completing his idea on operator-based nonlinear control system design and experimental work, and the author appreciates the many valuable discussions concerning the research approach with the above Ph.D. students. Also, thanks to Dr. Akira Yanou and Dr. Tomohiro Henmi for their support of some of this research. Compilation of this book involved the assistance of my graduate student, Mr. Toshihiro Kawashima. Finally, a special thanks to my wife, Yan Yu, and my children, Mengyan and Huili, for their constant encouragement and patience and my friends in China, the United Kingdom, the United States, and Canada for many interesting discussions during the writing of this book.

CHAPTER 2

Robust Right Coprime Factorization for Nonlinear Plants with Uncertainties

2.1 PRELIMINARIES

This chapter gives some basic definitions and notation needed throughout this book and some important remarks necessary in describing the problems to be investigated later.

2.1.1 Definition of Spaces

In mathematics, a space is a set with some added structures. There are two basic spaces: linear spaces (also called vector spaces) and topological spaces, where linear spaces are of algebraic nature and topological spaces are of analytic nature. There are three types of linear spaces; real linear spaces (over the field of real numbers), complex linear spaces (over the field of complex numbers), and more generally linear spaces over any field. The discussion in this book is based on linear spaces.

2.1.1.1 Normed Linear Space A space X of time functions is said to be a vector space if it is closed under addition and scalar multiplication. The space X is said to be *normed* if each element x in X is endowed with norm $\|\cdot\|_X$, which can be defined in any way so long as the following three properties are fulfilled:

1. $\|x\|$ is a real, positive number and is different from zero unless x is identically zero,
2. $\|ax\| = |a|\|x\|$, and
3. $\|x_1 + x_2\| \leq \|x_1\| + \|x_2\|$.

It should be mentioned that every normed space is a linear topological space.

2.1.1.2 Banach Space A Banach space is defined as a complete normed space. This means that a Banach space is a vector space X over the real or complex numbers

Operator-Based Nonlinear Control Systems: Design and Applications, First Edition. Mingcong Deng.

with a norm $\|\cdot\|$ such that every Cauchy sequence (with respect to the metric $d(x, y) = \|x - y\|$) in X has a limit in X. Many spaces of sequences or functions are infinite-dimensional Banach spaces.

2.1.1.3 *Extended Linear Space* Let Z be the family of real-valued measurable functions defined on $[0, \infty)$, which is a linear space. For each constant $T \in [0, \infty)$, let P_T be the projection operator mapping from Z to another linear space, Z_T, of measurable functions such that

$$f_T(t) := P_T(f)(t) = \begin{cases} f(t) & t \leq T \\ 0 & t > T \end{cases} \tag{2.1}$$

where $f_T(t) \in Z_T$ is called the truncation of $f(t)$ with respect to T. Then, for any given Banach space X of measurable functions, set

$$X^e = \{f \in Z : \|f_T\|_X < \infty \text{ for all } T < \infty\} \tag{2.2}$$

Obviously, X^e is a linear subspace of Z. The space so defined is called the extended linear space associated with the Banach space X.

It should be noted that the extended linear space is not complete in the norm in general and hence not a Banach space, but it is determined by a relative Banach space. The reason to use the extended linear space is that all the control signals have finite time duration in practice, and many useful techniques and results can be carried over from the standard Banach space X to the extended space X^e if the norm is suitably defined.

2.1.2 Definition of Operators

Let X and Y be linear spaces over the field of real numbers, and let X_s and Y_s be normed linear subspaces, called the stable subspaces of X and Y, respectively, defined suitably by two normed linear spaces under a certain norm $X_s = \{x \in X : \|x\| < \infty\}$ and $Y_s = \{y \in Y : \|y\| < \infty\}$.

2.1.2.1 *Linear and Nonlinear Operator* Let $Q : X \to Y$ be an operator mapping from X to Y, and denote by $\mathcal{D}(Q)$ and $\mathcal{R}(Q)$, respectively, the domain and range of Q. If the operator $Q : \mathcal{D}(Q) \to Y$ satisfies the addition rule and multiplication rule

$$Q : ax_1 + bx_2 \to aQ(x_1) + bQ(x_2)$$

for all $x_1, x_2 \in \mathcal{D}(Q)$ and all $a, b \in \mathcal{C}$, then Q is said to be linear. Otherwise, it is said to be nonlinear. Since linearity is a special case of nonlinearity, in what follows "nonlinear" will always mean "not necessarily linear" unless otherwise indicated.

2.1.2.2 *Bounded Input–Bounded Output (BIBO) Stability* Let Q be a nonlinear operator with its domain $\mathcal{D}(Q) \subseteq X^e$ and range $\mathcal{R}(Q) \subseteq Y^e$. If $Q(X) \subseteq Y$, Q is said to be input–output stable. If Q maps all input functions from X_s into the output space Y_s, that is, $Q(X_s) \subseteq Y_s$, then operator Q is said to be BIBO stable or simply stable. Otherwise, if Q maps some inputs from X_s to the set $Y^e \backslash Y_s$ (if not empty), then Q is said to be unstable. Any stable operators defined here and later in this book are BIBO stable.

2.1.2.3 *Invertible* An operator Q is said to be invertible if there exists an operator P such that

$$Q \circ P = P \circ Q = I \tag{2.3}$$

where P is the inverse of Q and is denoted by Q^{-1}, I is the identity operator, and $Q \circ P$ [or simply $Q(P(\cdot))$ or QP] is an operation satisfying

$$\mathcal{D}(Q \circ P) = P^{-1}(\mathcal{R}(P) \cap \mathcal{D}(Q)) \tag{2.4}$$

2.1.2.4 *Unimodular Operator* Let $\mathcal{S}(X, Y)$ be the set of stable operators mapping from X to Y. Then, $\mathcal{S}(X, Y)$ contains a subset defined by

$$\mathcal{U}(X, Y) = \{M : M \in \mathcal{S}(X, Y),\ M \text{ is invertible with } M^{-1} \in \mathcal{S}(Y, X)\} \tag{2.5}$$

Elements of $\mathcal{U}(X, Y)$ are called unimodular operators.

2.1.2.5 *Lipschitz Operator* For any subset $D \subseteq X$, let $\mathcal{F}(D, Y)$ be the family of nonlinear operators Q such that $\mathcal{D}(Q) = D$ and $\mathcal{R}(Q) \subseteq Y$. Introduce a (semi)-norm into (a subset of) $\mathcal{F}(D, Y)$ by

$$\|Q\| := \sup_{\substack{x, \tilde{x} \in D \\ x \neq \tilde{x}}} \frac{\|Q(x) - Q(\tilde{x})\|_Y}{\|x - \tilde{x}\|_X}$$

if it is finite. In general, it is a seminorm in the sense that $\|Q\| = 0$ does not necessarily imply $Q = 0$. In fact, it can be easily seen that $\|Q\| = 0$ if Q is a constant operator (need not to be zero) that maps all elements from D to the same element in Y.

Let $\text{Lip}(D, Y)$ be the subset of $\mathcal{F}(D, Y)$ with its all elements Q satisfying $\|Q\| < \infty$. Each $Q \in \text{Lip}(D, Y)$ is called a Lipschitz operator mapping from D to Y, and the number $\|Q\|$ is called the Lipschitz seminorm of the operator Q on D.

It is evident that a Lipschitz operator is both bounded and continuous on its domain. Next, a generalized Lipschitz operator is introduced which is defined on an extended linear space.

2.1.2.6 Generalized Lipschitz Operator Let X^e and Y^e be extended linear spaces associated respectively with two Banach spaces X and Y of measurable functions defined on the time domain $[0, \infty)$, and let D be a subset of X^e. A nonlinear operator $Q : D \to Y^e$ is called a generalized Lipschitz operator on D if there exists a constant L such that

$$\left\| [Q(x)]_T - [Q(\tilde{x})]_T \right\|_Y \leq L \|x_T - \tilde{x}_T\|_X \tag{2.6}$$

for all $x, \tilde{x} \in D$ and for all $T \in [0, \infty)$. Note that the least such constant L is given by the norm of Q with

$$\begin{aligned} \|Q\|_{\text{Lip}} &:= \|Q(x_0)\|_Y + \|Q\| \\ &= \|Q(x_0)\|_Y \\ &\quad + \sup_{T \in [0,\infty)} \sup_{\substack{x,\tilde{x} \in D \\ x_T \neq \tilde{x}_T}} \frac{\left\| [Q(x)]_T - [Q(\tilde{x})]_T \right\|_Y}{\|x_T - \tilde{x}_T\|_X} \end{aligned} \tag{2.7}$$

for any fixed $x_0 \in D$.

Based on (2.7), it follows immediately that for any $T \in [0, \infty)$

$$\begin{aligned} \left\| [Q(x)]_T - [Q(\tilde{x})]_T \right\|_Y &\leq \|Q\| \|x_T - \tilde{x}_T\|_X \\ &\leq \|Q\|_{\text{Lip}} \|x_T - \tilde{x}_T\|_X \end{aligned} \tag{2.8}$$

Lemma 2.1 [1] Let X^e and Y^e be extended linear spaces associated respectively with two Banach spaces X and Y, and let D be a subset of X^e. The following family of Lipschitz operators is a Banach space:

$$\text{Lip}(D, Y^e) = \left\{ Q : D \to Y^e \middle| \|Q\|_{\text{Lip}} < \infty \text{ on } D \right\} \tag{2.9}$$

Proof First, it is clear that $\text{Lip}(D, Y^e)$ is a normed linear space. Hence, it is sufficient to verify its completeness.

Let Q_n be a Cauchy sequence in $\text{Lip}(D, Y^e)$ such that $\|Q_m - Q_n\| \to 0$ as $m, n \to \infty$. We need to show that $\|Q_n - Q\| \to 0$ for some $Q \in \text{Lip}(D, Y^e)$ as $n \to \infty$.

Let $T \in [0, \infty)$ be fixed. For any $\tilde{x} \in D$, by definition of the Lipschitz norm with an $x_0 \in D$, we have

$$\begin{aligned} &\left\| [(Q_m - Q_n)(\tilde{x})]_T - [(Q_m - Q_n)(x_0)]_T \right\|_Y \\ &\leq \left\| Q_m - Q_n \right\|_{\text{Lip}} \left\| \tilde{x}_T - [x_0]_T \right\|_X \end{aligned} \tag{2.10}$$

so that

$$\begin{aligned}\left\|[Q_m(\tilde{x})]_T - [Q_n(\tilde{x})]_T\right\|_Y &= \left\|[(Q_m - Q_n)(\tilde{x})]_T\right\|_Y \\ &\leq \left\|[(Q_m - Q_n)(x_0)]_T\right\|_Y + \left\|Q_m - Q_n\right\|_{\text{Lip}}\left\|\tilde{x}_T - [x_0]_T\right\|_X \end{aligned} \tag{2.11}$$

Since the right-hand side of the above tends to zero as $m, n \to \infty$, it follows that the sequence $\{[Q_n(\tilde{x})]_T\}$ is Cauchy in the range Y^e (and in fact is uniformly Cauchy over each bounded subset of the domain D). Hence, for each fixed $T \in [0, \infty)$, $\tilde{v}_T := \lim[Q_n(\tilde{x})]_T$ exists in the range Y^e (and is uniform over bounded subsets of the domain D^T). Let v be a function such that $v_T = \tilde{v}_T$ for all $T \in [0, \infty)$, and define a nonlinear operator Q by $Q : \tilde{x} \to v$. Then, Q satisfies $[Q(\tilde{x})]_T = \tilde{v}_T$ for all $T \in [0, \infty)$. In the following, we will show that $Q \in \text{Lip}(D, Y^e)$. We first note that the operator Q so defined has domain D since in the above $\tilde{x} \in D$ is arbitrary. We then observe that Q is actually independent of T. Then, since $\|Q_m - Q_n\| \to 0$, we have $\lim \|Q_n\|_{\text{Lip}} = c$, a constant, so that, for any $\tilde{x}_1, \tilde{x}_2 \in D$,

$$\begin{aligned}\left\|[Q(\tilde{x}_1)]_T - [Q(\tilde{x}_2)]_T\right\|_Y &= \lim_{n\to\infty} \left\|[Q_n(\tilde{x}_1)]_T - [Q_n(\tilde{x}_2)]_T\right\|_Y \\ &\leq \lim_{n\to\infty} \left\|[Q_n]_T\right\|_{\text{Lip}}\left\|[\tilde{x}_1]_T - [\tilde{x}_2]_T\right\|_X \\ &= c\left\|[\tilde{x}_1]_T - [\tilde{x}_2]_T\right\|_X \end{aligned} \tag{2.12}$$

Therefore, taking the supremum over D and then the supremum over $[0, \infty)$ yields

$$\sup_{T\in[0,\infty)} \sup_{\substack{\tilde{x}_1,\tilde{x}_2\in D \\ [\tilde{x}_1]_T\neq[\tilde{x}_2]_T}} \frac{\|[Q(\tilde{x}_1)]_T - [Q(\tilde{x}_2)]_T\|_Y}{\|[\tilde{x}_1]_T - [\tilde{x}_2]_T\|_X} \leq c \tag{2.13}$$

which implies that $\|Q\| \leq c < \infty$, so that $Q \in \text{Lip}(D, Y^e)$.

We finally verify that $\|Q_n - Q\|_{\text{Lip}} \to 0$ as $n \to \infty$. Since the above explanation also proves (letting $\tilde{x} = x_0$ therein) that $\|[Q_n(x_0)]_T - [Q(x_0)]_T\|_Y \to 0$ as $n \to \infty$ for each $T \in [0, \infty)$, for $\epsilon > 0$ we can let N be such that $\|Q_m - Q_n\|_{\text{Lip}} \leq \epsilon/2$ and $\|[(Q_n - Q)(x_0)]_T\|_Y \leq \epsilon/2$ for $m, n \geq N$. Then, for any given $\tilde{x}_1, \tilde{x}_2 \in D$, we have

$$\begin{aligned}&\left\|[(Q - Q_n)(\tilde{x}_1)]_T - [(Q - Q_n)(\tilde{x}_2)]_T\right\|_Y \\ &= \lim_{k\to\infty} \left\|[(Q_k - Q_n)(\tilde{x}_1)]_T - [(Q_k - Q_n)(\tilde{x}_2)]_T\right\|_Y \\ &\leq \lim_{k\to\infty} \left\|Q_k - Q_n\right\|_{\text{Lip}}\left\|[\tilde{x}_1]_T - [\tilde{x}_2]_T\right\|_X \\ &= \epsilon/2\left\|[\tilde{x}_1]_T - [\tilde{x}_2]_T\right\|_X \end{aligned} \tag{2.14}$$

so that $\|Q_n - Q\|_{\text{Lip}} \leq \epsilon$ for $n \geq N$. This shows that $\|Q_n - Q\|_{\text{Lip}} \to 0$ as $n \to \infty$ and completes the proof of the lemma. ■

It should be remarked that the standard Lipschitz operator and the generalized Lipschitz operator are not comparable since they have different domains and ranges. The generalized Lipschitz operator has been proved more useful than the standard Lipschitz operator for nonlinear system control and engineering in the considerations of stability, robustness, and uniqueness of internal control signals. Operators defined throughout this book are assumed to be generalized Lipschitz operators. For simplicity, the Lipschitz operator is always the one defined in the generalized case in this book.

In this book, $\text{Lip}(D^e) = \text{Lip}(D^e, D^e)$. In the following, causality is introduced, a basic requirement for a physical system.

2.1.2.7 Causality Let X^e be the extended linear space associated with a given Banach space X, and let $Q : X^e \to X^e$ be a nonlinear operator describing a nonlinear control system. Then, Q is said to be causal if and only if

$$P_T Q P_T = P_T Q \tag{2.15}$$

for all $T \in [0, \infty)$, where P_T is the projection operator.

The physical meaning behind the definition of causality may be understood as follows. If the system outputs depend only on the present and past values of the corresponding system inputs, then we have $QP_T(u) = Q(u)$ for all input signals u in the domain of Q, so that $P_T Q P_T = P_T Q$. Conversely, if $P_T Q P_T = P_T Q$ for all $T \in [0, \infty)$, then we have $P_T Q(I - P_T)(u) = 0$ for all input u in the domain of Q, which implies that any future value of a system input, $(I - P_T)(u)$, does not affect the present and past values of the corresponding system output given by $P_T Q(\cdot)$, or in other words, system outputs depend only on the present and past values of the corresponding system inputs.

Lemma 2.2 [1] A nonlinear operator $Q : X^e \to X^e$ is causal if and only if, for any $x, y \in X^e$ and $T \in [0, \infty)$, $x_T = y_T$ implies $[Q(x)]_T = [Q(y)]_T$.

Proof Suppose that $Q : X^e \to X^e$ is causal. Then by definition we have that $P_T Q P_T = P_T Q$, so that if $x_T = y_T$, then

$$\begin{aligned}[Q(x)]_T &= P_T Q(x) = P_T Q P_T(x) = P_T Q(x_T) = P_T Q(y_T) \\ &= P_T Q P_T(y) = P_T Q(y) = [Q(y)]_T\end{aligned} \tag{2.16}$$

Conversely, suppose that $x_T = y_T$ implies $[Q(x)]_T = [Q(y)]_T$ for all $x, y \in X^e$ and all $T \in [0, \infty)$. Fix a $T \in [0, \infty)$, for any $x \in X^e$, let $y = x_T$, then $x_T = y_T$, so that $[Q(x)]_T = [Q(y)]_T$. Consequently, we have that

$$\begin{aligned}P_T Q P_T(x) &= P_T Q(x_T) = P_T Q(y) \\ &= [Q(y)]_T = [Q(x)]_T = P_T Q(x)\end{aligned} \tag{2.17}$$

Since $x \in X^e$ and $T \in [0, \infty)$ are arbitrary, it follows that $P_T Q P_T = P_T Q$ for all $T \in [0, \infty)$, which implies that Q is causal. ∎

Lemma 2.3 [1] If $Q : X^e \to X^e$ is a generalized Lipschitz operator, then Q is causal.

Proof Let

$$\|[Q(x)]_T - [Q(y)]_T\|_X \leq \|Q\|_{\text{Lip}} \|x_T - y_T\|_X \tag{2.18}$$

for all $x, y \in X^e$ and all $T \in [0, \infty)$. Hence, $x_T = y_T$ implies that $[Q(x)]_T = [Q(y)]_T$ for all $x, y \in X^e$ and all $T \in [0, \infty)$. ∎

Note that a nonlinear operator may produce nonunique outputs from an input, particularly for a set-valued mapping. However, in practice, the internal signals in the system are always required to be unique. It is clear from the definition of the generalized Lipschiz operator that it guarantees the uniqueness requirement.

Lemma 2.4 A nonlinear generalized Lipschitz operator produces a unique output from an input in the sense that if input x and output y are related by a generalized Lipschitz operator Q such that $y = Q(x)$, then $x_T = \tilde{x}_T$ implies that $y_T = \tilde{y}_T$ for all $T \in [0, \infty)$.

2.2 OPERATOR THEORY

A normal operator theory nonlinear feedback control system is shown in Figure 2.1, where U and V are used to denote the input and output spaces of a given plant operator P, that is, $P : U \to V$.

2.2.1 Right Coprime Factorization

2.2.1.1 Right Factorization The given plant operator $P : U \to V$ is said to have a right factorization if there exist a linear space W and two stable operators $D : W \to U$ and $N : W \to V$ such that D is invertible and $P = ND^{-1}$. Such

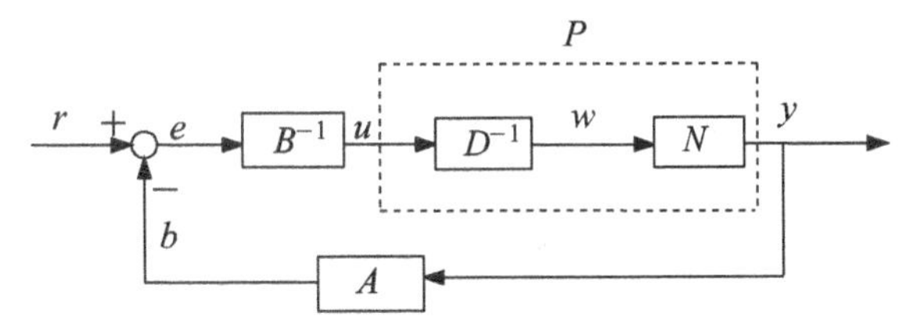

FIGURE 2.1 Nonlinear feedback system.

a factorization of P is denoted by (N, D) and space W is called a quasi-state space of P.

2.2.1.2 Right Coprime Factorization

Let (N, D) be a right factorization of P. The factorization is said to be coprime, or P is said to have a right coprime factorization, if there exist two stable operators $A : V \to U$ and $B : U \to U$ satisfying the Bezout identity

$$AN + BD = M \quad \text{for some } M \in \mathcal{U}(W, U) \tag{2.19}$$

where B is invertible. Usually, P is unstable and (N, D, A, B) are to be determined.

It is worth mentioning that the initial state should also be considered, that is, $AN(w_0, t_0) + BD(w_0, t_0) = M(w_0, t_0)$ should be satisfied. In this book, $t_0 = 0$ and $w_0 = 0$ are selected.

In general, researchers choose $W = U$ briefly, meaning U and W are the same linear space and $M = I$ and I is the identity operator. In this book, the arbitrary unimodular operator M is chosen.

2.2.1.3 Well-Posedness

The feedback control system shown in Figure 2.1 is said to be well-posed if for every input signal $r \in U$ all signals in the system (i.e., e, u, w, b, and y) are uniquely determined.

2.2.1.4 Overall Stable

The feedback control system shown in Figure 2.1 is said to be overall stable if $r \in U_s$ implies that $u \in U_s$, $y \in V_s$, $w \in W_s$, $e \in U_s$, and $b \in U_s$.

Lemma 2.5 [1] Assume that the system shown in Figure 2.1 is well-posed. If the system has a right factorization $P = ND^{-1}$, then the system is overall stable if and only if the operator M in (2.19) is a unimodular operator.

Proof *Sufficiency:* Since $M \in \mathcal{U}(W, U)$, for any $r \in U_s$, we have $r = (AN + BD)(w)$, that is, $w = M^{-1}r \in W_s$. Moreover, since $y = Nw$, $e = BDw$, and $b = Ay = ANw$, the stability of A, B, N, and D implies that $y \in V_s$, $e \in U_s$ and $b \in U_s$. Thus, the system is overall stable.

Necessity: First, it follows from the well-posedness and through the path of N and A that $M : W \to U$ is invertible. Then, it can be verified that both M and M^{-1} are stable. As a result, $M \in \mathcal{U}(W, U)$. ■

2.2.2 Robust Right Coprime Factorization

Generally speaking, for a nonlinear system, if the corresponding system with uncertainty remains stable, the system is said to have the robust stability property.

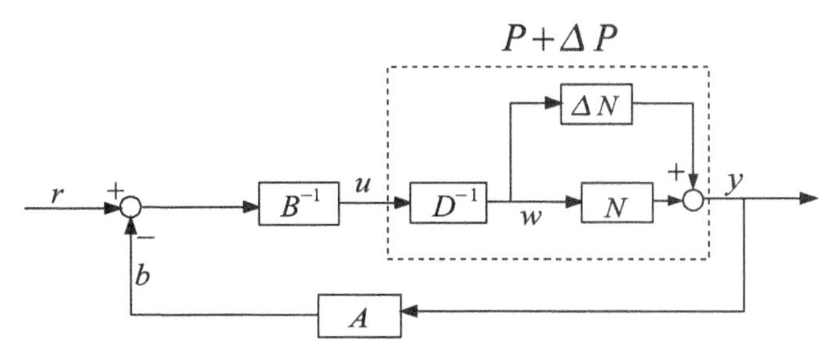

FIGURE 2.2 Nonlinear feedback system with uncertainty.

Concerning nonlinear feedback control systems with unknown bounded uncertainty, the robust condition of right coprime factorization was derived in [2, 3].

The operator theory based nonlinear feedback control system with uncertainty is given in Figure 2.2, where, the nominal plant and uncertainty are P and ΔP, respectively, and the overall plant $\tilde{P} = P + \Delta P$. The right factorization values of the nominal plant P and the overall plant $\tilde{P}$ are

$$P = ND^{-1} \tag{2.20}$$

and

$$P + \Delta P = (N + \Delta N)D^{-1} \tag{2.21}$$

where N, ΔN, and D are stable operators, D is invertible, ΔN is unknown, but the upper and lower bounds are known. According to (2.19) and Lemma 2.5, the BIBO stability of the nonlinear feedback control system with uncertainty can be guaranteed provided that

$$A(N + \Delta N) + BD = \tilde{M} \tag{2.22}$$

where $\tilde{M}$ is the unimodular operator. This Bezout identity is called the perturbed Bezout identity in this book. To obtain this condition, according to [3], if

$$A(N + \Delta N) = AN \tag{2.23}$$

that is, if $\mathcal{R}(\Delta N) \subseteq \mathcal{N}(A)$, where $\mathcal{N}(A)$ is a null set defined by

$$\mathcal{N}(A) = \{x : x \in \mathcal{D}(A) \text{ and } A(y + x) = Ay \text{ for all } y \in \mathcal{D}(A)\} \tag{2.24}$$

then the system is stable because

$$A(N + \Delta N) + BD = AN + BD = M \tag{2.25}$$

In fact, the condition is restrictive and design A does not easily satisfy (2.23) because ΔN is unknown. To improve this, an extended condition, the robust right coprime factorization condition, was proposed in [2].

To begin with, we recall a sufficient condition for judging an operator's invertibility.

Lemma 2.6 [1] Let X and Y be Banach spaces, $S \in \mathrm{Lip}(X, Y)$ be an invertible operator, and R be an operator in $\mathrm{Lip}(X, Y)$ such that $\|RS^{-1}\| < 1$, where $\mathrm{Lip}(X, Y) = \{S : X \to Y, \|S\|_{\mathrm{Lip}} < \infty\}$. Then, the operator $R + S$ is invertible in $\mathrm{Lip}(X, Y)$ and

$$\|(R + S)^{-1}\| \leq \|S^{-1}\|(1 - \|RS^{-1}\|)^{-1} \tag{2.26}$$

Proof We first prove the following assertion: If one operator $J \in \mathrm{Lip}(X)$ such that $\|J\| < 1$, then $I - J$ is invertible and

$$\|(I - J)^{-1}\| \leq (1 - \|J\|)^{-1} \tag{2.27}$$

In fact, for $x, y \in X$,

$$\begin{aligned} \|(I - J)x - (I - J)y\| &\geq \|x - y\| - \|Jx - Jy\| \\ &\geq (1 - \|J\|)\|x - y\| \end{aligned} \tag{2.28}$$

Thus, $I - J$ is injective. Next, we show that $I - J$ is surjective and $(I - J)^{-1} \in \mathrm{Lip}(X)$.

Define $Q_0 := I$ and $Q_n = I + JQ_{n-1}$ $\forall n = 1, 2, \ldots$. We can prove that for $x \in X$

$$\|Q_{n+1}(x) - Q_n(x)\| \leq \|J\|^n \|J(x)\|_X, \ n = 1, 2, \ldots \tag{2.29}$$

Then for any positive integer m, we have that

$$\begin{aligned} \|Q_{n+m}(x) - Q_n(x)\|_X &= \left\| \sum_{k=0}^{m-1} (Q_{n+k+1}(x) - Q_{n+k}(x)) \right\| \\ &\leq \sum_{k=0}^{m-1} \|J\|^{n+k} \|J(x)\|_X \leq \frac{\|J\|^n \|J(x)\|_X}{1 - \|J\|} \end{aligned} \tag{2.30}$$

Since $\|J\| < 1$ and X is a Banach space, then $Q(x) = \lim_{n\to\infty} Q_n(x)$ exists and

$$\|Q(x) - Q_n(x)\|_X = \lim_{n\to\infty} \|Q_{n+m}(x) - Q_n(x)\| \leq \frac{\|J\|^n \|J(x)\|_X}{1 - \|J\|} \tag{2.31}$$

Since J is Lipschitz and hence is continuous,

$$Q(x) = \lim_{n\to\infty} Q_n(x) = \lim_{n\to\infty} (I + JQ_{n-1})x = x + JQx \tag{2.32}$$

that is, $Q = I + JQ$, namely, $(I - J)Q = I$, which implies that $I - J$ is surjective in $\text{Lip}(X)$. Then, for $r, z \in \mathcal{R}(I - L)$,

$$\|(I - J)^{-1}r - (I - J)^{-1}z\| \leq (1 - \|J\|)^{-1}\|r - z\| \tag{2.33}$$

Thus, the assertion is proved. As a consequence, since $R + S = (I + RS^{-1})S$, then $\|RS^{-1}\| < 1$ follows that $(I + RS^{-1})^{-1}$ exists and $(R + S)^{-1} = S^{-1}(1 + RS^{-1})^{-1}$. This completes the proof of Lemma 2.6. ■

Lemma 2.7 [2] Let D^e be a linear subspace of the extended linear space U^e associated with a given Banach space U_B, and let $(A(N - \Delta N) + AN)M^{-1} \in \text{Lip}(D^e)$. Let the Bezout identity of the nominal plant and the exact plant be $AN + BD = M \in \mathcal{U}(W, U)$, $A(N + \Delta N) + BD = \tilde{M}$, respectively. If

$$\|(A(N + \Delta N) - AN)M^{-1}\| < 1 \tag{2.34}$$

then the system shown in Figure 2.2 is robust stable for ΔN.

Proof That M is a unimodular operator implies it is invertible. From

$$AN + BD = M \tag{2.35}$$

$$A(N + \Delta N) + BD = \tilde{M} \tag{2.36}$$

we have

$$\begin{aligned} \tilde{M} &= M + [A(N + \Delta N) - AN] \\ &= [I + (A(N + \Delta N) - AN)M^{-1}]M \end{aligned} \tag{2.37}$$

and $(A(N + \Delta N) - AN)M^{-1} \in \text{Lip}(D^e)$. Then $I + (A(N + \Delta N) - AN)M^{-1}$ is invertible based on Lemma 2.6, where I is the identity operator. Consequently

$$\tilde{M}^{-1} = M^{-1}[I + (A(N + \Delta N) - AN)M^{-1}]^{-1} \tag{2.38}$$

Meanwhile, since $(A(N + \Delta N) - AN)M^{-1} \in \text{Lip}(D^e)$ and $M \in \mathcal{U}(W, U)$, then $\tilde{M} \in \mathcal{U}(W, U)$ provided that the system shown in Figure 2.2 is well-posed. As a result, for any $r \in U_s$, $w = \tilde{M}^{-1}r \in W_s$. Further, since $y = (N + \Delta N)w$, $e = BDw$ and $b = A(N + \Delta N)w$, the stability of A, B, N, D, and ΔN implies that $y \in V_s$, $e \in U_s$, and $b \in U_s$. Then, the system is overall stable, namely robust stable for ΔN. ■

It is worth mentioning that the initial state should also be considered, that is, $AN(w_0, t_0) + BD(w_0, t_0) = M(w_0, t_0)$ should be satisfied.

This book studies the problems encountered to satisfy the extended condition, namely, Lemma 2.7. In the following section, the robust right prime factorization problem is explored by using the idea of isomorphism and passivity.

2.2.3 Isomorphism-Based Robust Right Prime Factorization

Factorization of the plant and the design of the controllers in the system design problem were discussed in [5]. Therein, the right factorization of the nonlinear plant is realized by means of isomorphism, and a quantitative design scheme of controllers S and R is proposed to guarantee the stability of the nonlinear feedback control system, which also means that the factorization of the perturbed plant is a robust right coprime factorization. Meanwhile, the plant output is guaranteed to track to the reference output. However, the factorization of the nonlinear plant is limited to some nonlinear systems, and for the case of unknown perturbations tracking is unsolved. Therefore, the right factorization of the plant will be realized by a combination of isomorphism and the passivity property. Then for the unknown perturbation, robust controllers will be designed to guarantee the stability of the nonlinear feedback systems as well as the plant output asymptotically tracking to the reference output. Meanwhile, the passivity of the whole nonlinear feedback system can be found to be guaranteed.

In this chapter, the relationship between robust right coprime factorization and passivity for passive systems is considered, where, according to the passivity principle, right factorization of the nonlinear perturbed plant is realized and guaranteed to be robust right coprime factorization for the nonlinear perturbed plant. Passive systems constitute an important class of dynamical systems where the energy exchanges with the environment. Simply speaking, in passive systems, the amount of stored energy cannot exceed that of the energy supplied by the outside—the difference being the dissipated energy [9]. Given this property, passivity has been taken as a building block for stabilization of nonlinear systems by an increasing number of researchers [10–14].

Lemma 2.8 [3] Suppose that the system shown in Figure 2.3 is well-posed if the unstable plant has a right coprime factorization, that is, the Bezout identity $SN + RD = M$ is satisfied, where M, S, and R^{-1} represent the unimodular operator and feedback and feedforward controllers, respectively. Then the system is overall stable.

Lemma 2.8 can be obtained by using the result in Section 2.2.1. Based on Lemma 2.8, the nonlinear feedback system shown in Figure 2.3 can be transformed to the system shown in Figure 2.4. Then the plant output will track to the reference output by satisfying $NM^{-1} = I$ [5].

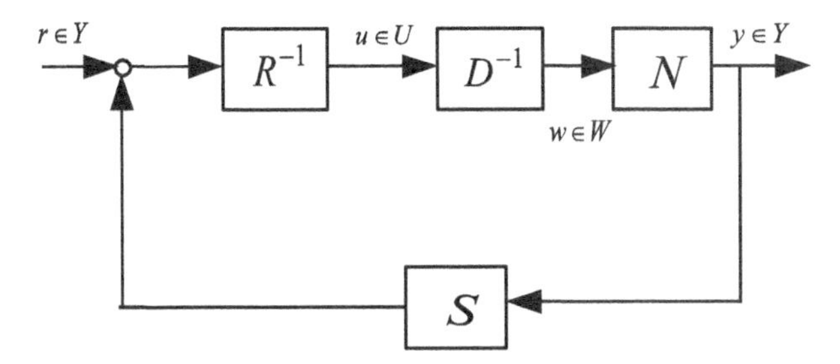

FIGURE 2.3 Nonlinear feedback system.

FIGURE 2.4 Equivalent system.

Lemma 2.9 [7]: Let $\mathbf{D}^e$ be a linear subspace of the extended linear space $\mathbf{U}^e$ associated with a given Banach space $\mathbf{U}_B$ and let $M^{-1}[S(N + \Delta N) - SN + R(D + \Delta D) - RD] \in \mathrm{Lip}(\mathbf{D}^e)$. Let the Bezout identities of the nominal plant and the exact plant be $SN + RD = M \in \mu(\mathbf{W}, \mathbf{U})$ and $S(N + \Delta N) + R(D + \Delta D) = \tilde{M}$, respectively. If the condition

$$\left\| [S(N + \Delta N) - SN + R(D + \Delta D) - RD]M^{-1} \right\| < 1 \tag{2.39}$$

is satisfied, the system shown in Figure 2.5 is stable, where $\| \cdot \|$ is the Lipschitz norm.

In the above results, what needs to be mentioned is that if the perturbed systems remain overall stable, then the systems are said to have robust stability properties and the perturbed plants are said to have a robust right coprime factorization.

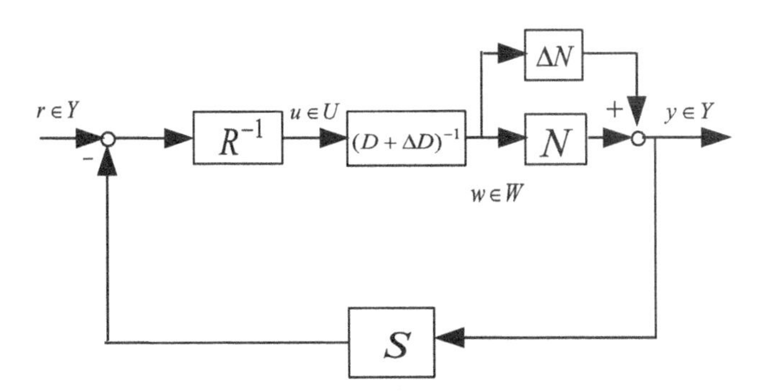

FIGURE 2.5 Perturbed nonlinear feedback system [7].

In the following, the definitions and notation of passivity are introduced, where we consider nonlinear systems described by the operator form

$$\Sigma : \quad \begin{cases} w(t) = f(u)(t) \\ y(t) = g(w)(t) \end{cases} \tag{2.40}$$

where $u \in U$, $w \in W$, and $y \in Y$ are the control input and the quasi-state and plant output, respectively.

Definition 2.1 The system Σ is said to be passive if there exists a nonnegative function $V : W \to R_+$, called the storage function, such that $\forall u \in U$, $w_0 \in W$, $t \geq 0$,

$$\underbrace{V(w) - V(w_0)}_{\textit{Stored energy}} \leq \underbrace{\int_0^t y(s)u(s)\,ds}_{\textit{Supplied energy}} \tag{2.41}$$

where $V(0) = 0$.

Moreover, if the stored energy is equal to the supplied energy,

$$V(w) - V(w_0) = \int_0^t y(s)u(s)\,ds \tag{2.42}$$

then the passive system Σ with storage function V is said to be lossless. For another case, a passive system Σ with storage function V is said to be strictly passive if there exists a number γ such that

$$V(w) - V(w_0) = \int_0^t y(s)u(s)\,ds - \gamma^2 \tag{2.43}$$

If the nonnegative function V is differentiable, then the passivity inequality [7] is equivalent to

$$\dot{V} \leq yu \quad \forall u \in \mathbf{U} \tag{2.44}$$

The fundamental problem—robust right coprime factorization of the nonlinear plant—has been realized by using the definition of isomorphism in [5]. But the concerned systems are limited to some special nonlinear systems and the tracking problem for the case of unknown perturbations is unsolved. Therefore, in the following, operator-based robust right coprime factorization is combined with the passivity property to deal with the robust control and tracking problem in nonlinear systems with unknown perturbations.

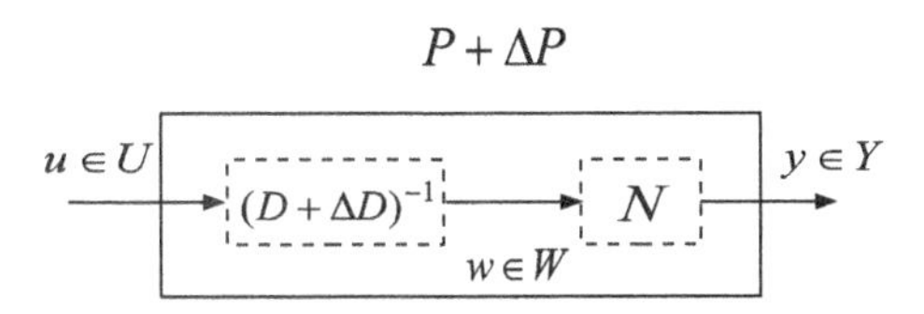

FIGURE 2.6 Nonlinear plant with unknown perturbation.

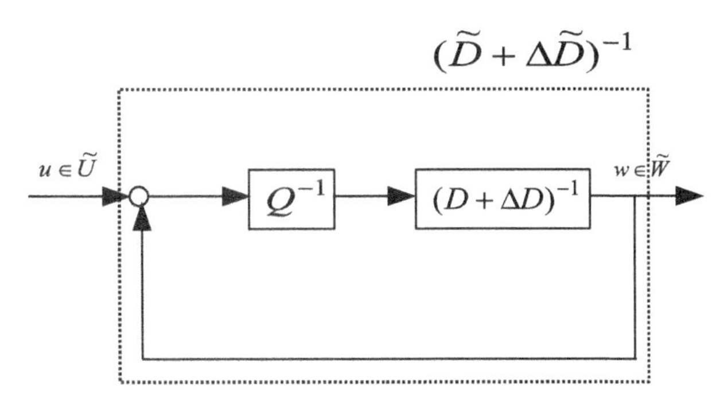

FIGURE 2.7 Construct of isomorphic subspace.

Lemma 2.10 Suppose that the system shown in Figure 2.6 is well-posed. The nonlinear plant with unknown perturbations can be right factorized into two parts N and $(D + \Delta D)^{-1}$, where $(\tilde{D} + \Delta\tilde{D} - I)^{-1}$ is designed to be passive and its inverse is also passive.

Proof First, by the passivity property, a feedback system with a compensator Q^{-1} is supposed to be constructed to guarantee that the constructed space is an isomorphic subspace $\tilde{\mathbf{W}}$ of the input space $\tilde{\mathbf{U}}$ ($\tilde{\mathbf{U}} \subseteq \mathbf{U}$, $\tilde{\mathbf{W}} \subseteq \tilde{\mathbf{U}}$, where $\tilde{\mathbf{U}}$ is a positive and stabilizable subspace of $\mathbf{U}$), which is shown in Figure 2.7. Its equivalent system $(\tilde{D} + \Delta\tilde{D})^{-1}$ is shown in Figure 2.8.

Since the space $\tilde{\mathbf{W}}$ is the isomorphic subspace of the input space $\tilde{\mathbf{U}}$, the isomorphism between $\tilde{\mathbf{U}}$ and $\tilde{\mathbf{W}}$ can be assumed to be ϕ, so for $w, u \in \tilde{\mathbf{U}}$, the following relationship is satisfied:

$$\phi(w \circ u) = \phi(w) * \phi(u) \tag{2.45}$$

where "$\circ$" is the quasi-inner product defined in space $\tilde{\mathbf{U}}$ and "$*$" is the operation defined in the operator theory [5]. Thus, $w = \phi(u)$ and the following equality is

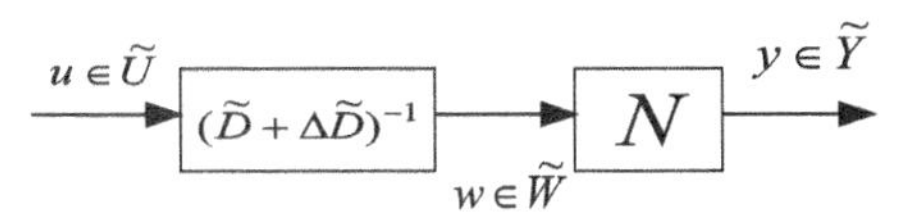

FIGURE 2.8 Factorized parts.

established:

$$\phi(\phi(u) \circ u) = \phi(\phi(u)) \tag{2.46}$$

With the injective property of isomorphism, relationship (2.47) can be obtained:

$$\phi(u) \circ u = \phi(u) \tag{2.47}$$

The operation result between $\phi(u)$ and u [the left side of (2.47)] is generally assumed to be $\Phi(\phi(u), u)$. Then

$$\Phi(\phi(u), u) = \phi(u) \tag{2.48}$$

Then, according to the elementary Gronwall equality, the passive solution of (2.48) $\phi(u)$ can be obtained, and the following two operators can be designed:

$$N(w)(t) = \phi(w)(t) \tag{2.49}$$

$$(\tilde{D} + \Delta\tilde{D})^{-1}(u)(t) = \phi(u)(t) + \Delta(t)u(t) \tag{2.50}$$

Then, $(D + \Delta D)^{-1}$ can be obtained by $P + \Delta P = N(D + \Delta D)^{-1}$. Since $(\tilde{D} + \Delta\tilde{D} - I)$ is designed to be passive and its inverse is also passive, then the stable Q^{-1} is obtained: $Q^{-1} = (D + \Delta D)(\tilde{D} + \Delta\tilde{D} - I)^{-1}$. Therefore, based on passivity, by constructing an isomorphic space of $\tilde{U}$, the right factorization of the plant with unknown perturbations can be realized. ■

In the above proof, based on passivity, the isomorphic subspace of the input space is constructed. Then the nonlinear plant with unknown perturbations is factored into two parts, N and $(D + \Delta D)^{-1}$, and the right factorization of the nonlinear plant with unknown perturbations is realized. Next two passive controllers $\tilde{S}$ and $\tilde{R}$ are designed to guarantee the robust stability of the nonlinear systems with unknown perturbations. Another problem here is that of the plant output tracking to the reference output. Since there exist unknown perturbations, perfect tracking cannot be realized, and we consider the case of asymptotic tracking. Therefore, in what follows, a robust control scheme will be given to guarantee the robust stability of the obtained nonlinear system with unknown perturbations as well as the plant output asymptotically tracking to the reference output.

Theorem 2.1 Suppose the obtained nonlinear feedback control system shown in Figure 2.9 is well-posed, where the plant output of the nonlinear system with a nominal plant tracks to the reference output, if the passive controller $\tilde{S} \nrightarrow I$ is designed to satisfy the condition

$$\tilde{S}N + (I - \tilde{S})N\tilde{D}^{-1}(\tilde{D} + \Delta\tilde{D}) \rightarrow N \quad as \quad t \rightarrow \infty \tag{2.51}$$

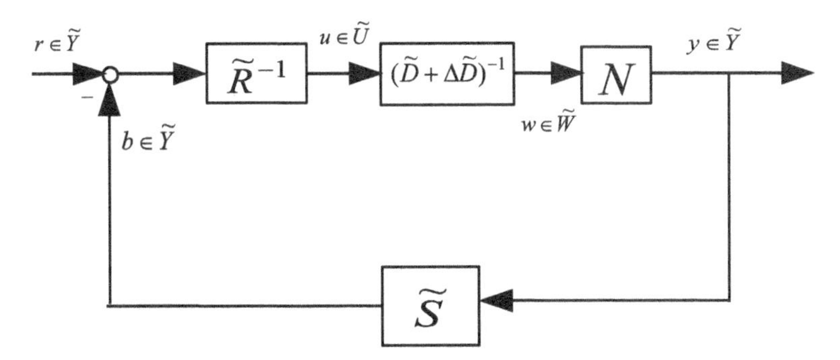

FIGURE 2.9 Obtained nonlinear feedback control system.

Then the robust stability as well as the plant output asymptotically tracking to the reference output can be guaranteed.

Proof According to the construct of the isomorphic subspace of the input space, the part N is passive. Since the plant output of nonlinear systems with a nominal plant tracks to the reference output, which means $\tilde{S}N + \tilde{R}\tilde{D} = N$, the inverse of N is also passive. Moreover, the controller $\tilde{R}$ can be presented by $\tilde{R}(u)(t) = (I - \tilde{S})N\tilde{D}^{-1}(u)(t)$. Further, by the property of the operator, we know that $[\tilde{D}^{-1}(\tilde{D} + \Delta\tilde{D})](z)$ is in a linear form of z. Therefore

$$\begin{aligned} \tilde{M} &= [\tilde{S}N + \tilde{R}(\tilde{D} + \Delta\tilde{D})] \\ &= [\tilde{S}N + (I - \tilde{S})N\tilde{D}^{-1}(\tilde{D} + \Delta\tilde{D})] \end{aligned} \tag{2.52}$$

can be found to be passive and its inverse is also passive (passivity of N and its inverse). Therefore, $\tilde{M}$ is a unimodular operator. Then the robust stability of the nonlinear feedback system is guaranteed.

Here $\tilde{M}$ is a unimodular operator, then the equivalent system can be obtained as shown in Figure 2.10, and

$$N\tilde{M}^{-1}(r)(t) = N[\tilde{S}N + (I - \tilde{S})N\tilde{D}^{-1}(\tilde{D} + \Delta\tilde{D})]^{-1}(r)(t) \tag{2.53}$$

Here $\tilde{S}$ is designed such that $\tilde{S}N + (I - \tilde{S})N\tilde{D}^{-1}(\tilde{D} + \Delta\tilde{D}) \to N$ as $t \to \infty$. Thus

$$y(t) = N\tilde{M}^{-1}(r)(t) \to I(r)(t) \quad t \to \infty \tag{2.54}$$

Then the plant output can asymptotically track to the reference output.

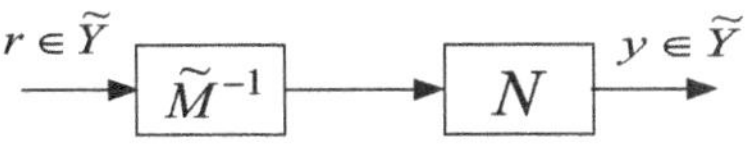

FIGURE 2.10 Equivalent perturbed nonlinear system.

Design $\tilde{S}$ so as to guarantee that the controller $\tilde{R}$ is a unimodular operator. Therefore, $\tilde{S}$ cannot be designed as $\tilde{S} \to I$, as $t \to \infty$. ■

In the above proof, the unimodular operator $\tilde{M}$ is guaranteed by the passivity of the operator which is an extension of the BIBO stability.

Corollary Considering the nonlinear feedback control system shown in Figure 2.9, where the system is well-posed, if the controllers are designed as in Theorem 2.1, the whole nonlinear feedback systems can be guaranteed to be passive.

Proof According to Theorem 2.1, $N\tilde{M}^{-1} \to I$. Then we can choose the storage function $V(w)(t)$ to be $V(w)(t) = \int_0^t N(w)(s)N(w)(s)ds$ [if $\tilde{M}(w)(t) \geq N(w)(t)$] or to be $V(w)(t) = \int_0^t \tilde{M}(w)(s)\tilde{M}(w)(s)ds$ [if $\tilde{M}(w)(t) < N(w)(t)$], and

$$\dot{V}(w)(t) = N(w)(t)N(w)(t) \quad \text{where } \tilde{M}(w)(t) \geq N(w)(t) \tag{2.55}$$

or

$$\dot{V}(w)(t) = \tilde{M}(w)(t)\tilde{M}(w)(t) \quad \text{where } \tilde{M}(w)(t) < N(w)(t) \tag{2.56}$$

Since the storage function is differentiable, then according to (2.44), we verify the passivity of the whole nonlinear feedback system shown in Figure 2.9 as

$$y(t)r(t) = N(w)(t)\tilde{M}(w)(t) \tag{2.57}$$

Then

$$\begin{aligned} y(t)r(t) \geq \dot{V}(w)(t) \quad &\text{where } \dot{V}(w)(t) \\ &= N(w)(t)N(w)(t),\ \tilde{M}(w)(t) \geq N(w)(t) \\ y(t)r(t) < \dot{V}(w)(t) \quad &\text{where } \dot{V}(w)(t) \\ &= \tilde{M}(w)(t)\tilde{M}(w)(t),\ \tilde{M}(w)(t) < N(w)(t) \end{aligned} \tag{2.58}$$

Thus the nonlinear feedback control system is guaranteed to be (strictly) passive when the robust controllers are designed as shown in Theorem 2.1. ■

The corollary shows the connection between passivity and robust right coprime factorization for the obtained nonlinear system. That is, by the design scheme based on robust right coprime factorization, the obtained nonlinear system can be guaranteed to be robust stable by satisfying the perturbed Bezout identity $\tilde{S}N + \tilde{R}(\tilde{D} + \Delta\tilde{D}) = \tilde{M}$. Moreover, from the corollary, the nonlinear feedback control system can be guaranteed to be passive. In the following, a numerical example is given to show the effectiveness of the method.

First, two spaces **U** and **Y** are given: $\mathbf{U} = \mathbf{C}^*_{[0,\infty)}$, $\mathbf{Y} = \{\beta(t)u^3(t)|u(t) \in \mathbf{U}\}$, where $\mathbf{C}^*$ is a positive and continuous set defined in $t \in [0, \infty)$ and $\beta(t)$ is also a continuous function.

Let $\|\cdot\|$ be the sup norm defined by $\|u\|_\infty = \sup_{t\in[0,\infty)} |u(t)|$. Define $\mathbf{U}_s = \{u(t)|u(t) \in \mathbf{U}, \|u\|_\infty < \infty\}$ and $\mathbf{Y}_s = \{y(t)|y(t) \in \mathbf{Y}, |\beta(t)|\|u^2\|_\infty < \infty\}$. It can be verified that both of the above spaces are linear normed, so they can be used as the stable subspaces of $\mathbf{U}$ and $\mathbf{Y}$, respectively.

Consider the system in Figure 2.6 in which the input space and output space are $\mathbf{U}$ and $\mathbf{Y}$, respectively. The given plant $P + \Delta P : \mathbf{U} \to \mathbf{Y}$ is defined by:

$$(P + \Delta P)(u)(t) = \frac{1}{8}\big(-e^t + 101 + \eta(t) + (\eta(t) + 100)e^{-t}\big)u^3(t) \quad \text{where } \eta(t) \to 0 \text{ as } t \to \infty$$

According to Lemma 2.10, the isomorphic space can be constructed where the mapping "$\circ$" is defined by

$$\phi(u)(t) \circ u(t) = -\frac{1}{8}u(t)\left(\int_0^t \frac{8\phi(u)(\tau) - u(\tau)}{u(\tau)}d\tau - 2\right) \tag{2.59}$$

The solution of equation (2.58) is $\phi(u)(t) = \frac{1}{8}(1 + e^{-t})u(t)$, so we can define the operators N and $(\tilde{D} + \Delta\tilde{D})^{-1}$ as

$$\begin{aligned} N(w)(t) &= \tfrac{1}{8}(1 + e^{-t})w(t) \\ (\tilde{D} + \Delta\tilde{D})^{-1}(u)(t) &= \left[\tfrac{1}{8}(1 + e^{-t}) + \eta(t)\right]u(t) \end{aligned} \tag{2.60}$$

According to the definition of right factorization

$$(D + \Delta D)^{-1}(u)(t) = [100 - e^t + \eta(t)]u^3(t) \tag{2.61}$$

The right factorization of the nonlinear plant with unknown perturbations is realized. Next, the passive controllers $\tilde{S}$ and $\tilde{R}$ can be designed according to Theorem 2.1:

$$\begin{aligned} \tilde{S}(y)(t) &= \frac{2}{3}y(t) \\ \tilde{R}(u)(t) &= (I - \tilde{S})N\tilde{D}^{-1}(u)(t) = \frac{1}{192}(1 + e^{-t})^2(u)(t) \end{aligned} \tag{2.62}$$

Then

$$\begin{aligned} \tilde{M}(w)(t) &= [\tilde{S}N + \tilde{R}(\tilde{D} + \Delta\tilde{D})](w)(t) \\ &= \frac{1}{24}(1 + e^{-t})\frac{3(1 + e^{-t}) + 8\eta(t)}{1 + e^{-t} + 8\eta(t)}w(t) \end{aligned} \tag{2.63}$$

The passive operator $\tilde{M}$ is found to be unimodular with

$$\tilde{M}^{-1}(r)(t) = \frac{24[1+e^{-t}+8\eta(t)]}{(1+e^{-t})[3(1+e^{-t})+8\eta(t)]}(r)(t)$$

which is also passive.

Since the Bezout identity for the nominal plant $\tilde{S}N + \tilde{R}\tilde{D} = N$ is satisfied, and the perturbed Bezout identity for the nonlinear plant with unknown perturbations is also verified to be guaranteed, then the designed controllers for the nonlinear feedback system with unknown perturbations are robust.

Moreover, from the assumption, (2.59) and (2.62), we can find $\tilde{M} \to N$ as $t \to \infty$ since $\eta(t) \to 0$. Therefore

$$y(t) = N\tilde{M}^{-1}(r)(t) \to I(r)(t) \tag{2.64}$$

and, by the robust control scheme, the plant output asymptotically tracks to the reference output.

Finally, we consider the passivity of the nonlinear feedback control by constructing the score function as $V(w)(t) = \int_0^t \tilde{M}(\tilde{M}(w))(s)ds$ $(\tilde{M}(w)(t) \le N(w)(t))$. Similar with Theorem 2.1, the passivity of the nonlinear feedback system can be guaranteed:

$$y(t)r(t) = N(w)(t)\tilde{M}(w)(t) \ge \dot{V}(w)(t) = \tilde{M}(w)(t)\tilde{M}(w)(t) \tag{2.65}$$

The simulation results are shown in Figures 2.11 and 2.12. It is easy to see that the plant output $y(t)$ asymptotically tracks to the reference output $r(t)$ while the reference

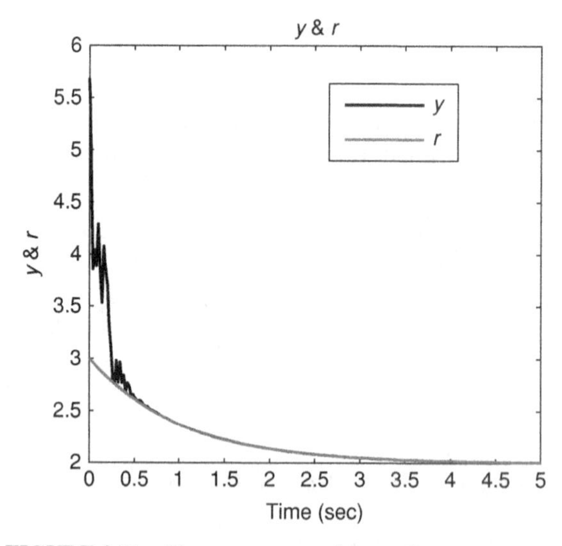

FIGURE 2.11 Plant output y tracks to reference output r.

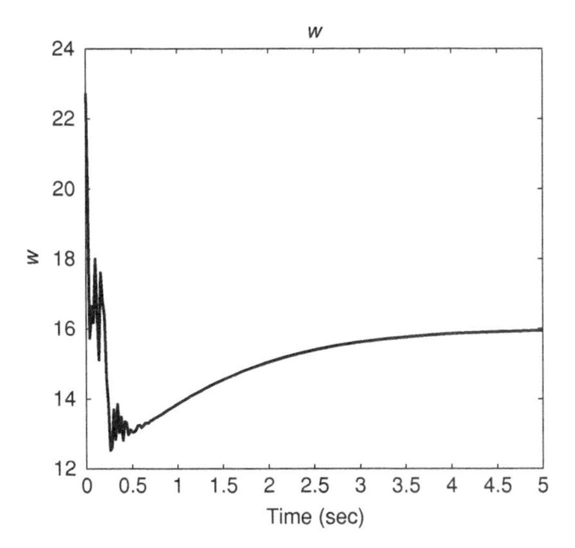

FIGURE 2.12 Quasi-state w.

output, quasi-state, and plant output are all stable, where the reference output is $r(t) = 2 + e^{-t}$, $\eta(t) = e^{-10t} * \sigma(t)$, and $\sigma(t)$ is a function that generates random noise between 0 and 1. Moreover, from Figures 2.13 and 2.14, we can find that the nonlinear feedback control system is guaranteed to be passive by the robust control scheme. That is, (2.65) is satisfied from Figure 2.14. The robust control problem for a nonlinear feedback control system with unknown perturbations is discussed here using operator-based robust right coprime factorization and the passivity property.

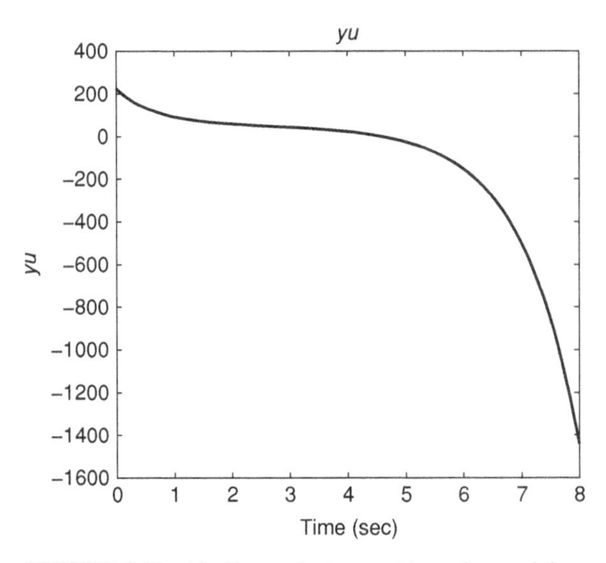

FIGURE 2.13 Nonlinear plant cannot keep the passivity.

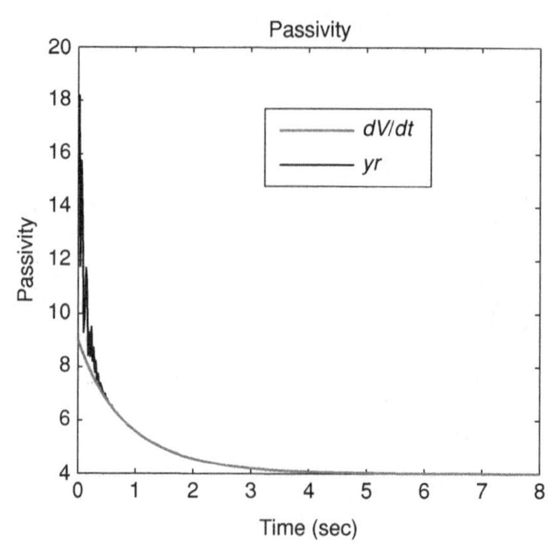

FIGURE 2.14 Obtained passive system by robust controllers.

After factorizing the nonlinear plant with unknown perturbations, a robust control scheme is shown to guarantee not only the robust stability of the nonlinear system but also the asymptotic tracking property. Meanwhile, the connection between passivity and robust right coprime factorization is also shown in the obtained equivalent system.

CHAPTER 3

Robust Stability of Operator-Based Nonlinear Control Systems

3.1 CONCEPT OF OPERATOR-BASED ROBUST STABILITY

So far, in Chapter 2 for the nonlinear system with uncertainties shown in Figure 2.2, if Lemma 2.7 is satisfied, the system is robust stable. Section 3.2 presents a detailed explanation of the lemma and gives a tracking control design scheme.

3.2 DESIGN METHODS OF NONLINEAR SYSTEMS WITH UNCERTAINTIES

3.2.1 Robust Right Coprime Factorization Condition

3.2.1.1 Introduction Nonlinear control system design problems have been considered by many researchers in different fields. One of the approaches is based on coprime factorization [3, 15–20]. The concept of coprime factorization is first considered in linear feedback control systems and provides a convenient framework for researching input–output stability problems of feedback control systems. The coprime factorization problem of nonlinear feedback control systems is also discussed for nonlinear analysis, design, stabilization, and control. Especially, right coprime factorization of nonlinear systems has attracted much attention due to its usefulness in the stabilization of nonlinear plants. The robust right coprime factorization of nonlinear plants under perturbation has been studied in [3]. The output tracking problem of perturbed nonlinear plants [21] has been considered by extending the design scheme given in [3] for the case in which the nonlinear plant output and the reference input share the same space and the Bezout identity is equal to the identity operator, where the perturbation should be known. The above robust right coprime factorization led to the robust stabilization of the entire feedback control system using an operator-theoretic approach, where the stability is based on the internal stability. However, the method only controls a class of nonlinear plants with bounded perturbations; the problem of checking the robust right coprime factorization condition for nonlinear

Operator-Based Nonlinear Control Systems: Design and Applications, First Edition. Mingcong Deng.

plants with unknown bounded perturbations might also be difficult in practice; the plant output tracking problem has not been considered for the case in which the nonlinear plant output and the reference input are different.

The purpose of this chapter is concerned with the robust stabilization and tracking performance of operator-based nonlinear feedback control systems using the robust right coprime factorization of Lemma 2.7. That is, the robust right coprime factorization condition is explained in detail and a tracking controller based on the generalized Lipschitz operator [1] is shown. The detailed explanation is given as follows. A condition for the robust right coprime factorization of nonlinear plants with unknown bounded perturbations is given as Lemma 2.7. The condition is obtained by using the generalized Lipschitz operator and the unimodular operator. Robust stabilization of the nonlinear feedback control system can be obtained by using the condition. In the plant output tracking problem, in general the spaces of the plant output and reference input are different. In this case a space change operator is designed, and we consider a tracking controller using the exponential iteration theorem [22], where the spaces of operators are defined using the generalized Lipschitz norm. As a result, a broader class of nonlinear plants with bounded perturbations can be robustly controlled and satisfactory tracking performance can be obtained.

3.2.1.2 Mathematical Preliminaries In this section, we recall several definitions of operators, right coprime factorization, and internal stability from Chapter 2.

Consider a space U of time functions, where U is said to be a vector space if it is closed under addition and scalar multiplication. The space U_s is said to be normed if each element x in U_s has a norm $\|x\|$ which can be defined in any way so long as the following three properties are fulfilled: (1) $\|x\|$ is a real, positive number that is different from zero unless x is identically zero, (2) $\|ax\| = |a|\|x\|$, and (3) $\|x_1 + x_2\| \leq \|x_1\| + \|x_2\| (x_1, x_2 \in U_s)$.

Let U_s and Y_s be two normed linear spaces over the field of complex numbers, endowed, respectively, with norms $\|\cdot\|_{U_s}$ and $\|\cdot\|_{Y_s}$. Let $A : U_s \to Y_s$ be an operator mapping from U_s to Y_s and denote by $\mathcal{D}(A)$ and $\mathcal{R}(A)$, respectively, the *domain* and *range* of A. As mentioned above, if the operator $A : \mathcal{D}(A) \to Y_s$ satisfies

$$A : ax_1 + bx_2 \to aA(x_1) + bA(x_2)$$

for all $x_1, x_2 \in \mathcal{D}(A)$ and all $a, b \in \mathcal{C}$, then A is said to be linear; otherwise it is said to be nonlinear. Let $\mathcal{N}(U_s, Y_s)$ be the family of all nonlinear operators mapping from $\mathcal{D}(A) \subseteq U_s$ into Y_s. Recall that $\mathcal{L}(U_s, Y_s)$ is used to denote the family of bounded linear operators from U_s to Y_s. Obviously, $\mathcal{L}(U_s, Y_s) \in \mathcal{N}(U_s, Y_s)$. In the case that $U_s = Y_s$, we use the notation $\mathcal{L}(U_s)$ and $\mathcal{N}(U_s)$, respectively, instead of $\mathcal{L}(U_s, U_s)$ and $\mathcal{N}(U_s, U_s)$ for simplicity.

Let D_s be a subset of U_s and $\mathcal{F}(D_s, Y_s)$ be the family of operators A in $\mathcal{N}(U_s, Y_s)$ with $\mathcal{D}(A) = D_s$. A (semi)-norm on (a subset of) $\mathcal{F}(D_s, Y_s)$ is denoted by

$$\|A\| := \sup_{\substack{x_1, x_2 \in D_s \\ x_1 \neq x_2}} \frac{\|A(x_1) - A(x_2)\|_{Y_s}}{\|x_1 - x_2\|_{U_s}} \tag{3.1}$$

if it is finite. In general, it is a seminorm in the sense that $\|A\| = 0$ does not necessarily imply $A = 0$. In fact, it can be easily seen that $\|A\| = 0$ if and only if A is a constant operator (need not be zero) that maps all elements from D_s to the same element in Y_s.

Definition 3.1 Let $\text{Lip}(D_s, Y_s)$ be the subset of $\mathcal{F}(D_s, Y_s)$ with each element A satisfying $\|A\| < \infty$. Each $A \in \text{Lip}(D_s, Y_s)$ is called a Lipschitz operator mapping from D_s to Y_s, and the number $\|A\|$ is called the Lipschitz seminorm of the operator A on D_s.

In this session, we assume operators are of Lipschitz type and use a seminorm of Lipschitz operators. It is clear that an element A of $\mathcal{F}(D_s, Y_s)$ is in $\text{Lip}(D_s, Y_s)$ if and only if there is a number $L \leq 0$ such that

$$\|A(x_1) - A(x_2)\|_{Y_s} \leq L\|x_1 - x_2\|_{D_s}$$

for all $x_1, x_2 \in D_s$. The norm $\|A\|$ is the least such number L. It is also evident that a Lipschitz operator is both bounded and continuous on its domain. Basic theories of nonlinear Lipschitz operators are given in [1].

Definition 3.2 Let U^e and Y^e be two extended linear spaces which are associated respectively with two given Banach spaces U_B and Y_B of measurable functions defined on the time domain $[0, \infty)$, where a Banach space is a complete vector space with a norm. Let D^e be a subset of U^e. A nonlinear operator $A : D^e \to Y^e$ is called a generalized Lipschitz operator on D^e if there exists a constant L such that

$$\|[A(x)]_T - [A(\tilde{x})]_T\|_{Y_B} \leq L\|x_T - \tilde{x}_T\|_{U_B} \tag{3.2}$$

for all $x, \tilde{x} \in D^e$ and for all $T \in [0, \infty)$.

Note that the least such constant L is given by

$$\|A\| := \sup_{T \in [0,\infty)} \sup_{\substack{x,\tilde{x} \in D^e \\ x_T \neq \tilde{x}_T}} \frac{\|[Ax]_T - [A\tilde{x}]_T\|_{Y_B}}{\|x_T - \tilde{x}_T\|_{U_B}} \tag{3.3}$$

which is a seminorm for general nonlinear operators and is the actual norm for linear A. The actual norm for a nonlinear operator A is given by

$$\|A\|_{\text{Lip}} = \|A(x_0)\|_Y + \|A\| = \|A(x_0)\|_{Y_B} + \sup_{T \in [0,\infty]} \sup_{\substack{x,\tilde{x} \in D \\ x_T \neq \tilde{x}_T}} \frac{\|[Ax]_T - [A\tilde{x}]_T\|_{Y_B}}{\|x_T - \tilde{x}_T\|_{U_B}} \tag{3.4}$$

for any fixed $x_0 \in D^e$.

Here, it follows that for any $T \in [0, \infty)$

$$\|[A(x)]_T - [A(\tilde{x})]_T\|_{Y_B} \leq \|A\| \|x_T - \tilde{x}_T\|_U \leq \|A\|_{\text{Lip}} \|x_T - \tilde{x}_T\|_{U_B}$$

and let $\text{Lip}(D^e)$ denote the family of nonlinear generalized Lipschitz operators that map D^e to itself.

Remark 3.1 The family of standard Lipschitz operators and the family of generalized Lipschitz operators are not comparable since they have different domains and ranges. However, it can be easily verified that when the extended linear spaces become standard with subscript T dropped, generalized Lipschitz operators become standard ones. It can also be verified that many standard Lipschitz operators are also extended Lipschitz. In Section 3.2, we assume that $U_s = U^e$ and $Y_s = Y^e$.

Let U and Y be linear spaces over the field of complex numbers. The normed linear subspaces U_s and Y_s are also called the stable subspaces of linear spaces U and Y, respectively. Let $Q : U \to Y$ be an operator. All defined operators are causal but not necessarily linear or bounded (with a finite operator norm). We always assume that $\mathcal{D}(Q) = U$ with $\mathcal{R}(Q) \subseteq Y$. In this chapter, an operator $Q : U \to Y$ is said to be BIBO stable or simply stable if $Q(U_s) \subseteq Y_s$.

For brevity, we omit the definition of internal stabilization given in [3]. We can use the definition of right coprime factorization in Section 2.2.1 and denote this as rcf.

In the following, the robust rcf condition in Lemma 2.7 is explained. Consider the nonlinear feedback system shown in Figure 3.1, which is assumed to be well-posed.

Let the nominal plant and the plant perturbation be P and ΔP, respectively. The overall plant $\tilde{P}$ is given as

$$\tilde{P} = P + \Delta P \tag{3.5}$$

where $\tilde{P}$ and P are nonlinear and unstable operators. The rcf's of the nominal plant P and $P + \Delta P$ are

$$P = ND^{-1} \qquad P + \Delta P = (N + \Delta N)D^{-1} \tag{3.6}$$

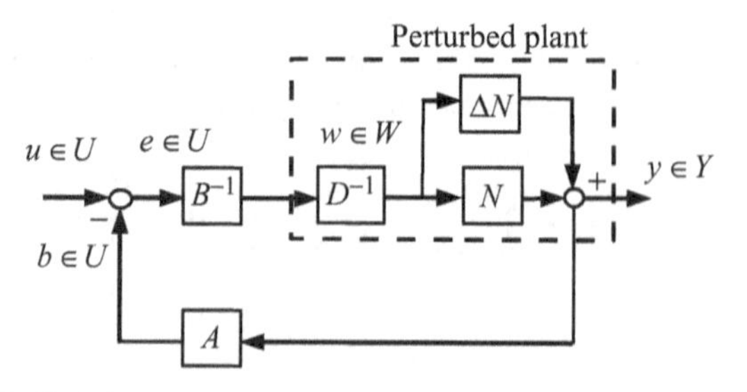

FIGURE 3.1 Nonlinear feedback system with perturbation.

where N, ΔN, and D are stable operators and D is invertible. We assume that ΔN is unkown but the upper and lower bounds of ΔN are known.

Let the input space, output space, and quasi-state space be U, Y, and W, respectively, and N, ΔN, and D be $N : W \to Y$, $\Delta N : W \to Y$, and $D : W \to U$, respectively. Assume A, B are the controller and stable operators and B is invertible. We can choose $W = U$, meaning U and W are the same linear space.

Then, we can get the Bezout identity,

$$AN + BD = M \quad \text{for some } M \in \mathcal{U}(W, U) \tag{3.7}$$

where $\mathcal{U}(W, U)$ is the set of unimodular operators. So, when M is a unimodular operators, it is said that operators A, N, B, D satisfy the Bezout identity. The operator M of the Bezout identity is equal to the operator $M : W \to U$ of the overall system. If M is a unimodular operator, the operator $M^{-1} : U \to W$ of the overall system is stable, and N is a stable operator because of rcf. Finally, the system is internally stable because all signals are bounded by the fact that input signal $u \in U$. Here, $N + \Delta N$ is a nonlinear and stable operator, but the stability of the feedback system shown in Figure 3.1 is unknown. We provide the following condition. The equation of the system with the perturbation ΔN is

$$A(N + \Delta N) + BD = M \qquad M \text{ is unimodular} \tag{3.8}$$

When N, D, A, and B satisfy (3.8) and (3.7) is the Bezout identity of the nominal plant P, the system shown in Figure 3.1 is stable [3]. It means $\mathcal{R}(\Delta N)$, the range of ΔN, is included in $\mathcal{N}(A)$, the null set of A, where ΔP is the perturbation of the plant which can represent only ΔN. The reason is that ΔP is an additive uncertainty. However, it is difficult to check (3.8) if ΔN is unknown. Also, in some cases, (3.8) is not satisfied.

In the following, consider the nonlinear system in Figure 3.2. Recalling Lemma 2.7 we can guarantee the stability of the nonlinear feedback control system with perturbation. We call the following a robust condition:

$$\|[A(N + \Delta N) - AN]M^{-1}\| < 1 \tag{3.9}$$

Then the system shown in Figure 3.1 is robust stable, where $\|\cdot\|$ is defined in (3.3).

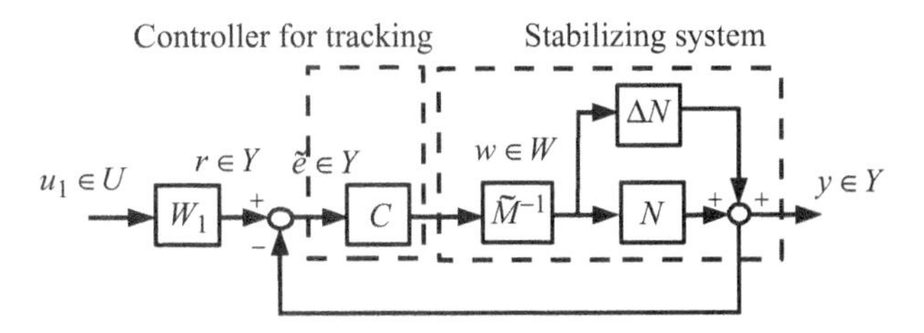

FIGURE 3.2 Nonlinear feedback system for tracking.

The main difference between the above condition and the condition in [3] is that the one above in an inequality and the one in [3] is $A(N + \Delta N) - AN = 0$ for $A(N + \Delta N) + BD = AN + BD$. This shows that the one above includes more sets for designing controllers. That is, (3.9) includes the condition $A(N + \Delta N) - AN = 0$. Also, if $\|[A(N + \Delta N) - AN]M^{-1}\|$ of (3.9) can be obtained by using bounded information of ΔN, the detailed ΔN is not necessary. It is also worth mentioning that the initial state of the system should be considered, that is, $AN(\omega_0, t_0) + BD(\omega_0, t_0) = M(\omega_0, t_0)$ should be satisfied. In the following, plant output tracking performance will be considered.

3.2.2 Tracking Control Design Scheme

It has been shown that the system in Figure 3.1 is stable, but we have not considered the plant output tracking performance yet. In this section, we discuss the plant output tracking problems for the stabilizing system described in Figure 3.1, where we assume that the spaces of the nonlinear plant output and reference input are different, namely, $U \neq Y$. First, a space change operator is designed. Next, we consider a tracking controller based on the exponential iteration theorem.

Consider the nonlinear feedback system shown in Figure 3.1. We design a tracking control system as given in Figure 3.2. The stabilizing system in Figure 3.2 is the same as the system in Figure 3.1 stabilized by Lemma 2.7. Here, $u_1 \in U$ is the reference input, W_1 is the space change operator to transform the reference input signal $u_1 \in U$ into the real reference input signal $r \in Y$, and C is the designed tracking controller.

First, we design a space change operator W_1 for the real reference input signal r in space Y so that one of the conditions of the exponential iteration theorem is satisfied. That is, the spaces of r and y are the same. In general, if $r \neq u_1$, W_1 is designed such that $W_1(u_1) - u_1$ can be made arbitrarily small by making T large enough. Figure 3.2 can be further reexpressed in the form of Figure 3.3.

The error signal $\tilde{e}$ is given as

$$\tilde{e} = (I + \tilde{P}C)^{-1}(r) \tag{3.10}$$

It is obvious that the operator $(I + \tilde{P}C)^{-1}$ is mapping Y to Y from Figure 3.2. Hence, the reference signal r and the error signal $\tilde{e}$ are in linear space. Then, one of the conditions of the exponential iteration theorem is also satisfied, namely, the spaces of $\tilde{e}$ and y are the same.

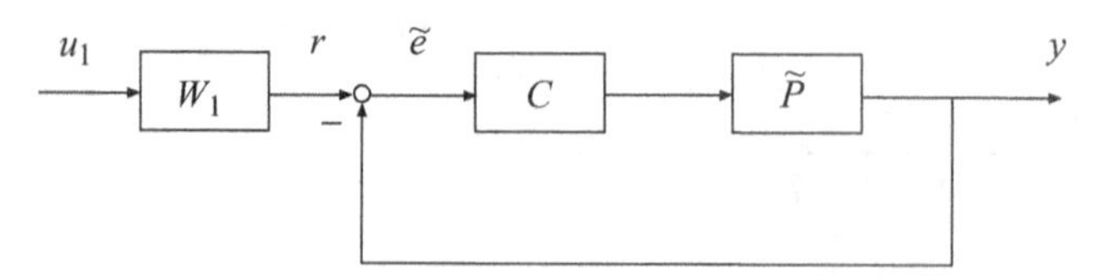

FIGURE 3.3 Equivalent block diagram of Figure 3.2.

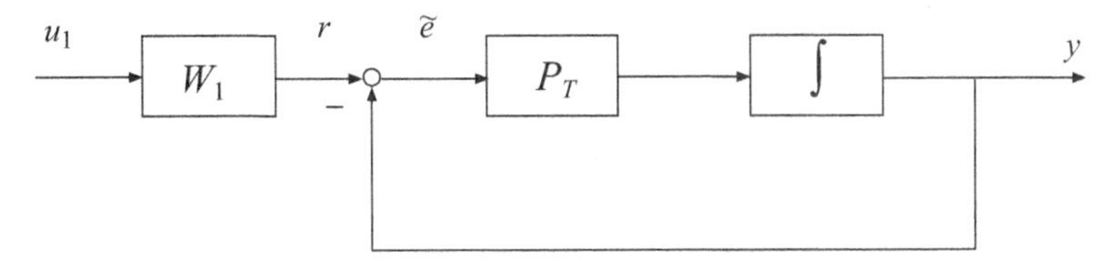

FIGURE 3.4 Equivalent block diagram of Figure 3.3.

Next, the controller C is designed so that the open loop $\tilde{P}C$ of the feedback system in Figure 3.3 consists of an integrator in cascade with a system P_T (Figure 3.4) and satisfies the following conditions:

1. For all t in $[0, T]$, C is stable, and $P_T(r) \geq K_1 > 0$ as $T \geq t \geq t_1 \geq 0, r > 0$;
2. $\tilde{P}C(0) = 0$;
3. $\|\tilde{P}C(x) - \tilde{P}C(y), t\| \leq h \int_0^t \|x - y, t_1\| \, dt_1$ for all x, y in Y_s and for all t in $[0, T]$, where h is any constant and is the gain of P_T in the first norm and the norm of x restricted to any interval $[0, T]$ will be denoted by $\|x, t\|$.

In this session, the gain of P_T is the generalized Lipschitz operator norm defined in Definition 3.2. Since C and $\tilde{P}$ are stable, the existence of h is ensured. Defining an operator from r to y as $\tilde{G}$, we have $\tilde{G} = PC * (I - \tilde{G})$ as the feedback equation, where the cascade $PC * (I - \tilde{G})$ means the operator PC after the operator $I - \tilde{G}$. Then we summarized the exponential iteration theorem in Lemma 3.1.

Lemma 3.1 *(Exponential Iteration Theorem [22])* The feedback equation $\tilde{G} = PC * (I - \tilde{G})$, in which all operators map the Banach space Y_B into itself, has a unique solution for $\tilde{G}$ which converges uniformly on $[0, T]$ provided that conditions 2 and 3 are satisfied. The plant output is bounded.

Lemma 3.1 means that, since Y_s is complete, the sequence is uniformly convergent on $[0, T]$. It may be established that $\tilde{G} - PC * (I - \tilde{G}) = 0$ and it is unique. Then the plant output is bounded [22]. Further, $(I + \tilde{P}C)^{-1}(r)(t)$ exists.

Theorem 3.1 The error signal $\tilde{e}$ with the controller C can be made arbitrarily small. That is, $y(t) - r(t)$ can be made arbitrarily small by $t \leq T$ large enough.

Proof From Figures 3.3 and 3.4, we have

$$y(t) = r(t) - \tilde{e} \tag{3.11}$$

where

$$\tilde{P}C = \int_0^t P_T \, dt_2 \tag{3.12}$$

From (3.10) and (3.11), we have

$$y(t) = r(t) - (I + \tilde{P}C)^{-1}(r)(t) \tag{3.13}$$

Since I is the identity operator, namely, $I(r) = r$ [3,22], from (3.12)

$$y(t) = r(t) - [r(t) + \tilde{P}C(r(t))]^{-1} = r(t) - \left(r(t) + \int_0^t P_T[r(t)]\,dt\right)^{-1}$$

Considering condition 1 of the controller design, namely, $P_T(r) \geq K_1 > 0$ as $T \geq t \geq t_1 \geq 0$, we obtain

$$\int_0^t P_T[r(t)]\,dt_2 \geq \int_0^{t_1} P_T[r(t)]\,dt_2 + K_1 \int_{t_1}^t dt_2 \tag{3.14}$$

where $K_1 \int_{t_1}^t dt_2$ can be made arbitrarily large by making $t \leq T$ large enough. Then, $\{r(t) + \int_0^t P_T[r(t)]\,dt_2\}^{-1}$ becomes arbitrarily small. From (3.14), $y(t) - r(t)$ can be made arbitrarily small. This leads to the desired result. ■

It is noted that when the spaces of plant output and reference input are the same, instead of a space change filter, a linear filter is required and the tracking controller shown in this session still works. Also, the method in [21] is difficult to use for the case in this chapter, because the method requires a detailed information of ΔN.

We now illustrate the design scheme developed and analyzed above through several simulations.

Let the given plant operators $P : U \to Y$ and $\hat{P} : U \to Y$[3] be defined by

$$P(u)(t) = \int_0^t u^{\frac{1}{3}}(\tau)\,d\tau + e^{\frac{t}{3}} u^{\frac{1}{3}}(t) \tag{3.15}$$

$$\begin{aligned}\hat{P}(u)(t) &= (1 + \Delta_1) \int_0^t u^{\frac{1}{3}}(\tau)\,d\tau + (e^{\frac{t}{3}} + \Delta_2) u^{\frac{1}{3}}(t) \\ &= [P + \Delta P](u)(t)\end{aligned} \tag{3.16}$$

The rcf's N, D, and ΔN are given as

$$N(w)(t) = \int_0^t e^{-\frac{\tau}{3}} w^{\frac{1}{3}}(\tau)\,d\tau + w^{\frac{1}{3}}(t) \tag{3.17}$$

$$D(\omega)(t) = e^{-t} w(t) \tag{3.18}$$

$$\Delta N(\omega)(t) = \Delta_1 \int_0^t e^{-\frac{\tau}{3}} w^{\frac{1}{3}}(\tau)\,d\tau + \Delta_2 e^{-\frac{t}{3}} w^{\frac{1}{3}}(t) \tag{3.19}$$

where Δ_1, Δ_2 are random values:

$$\Delta_1 \leq 0.1 \qquad \Delta_2 \leq 0.1 \tag{3.20}$$

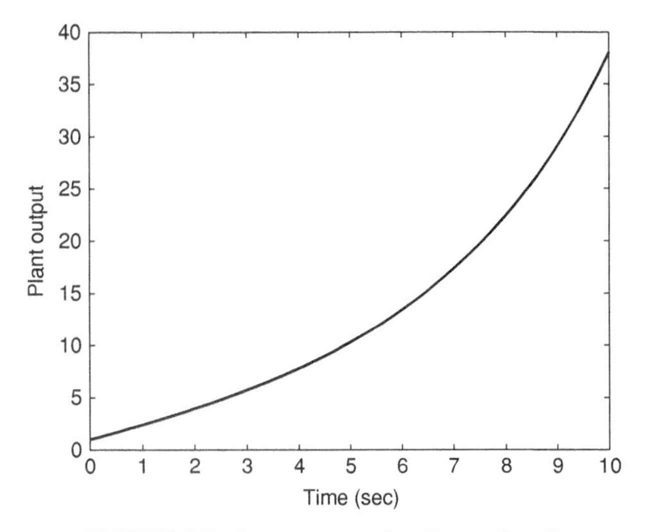

FIGURE 3.5 Step response of nonlinear plant P.

Here we assume that the random values are unknown but the upper and lower bounds of the values are known. We show the step response of the plant operator P in Figure 3.5. It is clear that the plant operator P is unstable and $P + \Delta P$ is as well.

Based on the design scheme in Section 3.2.1, we design controllers A, B as follows:

$$A(y)(t) = [(e^{\frac{t}{3}} + \Delta_2)^3 - (1 + \Delta_1)^3][h(t)]^3 \tag{3.21}$$

$$B(u)(t) = I(u)(t) \tag{3.22}$$

where $h(t) = e^{-\frac{t}{3}} w^{\frac{1}{3}}(t)$ and $I(\cdot)$ is an identity operator on U. We can see A, B to satisfy the Bezout identity for the nominal plant defined in (3.7). That is,

$$AN + BD(w) = \begin{cases} [(e^{\frac{t}{3}})^3 - (1)^3][e^{-\frac{t}{3}} w^{\frac{1}{3}}(t)]^3 + e^{-t} w(t) & (3.23) \\ [e^t - 1][e^{-t} w(t)] + e^{-t} w(t) & (3.24) \\ w(t) & (3.25) \\ I(w)(t) & (3.26) \end{cases}$$

Define Δ_s as

$$\Delta_s = \begin{cases} \tilde{M} - M & (3.27) \\ [A(N + \Delta N)](w)(t) - AN(w)(t) & (3.28) \\ [(e^{\frac{t}{3}} + \Delta_2)^3 - e^t - (1 + \Delta_1)^3 + 1][h(t)]^3 & (3.29) \\ [3\Delta_1 + 3(\Delta_1)^2 + (\Delta_1)^3 - 3e^{\frac{2}{3}t}\Delta_2 - 3e^{\frac{t}{3}}(\Delta_2)^2 - (\Delta_2)^3][e^{-t} w(t)] & (3.30) \end{cases}$$

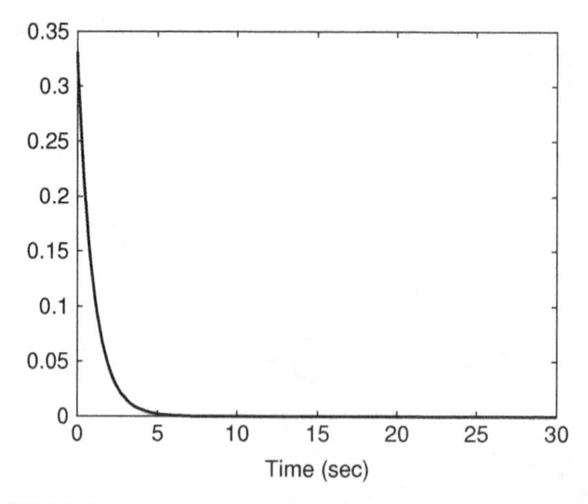

FIGURE 3.6 Time response for verification of $\|\Delta_s w^{-1}(t)\| < 1$.

Then, by using the method in [3], the system can be stabilized if and only if

$$\Delta_s = 0 \tag{3.31}$$

On the other hand, the condition for satisfying Lemma 2.7 is given as

$$\|\Delta_s w^{-1}(t)\| < 1 \tag{3.32}$$

In the simulation, selecting $\Delta_1 = 0.1$ rand(r_1) and $\Delta_2 = 0.1$ rand(r_2), the condition in (3.32) space (see Figure 3.6) is satisfied, where the reference input signal $u = 1$ and rand(r_i) is a function to generate random noise with the initial value r_i. In the calculation, we use the bounded information of 0.1 rand(r_i). Figure 3.7 shows the

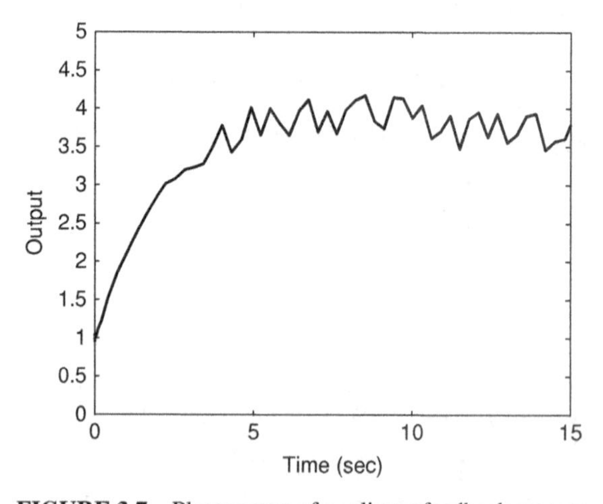

FIGURE 3.7 Plant output of nonlinear feedback system.

nonlinear plant output. From Figure 3.7, the system maintains its stability because the plant output signal is bounded by a reference input u_1 but tracking performance is unsatisfactory.

In this section, two examples are given to show the plant output tracking performance by using the presented design scheme.

First, the nominal plant P without perturbation is considered. Recall the nominal plant in (3.15),

$$P(u)(t) = \int_0^t u^{\frac{1}{3}}(\tau)\,d\tau + e^{\frac{t}{3}}u^{\frac{1}{3}}(t)$$

The controllers A, B for stabilizing the system are chosen by the method discussed in this section. The operator M^{-1} can be set as the identity operator since the Bezout identity $AN + BD = I$ in the above section. So, the system $P(u)(t)$ for the tracking problem is equivalent to Figure 3.2 for $\Delta N = 0$. Recalling the operator N,

$$N(w)(t) = \int_0^t e^{-\frac{\tau}{3}}w^{\frac{1}{3}}(\tau)\,d\tau + w^{\frac{1}{3}}(t)$$

it is obvious that N is a nonlinear operator. Based on the design condition given in Section 3.2.2, the controller C is defined as

$$C(\tilde{e}) = k\left[\int_0^t \tilde{e}(\tau)\,d\tau\right]^3 \tag{3.33}$$

where k is a constant. It is obvious that the open loop $\tilde{P}C$ of the feedback system consists of an integrator. That is, from (3.14) and (3.33), we have

$$\begin{aligned} y(t) &= r(t) - \left(r(t) + \int_0^t e^{-\frac{\tau}{3}}C^{\frac{1}{3}}(r(\tau))\,d\tau + C^{\frac{1}{3}}[r(t)]\right)^{-1} \\ &= r(t) - \left(r(t) + 4k^{\frac{1}{3}}\int_0^t r(\tau)\,d\tau - 3k^{\frac{1}{3}}\int_0^t e^{-\frac{\tau}{3}}r(\tau)\,d\tau\right)^{-1} \end{aligned}$$

Meanwhile, the space change filter W_1 is defined as

$$W_1 = \frac{1}{1.03}\left[0.01\int_0^t e^{-\frac{\tau}{3}}u_1^{\frac{1}{3}}(\tau)\,d\tau + u_1^{\frac{1}{3}}(t)\right] \tag{3.34}$$

where $u_1 = 1$ and $W_1(u_1) - u_1$ can be made arbitrarily small by making $t \le T$ large enough. Then, we have $r = W_1(u_1) = \frac{1}{1.03}(1.03 - 0.03e^{-\frac{t}{3}})$. When $t \le T$ is large enough, $r(t) \to 1$, we obtain $y(t) \to r(t) - [r(t) + 4k^{\frac{1}{3}}\int_0^t d\tau - 9k^{\frac{1}{3}}(1 - e^{-\frac{t}{3}})]^{-1} \to 1$ based on Theorem 3.1.

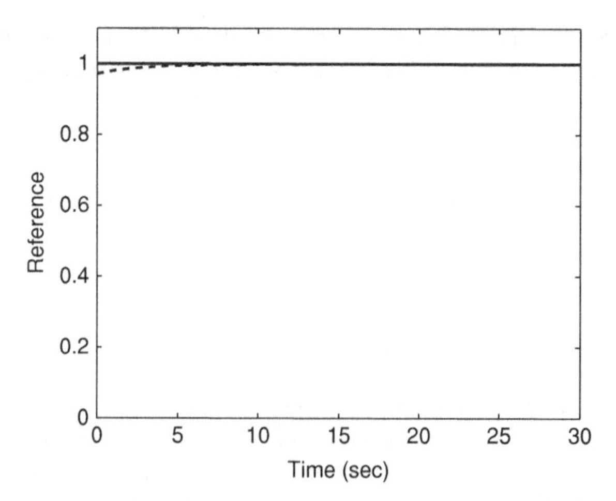

FIGURE 3.8 Simulated results of reference input signal u and real reference signal r.

The simulation results are shown in Figures 3.8–3.10 ($k = 1$). Figure 3.8 shows the result of the real reference signal r (solid line) and the reference input signal u_1 (dashed line). The signal r is the output signal of the regulator W_1. It is shown that r is probably equal to the input signal u, but the spaces of each signal are different, namely, $u \in U$, $r \in Y$. Figure 3.9 shows the result of the output signal y (solid line) and the real reference signal r (dashed line). It can be seen that the output signal y of the system tracks the reference signal r when t is large enough. In Figure 3.10, the error signal $\tilde{e}$ converges to zero. The design parameters are the same in the above simulation, but $k = \frac{1}{900}$. Figure 3.11 gives the results.

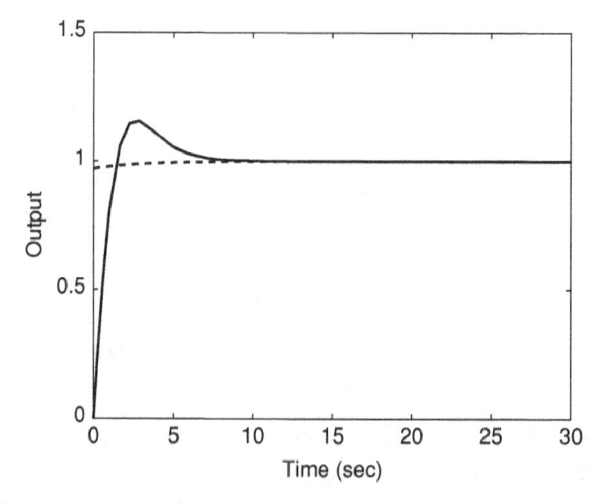

FIGURE 3.9 Simulated results of real reference signal r and plant output signal y.

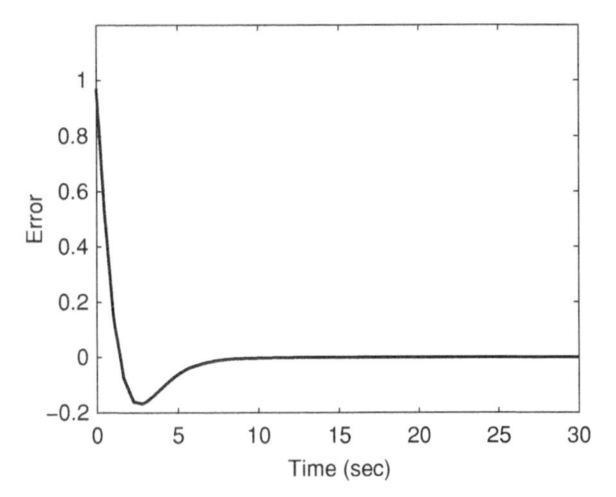

FIGURE 3.10 Time response of error signal $\tilde{e}$.

Comparing Figure 3.11 with Figure 3.9, there is no overshoot in Figure 3.11 but the response characteristic of Figure 3.9 is better. Hence the output response of the plant can be changed by choosing the controller gain k, but the effect is determined by the relation of the spaces U and Y.

Next, we consider the plant $\hat{P}$ with perturbations:

$$\hat{P}(u)(t) = (1 + \Delta_1) \int_0^t u^{\frac{1}{3}}(\tau)\, d\tau + (e^{\frac{t}{3}} + \Delta_2) u^{\frac{1}{3}}(t)$$

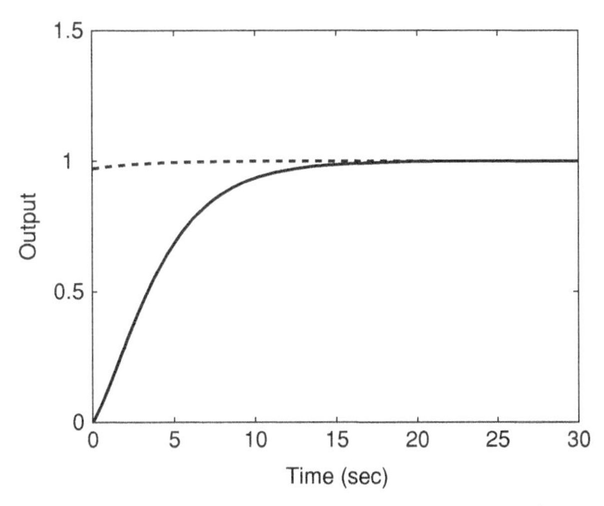

FIGURE 3.11 Simulated results of r and y $\left(k = \frac{1}{900}\right)$.

The plant $\hat{P}$ has a coprime factorization $\hat{P} = (N + \Delta N)D^{-1}$, where N, ΔN, D are given as

$$N(w)(t) = \int_0^t e^{-\frac{\tau}{3}} w^{\frac{1}{3}}(\tau)\, d\tau + w^{\frac{1}{3}}(t)$$

$$\Delta N(\omega)(t) = \Delta_1 \int_0^t e^{-\frac{\tau}{3}} w^{\frac{1}{3}}(\tau)\, d\tau + \Delta_2 e^{-\frac{t}{3}} w^{\frac{1}{3}}(t)$$

$$D(\omega)(t) = e^{-t} w(t)$$

where Δ_1 and Δ_2 are bounded and unknown perturbations. Consider the system in Figure 3.2. The controller C and the regulator W_1 are the same as in the above example,

$$C(\tilde{e}) = k \left[\int_0^t \tilde{e}(\tau)\, d\tau \right]^3$$

$$W_1 = \frac{1}{1.03} \left[0.01 \int_0^t e^{-\frac{\tau}{3}} u^{\frac{1}{3}}(\tau)\, d\tau + u^{\frac{1}{3}}(t) \right]$$

but Δ_1 and Δ_2 are chosen as in Section 5.1. The robust condition in (3.9) is satisfied. In the simulation, for tracking analysis, the perturbation of ΔN is shown in Figure 3.12. The plant output signal y and the real reference signal r are shown in Figure 3.13. From Figure 3.12, there is a large perturbation when the time is 15 sec. Then the output response has a little variation, but the plant output tracks the real reference signal quickly (Figure 3.13). This example does not show a sufficient tracking performance when perturbations vary quickly because of the effect of the integration in the loop. However, the desired tracking result has been obtained when t is large enough.

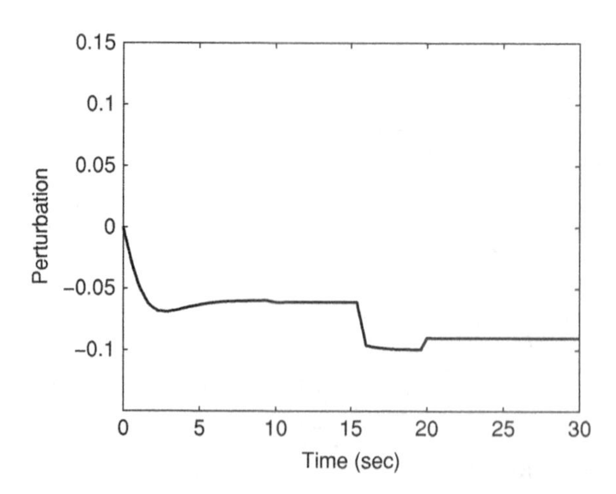

FIGURE 3.12 Perturbation ΔN of the plant.

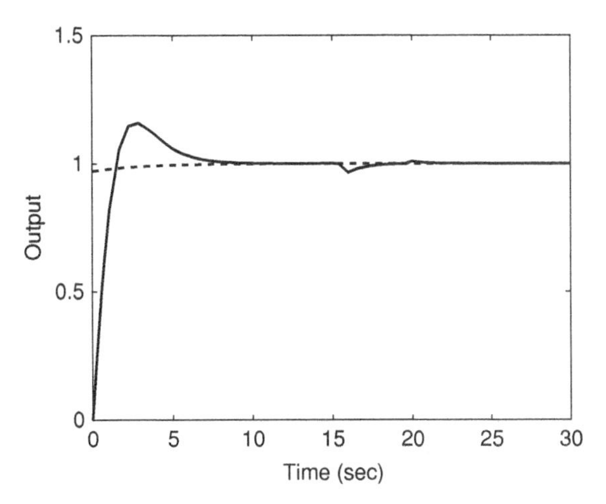

FIGURE 3.13 Simulation result of plant output signal y and real reference signal r for plant with perturbation.

3.3 OPERATOR-BASED ROBUST ANTI-WINDUP NONLINEAR FEEDBACK CONTROL SYSTEMS DESIGN

3.3.1 Introduction

In applying the control scheme to real plants in industry, the control systems must deal with some constraints in variables such as the pressure and temperature limit, and the control system must also avoid the unsafe operating regimes in [23]. In particular, constraints on the input variables of unstable plants and the unsafe problem of the controller are crucial in the control of unstable plants. So far, the input constraints of continuous-time unstable plants have been mainly taken into account in three ways. In the first case, for sinusoidal reference signals, stable setpoint tracking for linear open-loop unstable plants with saturated control input was studied in [24]. In the second case, the anti-windup controller design problem for exponentially unstable linear plants with input saturation was discussed in [25]. The third method is concerned with the anti-windup for the local stability of unstable plants with input saturation in [26]; namely, the controller should be unstable.

Section 3.3 considers a design problem of continuous-time unstable plants with input constraints and tracking characteristics from different viewpoint. Unlike the above methods, the features of this method does not need to consider the reference signal limitation and plants that are exponentially unstable. Since the stable controller of a system is necessary for systems with control input constraints to work as open and closed loops (see [27]), a stable feedback controller is designed using the coprime factorization approach. In coprime factorization, design control system techniques have been studied for decades, as can be seen in [28, 29] for linear systems and [17, 51] and Chapter 2 for nonlinear systems. The link between factorization theory and control problems is made by Youla–Kucera parameterization. Especially, coprime factorization for nonlinear control systems has been a promising approach for nonlinear systems analysis, design, and stabilization. Currently, more studies are needed of unstable plants with input constraints for which traditional control methods are insufficient. To address this, the research addressed here fuses continuous-time generalized predictive control and operator-based robust right coprime factorization to include input–output stability properties, namely overall stability (Lemma 2.5) and the desired output tracking performance.

The design scheme has three features, the first one is for unstable plants, the second one is for plants with input constraints, and the third one is for improved tracking performance. Corresponding to these features, the design scheme consists of three steps:

Step 1: To make an unstable plant a stable one, right coprime factorization of the plant is considered and a stable feedback controller is designed using continuous-time generalized predictive control (CGPC) (see [30, 31]) for the coprime factorization of the pole of an unstable plant. The stable controller ensures that the closed-loop system of the pole coprime factorization is stable.

Step 2: To design a strongly stable controller for a plant with input constraints, the control system should be strongly stable, that is, the system is stable in both of open and closed loops (see [31]), where open-loop stability is guaranteed provided that the closed-loop system of the pole coprime factorization is stable and closed-loop stability is guaranteed using a result on robust stability of the perturbed nonlinear plant in Chapter 2.

Step 3: A tracking operator is designed to obtain the desired output performance.

The following notation is used in this section: When A is a function of $*$, $A[*]$ means a polynomial function of $*$, whereas $A(*)$ is a rational function of $*$. When $A : B \to C$ is an operator, $A[*]$ means that it is causal, but not necessarily linear or bounded, the domain $\bar{D}(A) = B$ having the range $\bar{R}(A) \subseteq C$.

3.3.2 Design Description

Consider a single-input–single-output time-invariant linear plant described by the transfer function

$$P(p) = \frac{B[p]}{A[p]} \tag{3.35}$$

where $A[p]$ are $B[p]$ are polynomials in the differential operator p. Assume $A[p]$, $B[p]$ are nth unstable and mth stable coprime polynomials, respectively. The control input $u(t)$ of the plant is subject to the constraints

$$u(t) = \sigma[u_1(t)] \tag{3.36}$$

$$\sigma(v) = \begin{cases} u_{\max} & \text{if } v > u_{\max} \geq 0 \\ v & \text{if } u_{\min} \leq v \leq u_{\max} \\ u_{\min} & \text{if } v < u_{\min} \leq 0 \end{cases} \tag{3.37}$$

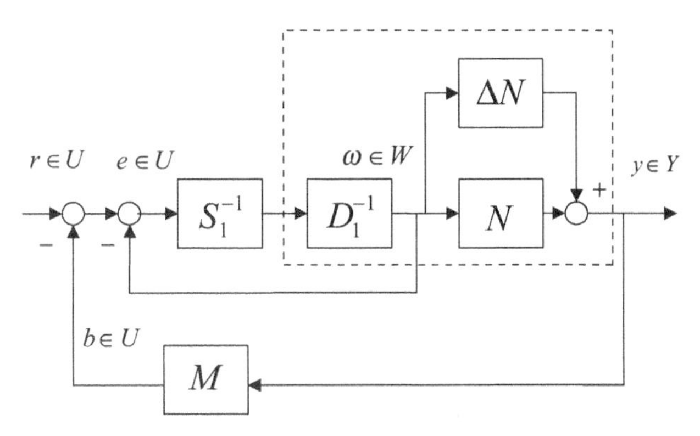

FIGURE 3.14 Control system.

where, $B_\sigma[p]u_1(t) = B[p]\sigma[u_1(t)]$, namely, $B_\sigma[p]$ is with input constraints. The coprime, factorization presentation of the plant (3.35) can be chosen as follows (Figure 3.14):

$$\begin{aligned} P(p) &= \frac{N(p)}{D_1(p)} \qquad N(p) = \frac{B[p]}{D_d[p]} \in RH_\infty \qquad D_1(p) \in RH_\infty \\ D_1(p) &= \frac{A[p]}{D_d[p]} \qquad N(p) + \Delta N = \frac{B_\sigma[p]}{D_d[p]} \end{aligned} \tag{3.38}$$

where, $D_d[p](\deg[D_d[p]] = n)$ is stable and is selected by the designer. The effect of $\sigma(t)$ is considered as ΔN.

The objective in this session is to design a stable robust control system for the above unstable plant with input constraints and to consider the plant output performance.

The controller for ensuring robust stability is given by using Youla–Kucera parametrization and robust right coprime factorization (Figures 3.14 and 3.15), where r is the reference input. This is related to Steps 1 and 2 given in Section 3.3.1. The

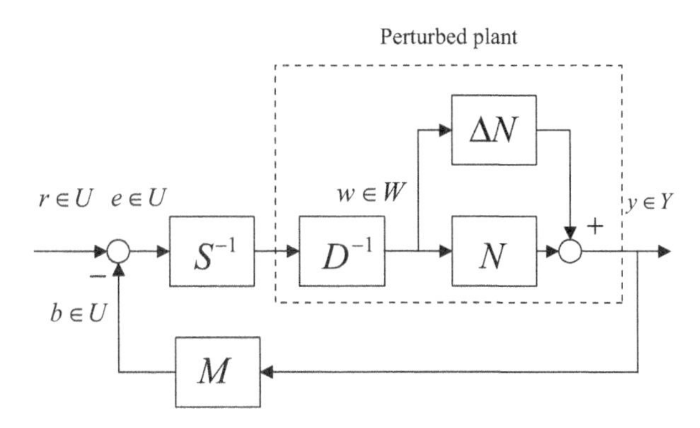

FIGURE 3.15 Equivalent diagram of control system.

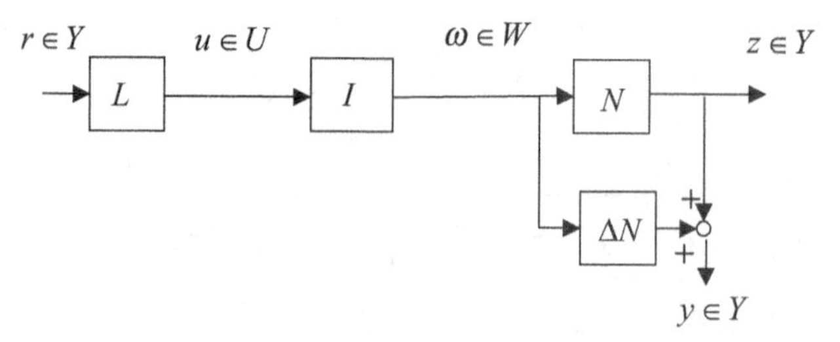

FIGURE 3.16 Equivalent diagram of Figure 3.15 with tracking operator.

plant output tracking performance is discussed using a tracking operator-based design method (Figure 3.16). This is related to Step 3. In the following, we summarize the three steps.

From (3.38), we know that D_1^{-1} is unstable. For the unstable part, stable CGPC is designed using the result in [31]. Using the result, S_1 is ensured to be always stable and the closed loop of $S_1^{-1}D_1^{-1}$ is stable (Figure 3.14), namely, $S_1(p)$ is a strong stable CGPC controller for $D_1(p)$,

$$S_1(p) = X(p) + Q(p)\bar{D}(p) \tag{3.39}$$

where $Q(p) \in RH_\infty$ is a design parameter for ensuring a stable controller $S_1(p)$. Introducing two design polynomials $u_n[p]$, $u_d[p]$ for $Q(p)$, we have

$$Q(p) = \frac{u_n[p]}{u_d[p]} \tag{3.40}$$

Assume $X(p) \in RH_\infty$ satisfies the Bezout identity

$$X(p)\bar{N}(p) + Y(p)\bar{D}(p) = 1 \tag{3.41}$$

where the coprime factorization presentation $\bar{N}(p)$ and $\bar{D}(p)$ of $D_1(p)$ can be chosen as

$$\bar{N}(p) = \frac{A[p]}{T_0[p]} \qquad \bar{D}(p) = \frac{D_d[p]}{T_0[p]} \tag{3.42}$$

where $T_0[p] = D_d[p] + L_0[p] + gA[p]$ is a stable closed-loop characteristic polynomial of unit output feedback of $S_1^{-1}D_1^{-1}$. Selecting the stable polynomial $C[p]$ so that $\deg[C[p]] = n - 1$, we obtain

$$X(p) = \frac{gC[p] + F_0[p]}{C[p]} \qquad Y(p) = \frac{C[p] + G_0[p]}{C(p)} \tag{3.43}$$

Using the results in [30, 31], $F_0(p)$, $G_0(p)$, $L_0(p)$, T_0, and g can be obtained.

Proof Consider the constraint of $\sigma(t)$ described by ΔN from Figure 3.14. The predicted output $\bar{\zeta}(t)$ (between S_1^{-1} ans D_1^{-1}) of $D_1(p)$ should belong to $[u_{\min}, u_{\max}]$. The approximation $\bar{\zeta}^*(t+T)$ of the predicted output at $t+T$ can be given by writing the appropriate Maclaurin expansion of $\bar{\zeta}(t+T)$ about t and truncating this after $N_{\bar{\zeta}}$ terms, where $N_{\bar{\zeta}}$ is predictor order. Further, we can replace the values of the derivatives of $\bar{\zeta}(t)$ by their emulated values $\bar{\zeta}_k^*(t)$. Multiplying $D_1(p)$ by s^k and decomposing $C[p]/D_d[p]$ and $A[p]E_k[p]/C[p]$ into their strict proper part and remainder, we obtain

$$\frac{s^k C[p]}{D_d[p]} = \frac{F_k[p]}{D_d[p]} + E_k[p] \qquad \frac{A[p]E_k[p]}{C[p]} = \frac{G_k[p]}{C[p]} + H_k[p] \tag{3.44}$$

where the orders of $F_k[p]$, $E_k[p]$, $G_k[p]$, and $H_k(p)$ are $n-1, k-1, n-2$, and $k-\rho(\rho = n-m)$. Then

$$k = \begin{bmatrix} k_0 \; k_1 \; \cdots \; k_{N_{\bar{\zeta}}} \end{bmatrix} r \tag{3.45}$$

$$G_0[p] = \sum_{i=1}^{N_{\bar{\zeta}}} k_i G_i[p], \; F_0[p] = \sum_{i=1}^{N_{\bar{\zeta}}} k_i F_i[p] \tag{3.46}$$

where

$$L_0[p] = \sum_{i=1}^{N_{\bar{\zeta}}} k_i L_i[p] \qquad \frac{s^k D_d[p]}{D_d[p]} = H_k[p] + \frac{L_k[p]}{D_d[p]} \tag{3.47}$$

The closed-loop transfer function $G_{ew}(p)$ can be obtained as (Figure 3.15)

$$G_{ew}(p) = \frac{gA[p]}{D_d[p] + L_0[p] + gA[p]} \tag{3.48}$$

We can select λ so that $T_0[p] = D_d[p] + L_0[p] + gA[p]$ is stable. ■

In the design, the output signal $w(t)$ of $D_1^{-1}(p)$ needs to be observable. Then $w(t)$ can be obtained as (see [1,32])

$$w(t) = \bar{X}(p)u_1(t) + \bar{Y}(p)y(t) \tag{3.49}$$

where $\bar{X}(p), \bar{Y}(p) \in RH_\infty$ satisfy the Bezout identity $\bar{X}(p)[N + \Delta N(p)] + \bar{Y}(p)D_1(p) = 1$. That is, the observer can be designed using the dynamic right factorization approach given in [32].

In the following, we consider the effect of input constraints. That is, the design of the nonlinear operator $M(y)$ in Figure 3.14 will be shown. Based on the robust right coprime factorization in Section 2.2.2, the nonlinear operator-based controller is designed according to the coprime factorization presentation of $N + \Delta N$. The nonlinear system depicted in Figure 3.14 can be regarded as a system with perturbation depicted in Figure 3.15, and $S^{-1}D^{-1}$ is a transfer function of unit output feedback of $S_1^{-1}D_1^{-1}$. Then, we have that

$$SD = 1 + S_1 D_1 \tag{3.50}$$

where $1 + S_1 D_1$ is invertible and from (3.38) D_1 is stable. From Figure 3.15, for $N \to N + \Delta N$, the satisfying design scheme for robust stabilization was given in [3]. That is, there exists a stable operator M satisfying the perturbed Bezout identity

$$M(N + \Delta N) + SD = MN + SD = I \tag{3.51}$$

where from (3.38) the operator $N + \Delta N$ is stable. Therefore, with the above presentation, the perturbed plant retains a robust right comprime factorization and it is robust stable. Then, we can design operator M so as to satisfy the following equations:

$$I(y) = \begin{cases} SD(y) + MN(u_1) & \text{if } u_{\min} \le u_1 \le u_{\max} \\ SD(y) + M(N + \Delta N)(\sigma) & \text{otherwise} \end{cases} \tag{3.52}$$

Note that, from (3.50), $S_1(p)$ must be stable so that S is stable, because a stable S is necessary to satisfy the Bezout identify in (3.51). That is, the stable controller shown in Section 3.3.2 is necessary.

The closed-loop system described in Figure 3.15 is robust stable in the presence of ΔN provided that $r(t)$ is bounded. Adding a tracking operator L for reference input $r(t)$, the nonlinear system depicted in Figure 3.15 can be regarded as the system with perturbation depicted in Figure 3.16. In this section, the objective is to design a tracking operator L for reference signal $r(t)$ such that plant output $y(t)$ tracks to the reference signal $r(t)$ under the perturbation of ΔN. Here, we assume that input signal u in Figure 3.16 is in a subset U^* of U and output y is in a subset Y^* of Y, where U^* and Y^* are the normed subspaces of U and Y, respectively, called the stable subspace in Section 3.1. Here, the linear input subspace $\bar{U}^*$ and the linear output subspace $\bar{Y}^*$ are designed based on the control objective. Namely, in some cases we can select that $\bar{Y}^*$ does not belong to $\bar{Y}$ for satisfying the desired tracking performance.

Lemma 3.2 Suppose that the following operator design condition is satisfied for the stable design operator L:

$$(N + \Delta N)L(r)(t) = I(r)(t) \tag{3.53}$$

where $r(t)$ is the reference input and $r \in \bar{Y}^*$. Then, the output $y(t)$ tracks to the reference input $r(t)$.

Proof From Figure 3.16, we have $u(t) = L(r)(t)$. Then, we can obtain $N(u)(t)$. From condition (3.53), we have

$$(N + \Delta N)L(r)(t) = I(r)(t) \tag{3.54}$$

This fact leads to the desired result under $r(t) \in \bar{Y}^*$, $y(t) \in \bar{Y}^*$. That is, $y(t) = I(r)(t) = r(t)$.

Based on Lemma 3.2, the detailed procedure of checking the design conditions is as follows. To apply the system to the reference signal $r(t) \in \bar{Y}^*$, we will design the operator $L(r)$ to satisfy

$$(N + \Delta N)L(r)(t) = I(r)(t) = r(t) \in \bar{Y}^* \tag{3.55}$$

Then we have $y(t) = r(t)$. Meanwhile, when the control constraints are not active, we use tracking operator $L(r)$ satisfying $NL(r)(t) = I(r)(t)$.

The merit of the design method described in this section is that the error signal $e(t)$ is not affected by the perturbed signal. That is, the perturbed signal $\Delta N w$ cannot be transmitted back to the error signal $e(t)$ (Figure 3.16). In practice, in this sense we can avoid the undesired influence from uncertainties, output disturbances, and feedback sensor error. Furthermore, when the lemma is not satisfied, we can design the operator $L(r)$ based on the steady-state tracking performance. ■

Remark 3.2 It is worth mentioning that the existing anti-windup methods (e.g., [27, 31]) can be utilized in the case of stable plants with input constraints. However, for unstable plants, when the input constraints are active, the closed loop may become unstable caused by the control limitation. Furthermore, other approaches, such as the state space approach, can also be employed if the state variable is observable (see [33]). The merit of this chapter lies in the fact that we used the concept of the operator, which is based on the input and output nature of the nonlinear plant.

3.3.3 Illustrative Examples

Next we present two examples to illustrate the efficacy of the design scheme for unstable plants with input constraints. The first example is for an unstable plant. The second is for an unstable plant with non–minimum phase.

The first simulation study is conducted using the following unstable plant:

$$A[p] = p^2 - p + 1 \qquad B[p] = p^2 + p + 1 \tag{3.56}$$

The design polynomial $C[p]$ is given as $C[p] = p + 0.45$. Three cases of input constraint are considered, namely, $u_{\max} = |u_{\min}| = 0.5, 1.0, 2.0$. The related design parameters of the method are shown in Table 3.1. From (3.43), we have

$$X(p) = \frac{0.1262p + 0.1437}{p + 0.45} \qquad Y(p) = \frac{p - 0.0245}{p + 0.45} \tag{3.57}$$

From (3.39), the stable controller is

$$S_1(p) = \frac{-0.1079p^3 - 0.2633p^2 - 0.2281p - 0.07447}{1.376p^3 + 1.321p^2 + 0.3658p + 0.106}$$

TABLE 3.1 Design Parameters for Example (3.56)

CGPC	Reference input $w_r = 1$; Predictor order $N_{\tilde{\xi}} = 6$ Control horizon $N_{\hat{u}} = 1$; Control weighting $\lambda = 0.1$ Minimum prediction horizon $T_1 = 0$; Maximum prediction horizon $T_2 = 10$ Design parameters $u_n = -0.25$; $u_d = 1.0$
Parameters	$T_0(p) : 1.1262s^2 + 1.4861s + 0.2647$ $D_d : s^2 + 2s + 1$

From (3.52),

$$M(y) = \begin{cases} \dfrac{0.054p^5 + 0.0777p^4 + 0.0364p^3 + 0.0548p^2 + 0.0768p + 0.0373}{1.3760p^5 + 2.6970p^4 + 3.0628p^3 + 1.7928p^2 + 0.4718p + 0.1060} & \text{if } y = N(u_1) \\ \dfrac{0.1079p^5 + 0.1553p^4 + 0.0728p^3 + 0.1096p^2 + 0.1536p + 0.0745}{1.3760p^5 + 2.6970p^4 + 3.0628p^3 + 1.7928p^2 + 0.4718p + 0.1060} & \text{otherwise} \end{cases}$$

and from (3.53) the tracking operator is

$$L = \frac{p^2 + 2p + 1}{p^2 + p + 1} \tag{3.58}$$

In the simulation, for simplicity we assume that $w(t)$ is known. In the case of $u_{\max} = |u_{\min}| = 2.0$, Figures 3.17–3.19 show the simulation results. It can be seen that the desired tracking performance is obtained for plant input $[u_1(t)]$ prior to the input constraint between the constraint ranges. The design parameters are the same as the above simulation, but $u_{\max} = |u_{\min}| = 1.0$. The simulation results are shown in Figures 3.20–3.22. It can be seen that the desired tracking performance in the above simulation is obtained when the transient of $u_1(t)$ exceeds the constraint ranges but

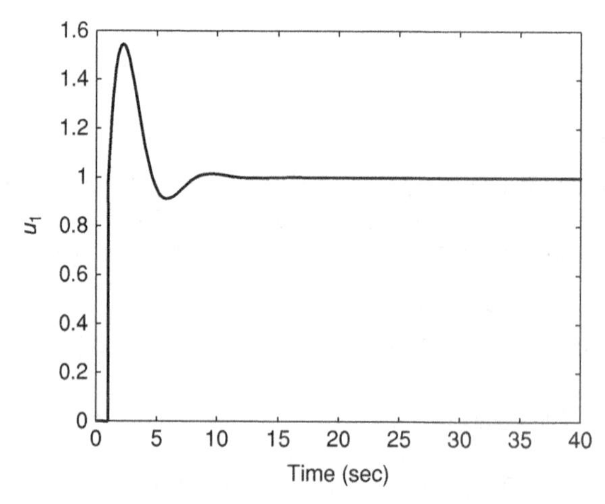

FIGURE 3.17 Plant input prior to input constraint ($u_{\max} = |u_{\min}| = 2.0$).

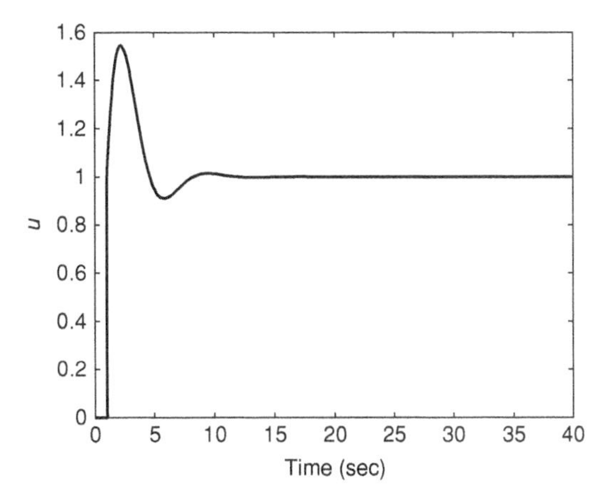

FIGURE 3.18 Plant input ($u_{\max} = |u_{\min}| = 2.0$).

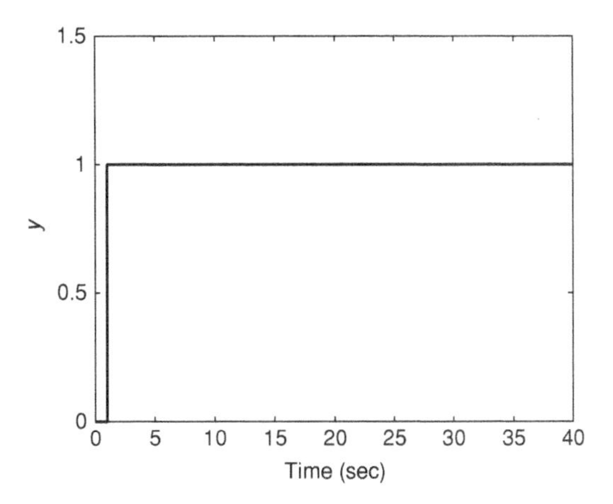

FIGURE 3.19 Plant output ($u_{\max} = |u_{\min}| = 2.0$).

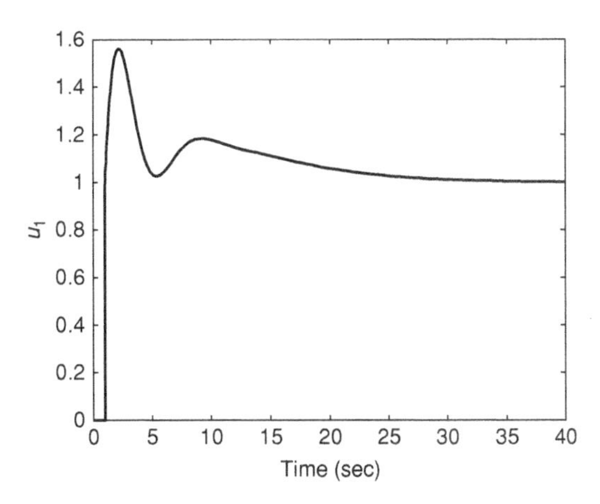

FIGURE 3.20 Plant input prior to input constraint ($u_{\max} = |u_{\min}| = 1.0$).

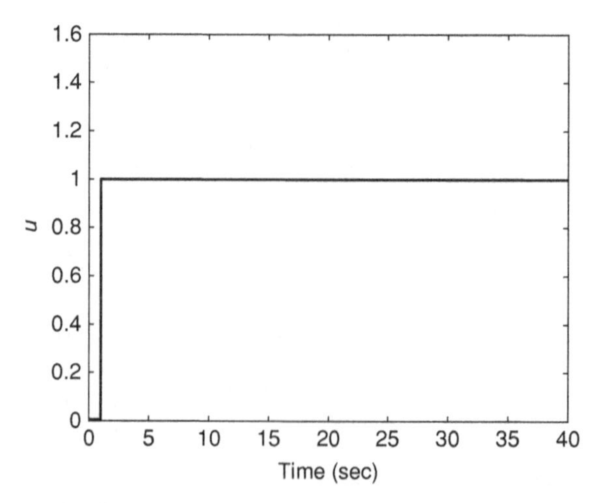

FIGURE 3.21 Plant input ($u_{\max} = |u_{\min}| = 1.0$).

the steady state of $u_1(t)$ is between the constraint ranges. For the plant with strict input constraint, $u_{\max} = |u_{\min}| = 0.5$. Figures 3.23–3.25 show the simulation results. It can be seen that the bounded tracking performance is obtained with the input constraint. It is also shown that the three simulations demonstrate the excellent robust stability characteristics of the closed-loop system in the presence of different input constraints. For the simulations $u_{\max} = |u_{\min}| = 2.0$ and $u_{\max} = |u_{\min}| = 1.0$, the main reason for the desired steady tracking performance is the condition of Lemma 3.2 is satisfied. For $u_{\max} = |u_{\min}| = 0.5$, since the tracking operator does not satisfy the lemma, bounded output tracking is obtained.

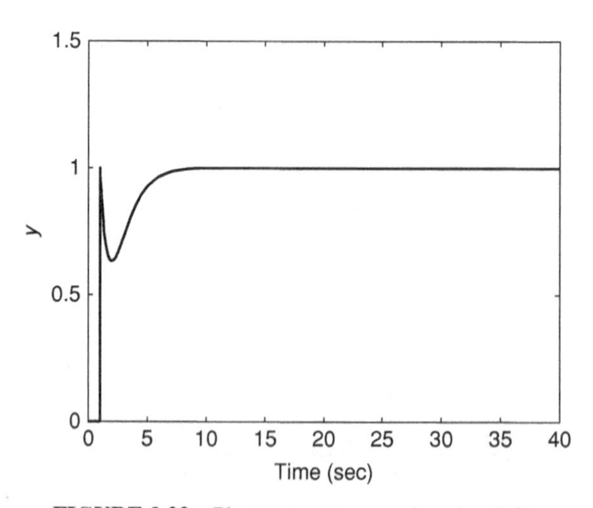

FIGURE 3.22 Plant output ($u_{\max} = |u_{\min}| = 1.0$).

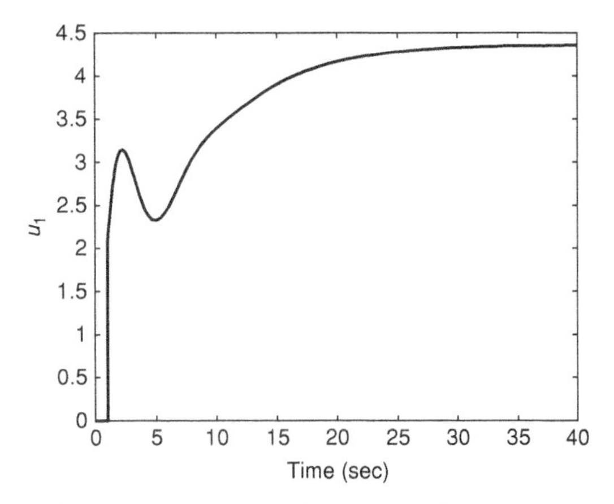

FIGURE 3.23 Plant input prior to input constraint ($u_{\max} = |u_{\min}| = 0.5$).

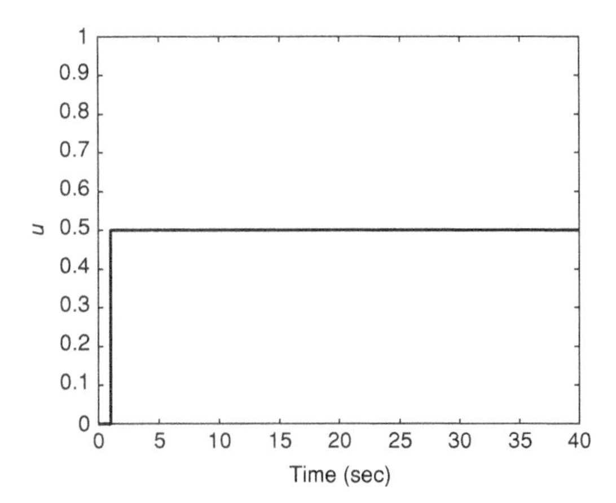

FIGURE 3.24 Plant input ($u_{\max} = |u_{\min}| = 0.5$).

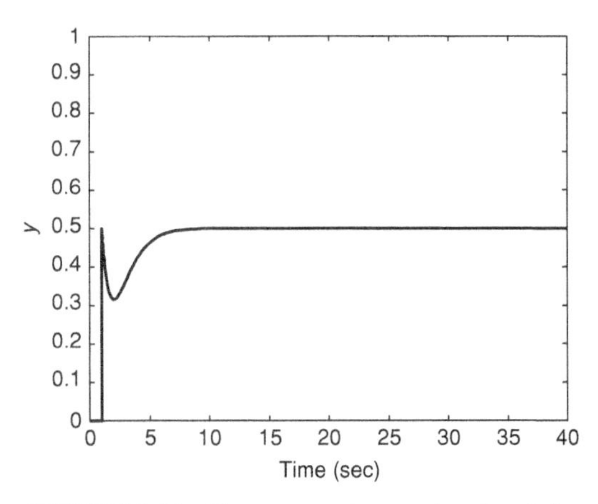

FIGURE 3.25 Plant output ($u_{\max} = |u_{\min}| = 0.5$).

The second example is considered as follows:

$$A[p] = p + 0.05,\ B[p] = -p + 0.8 \tag{3.59}$$

The design procedure is the same as Example 1, but the non–minimum phase is considered.

Two cases of input constraint are considered, namely, $u_{max} = |u_{min}| = 0.8, 1.0$, where $r(t) = 0.8$. The design parameters are given in Table 3.1, but $u_n = 0.01$, $u_d = -1.0$. Based on the method, we have

$$
\begin{aligned}
C[p] &= p + 0.45 \\
X(p) &= \frac{0.0.9942p - 0.416}{p + 200},\ Y(p) = \frac{p + 0.1559}{p + 200} \\
S_1(p) &= \frac{-1.214p^3 - 0.0002422p^2 - 0.00003287p - 0.031233}{0.01583p^3 + 1.997p^2 + 0.6501p + 0.117} \\
M(y) &= \begin{cases} \dfrac{p - 0.05}{p + 0.8} S_1 u_1 & \text{if } y = N(u_1) \\ \dfrac{p - 0.05}{p + 0.8} S_1 u_{max} & \text{otherwise} \end{cases}
\end{aligned}
$$

The tracking operator is designed by considering the condition of $t \to \infty$ in Lemma 3.2 as follows:

$$L = \frac{p + 1}{p + 0.8} \tag{3.60}$$

In the case of $u_{max} = |u_{min}| = 0.8$, Figures 3.26, 3.27, and 3.28 show the simulation results. Even though there are input limitations and output basis, the resulting

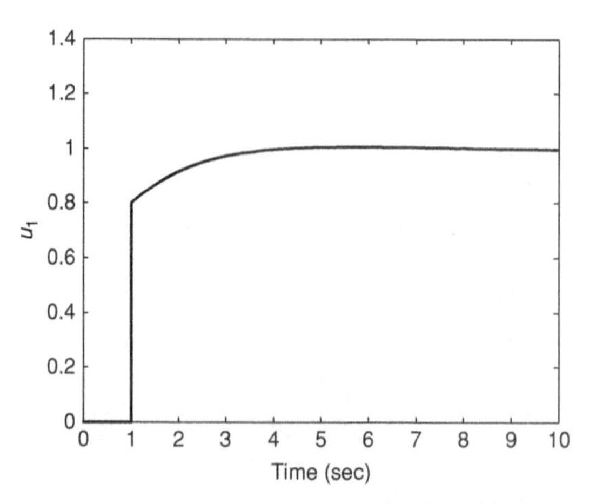

FIGURE 3.26 Plant input prior to input constraint ($u_{max} = |u_{min}| = 0.8$).

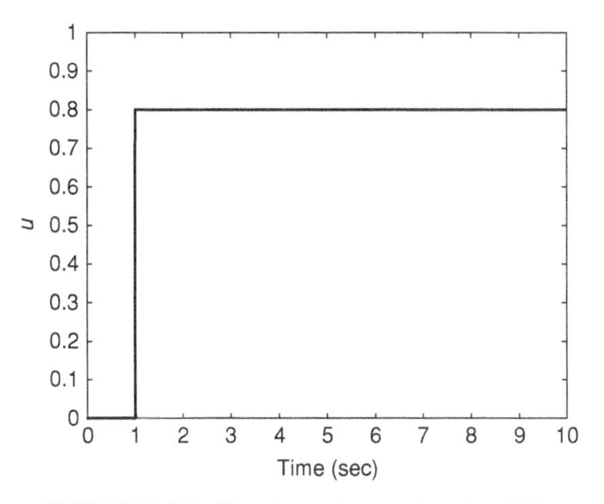

FIGURE 3.27 Plant input ($u_{\max} = |u_{\min}| = 0.8$).

closed loop is robustly stable. In this case, since the plant is of non–minimum phase, plant output is less than zero from about 1 to 2 sec. The design parameters are the same as Example 1, but with $u_{\max} = |u_{\min}| = 1.0$. The simulation results are shown in Figures 3.29, 3.30, and 3.31. Figure 3.31 shows that the plant output tracks the reference signal, because $u_1(t)$ lies between the constraint ranges (see Figures 3.29 and 3.30).

From the viewpoint of the robustness of right coprime factorization, a design problem of a stable robust feedback control system for unstable plants with input constraints was considered. Under the existence of input constraints and unstable open-loop poles, the closed-loop strong stability of the system was ensured. The tracking operator design method for plant output to track a reference input is given. Simulation results are presented to confirm the theoretical analysis.

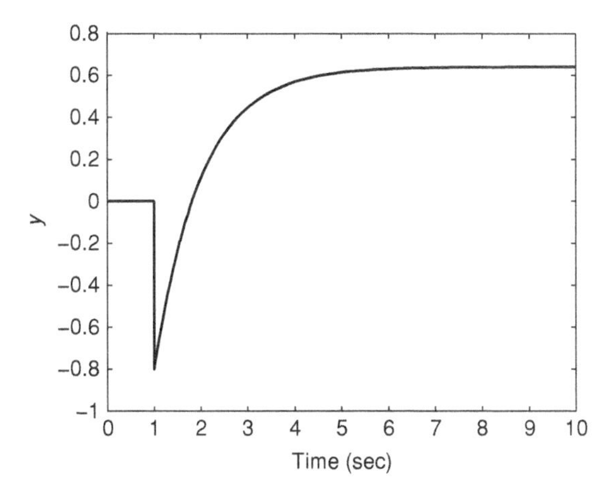

FIGURE 3.28 Plant output ($u_{\max} = |u_{\min}| = 0.8$).

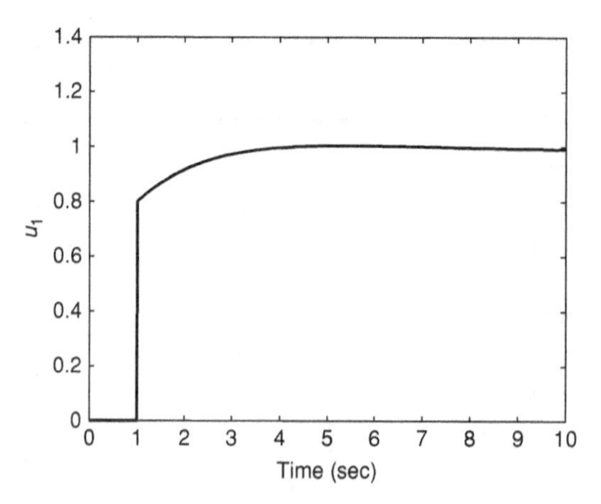

FIGURE 3.29 Plant input prior to input constraint ($u_{\max} = |u_{\min}| = 1.0$).

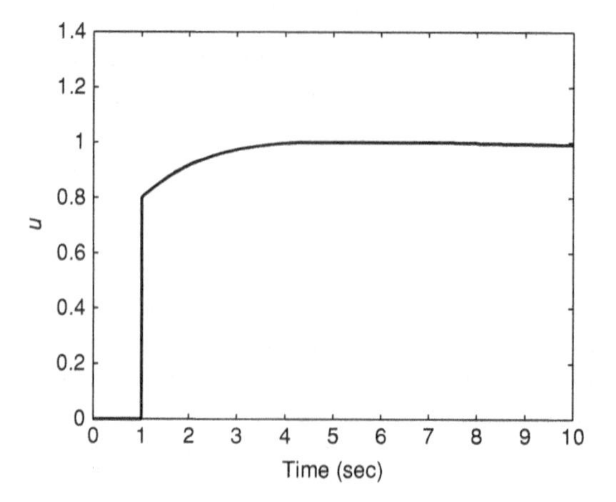

FIGURE 3.30 Plant input ($u_{\max} = |u_{\min}| = 1.0$).

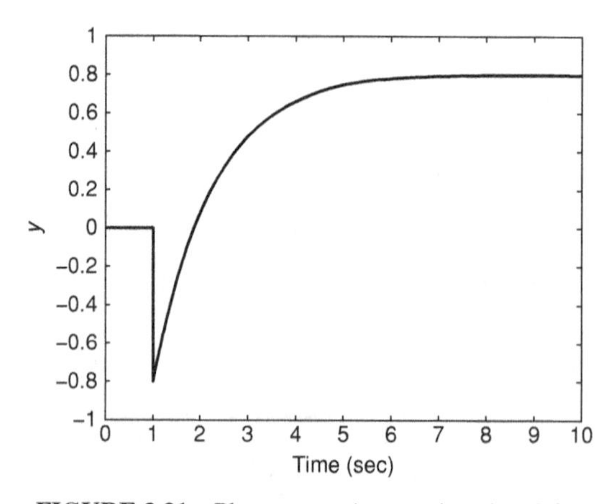

FIGURE 3.31 Plant output ($u_{\max} = |u_{\min}| = 1.0$).

3.3.4 Discussion

As a countermeasure to the design using a predictive control scheme, some results employing a parallel feedforward compensator (PFC) [34–37], also called a robust parallel compensator (RPC) [38–41], to the controlled plants have been considered in [42–44], where it is possible to apply the above-mentioned compensator to plants with non-minimum-phase effects. However, it is not clear whether the operator-based plant presentation can be made robust stable by this kind of compensator. Thus, somewhat serious problems remain with regard to the control system design for plants with operator descriptions and will be future work.

3.4 OPERATOR-BASED MULTI-INPUT–MULTI-OUTPUT NONLINEAR FEEDBACK CONTROL SYSTEMS DESIGN

3.4.1 Introduction

Nonlinear systems play an important role in control engineering owing to the complex structures and nonlinear characteristics. Another important issue with the control problem is to deal with the uncertainties. Generally speaking, except for the external disturbances, there two main types of uncertainties in the system: structured and unstructured. Structured uncertainties usually refer to parametric variations of the plant model due to inaccurate gains, poles, or zeros. Unstructured uncertainties usually represent unmodeled dynamics. In this book, uncertainties refer to the unmodeled ones for operator-based plant presentations.

Robust stability and output tracking performance are important issues in control engineering. Section 3.2 has discussed both issues, using various design methods and achieves desired results on single-input–single-output (SISO) system control design. However, the results cannot be extended to multi-input–multi-output (MIMO) systems mainly because there exist difficulties in dealing with the coupling effects. Therefore, control design for MIMO nonlinear systems with uncertainties is a challenging topic which has been attracting more and more attention.

As has been stated, the operator theory–based method has many advantages and has been proven effective in dealing with SISO nonlinear systems with uncertainties in Section 3.2. Thus, this chapter will be devoted to robust nonlinear control of MIMO nonlinear systems with coupling effects and uncertainties on the basis of robust right coprime factorization. Robust BIBO stability and the desired output tracking performance are considered in three ways using robust right coprime factorization.

In Section 3.4.3, differentiable operator-based robust nonlinear control for MIMO nonlinear systems is given by controller factorization. To begin with, an operator theory–based two-input, two-output feedback control system is described, where the interaction in the MIMO system is summarized as the internal operator in relation to the effect from the input of one subsystem to the input of another subsystem. Also, the most possible types of internal operators relating to the interaction are discussed. To deal with the coupling effects, the control operator is factorized using Taylor expansion. Using the factorized operator and robust right coprime factorization, robust BIBO stability and the desired output tracking performance are considered. Moreover, a numerical example is given to show the design scheme.

In Section 3.4.4, robust nonlinear control for MIMO nonlinear systems is considered by regarding the coupling effects as the uncertainties pertaining to the plants. Using robust right coprime factorization, the condition for guaranteeing the robust BIBO stability and for realizing the desired output tracking performance is derived. To demonstrate the control design scheme, a numerical example is given and the simulation results are shown.

In Section 3.4.5, operator theory–based robust nonlinear control for MIMO nonlinear systems is given by right factorizing the operators relating to the coupling effects. The design scheme is more accurate than the design scheme of Section 3.4.4, which is confirmed by analyzing the robustness of a given two-input, two-output nonlinear system. A discussion of the uncertainty for realizing the desired perfect tracking performance is given.

3.4.2 Definitions and Notation

In Section 3.4.2, some definitions of differentiable and stabilizable operators are introduced. Recall from Section 2.1 that a Lipschitz operator is always continuous on its domain. However, it may not be differentiable. As is known from real analysis, the differentiable operator has more elegant properties than the continuous one. For this reason, the differentiable operator and its properties are introduced.

Definition 3.3 Let X and Y be normed linear spaces and D be an open subset of X. A nonlinear operator $Q : D \to Y$ is said to be differentiable at a point $x_0 \in D$ in the sense of Fréchet if there exists a bounded linear operator $L(x_0) : X \to Y$, depending on x_0, such that, for $h \in X$ and $x_0 + h \in D$,

$$\lim_{\|h\|_X \to 0} \frac{\|Q(x_0 + h) - Q(x_0) - L(x_0)(h)\|_Y}{\|h\|_X} = 0 \tag{3.61}$$

In this case, $L(x_0) := L(x_0)(\cdot)$ is called the first Fréchet derivative of Q at x_0, which is denoted by $L(x_0) = Q'(x_0)$, and $dQ(x_0, h) := L(x_0)(h)$ is called the Fréchet differential (or simply, F-differentiable) of Q at x_0. Moreover, if $Q : D \to Y$ is differentiable at every point $x \in D$, then $L(\cdot) = Q'(\cdot)$ is a mapping from D to $\mathcal{L}(X, Y)$, the family of bounded linear operators defined on X with values in Y, and is called the derivative mapping of Q.

By considering Definition 3.3, although the domain of Q is D (or a subset contained in D), the domain of $L(x_0) = Q'(x_0)$ is the entire space X for any $x_0 \in D$. Also, although both Q and the derivative mapping $L = Q'$ have the common domain D, their ranges are in entirely different spaces. Of course, for each fixed $x_0 \in D$, Q and its derivative $L(x_0) = Q'(x_0)$ have their ranges in the same space Y.

It should be noted that for fixed $x_0 \in D$, Q' is a bounded linear operator. However, Q' is nonlinear in general if x_0 is variable. Taking the following operator as an example, define

$$Q(x) := Q(x)(t) = x(t) \int_0^1 K(t, \tau)x(\tau)\,d\tau$$

where $Q : C[0, 1] \to C[0, 1]$ is a nonlinear integral operator, $C[0, 1]$ is the family of continuous functions defined on $[0, 1]$, then for any $\tilde{x} \in C[0, 1]$

$$Q'(x_0)(\tilde{x}) = x_0(t) \int_0^1 K(t, \tau)\tilde{x}(\tau)\,d\tau + \tilde{x}(t) \int_0^1 K(t, \tau)x_0(\tau)\,d\tau$$

is nonliear if both $\tilde{x}$ and x_0 are variables.

Lemma 3.3 Let X, Y, Z be normed linear spaces and $D \subseteq X$ be an open set.

1. Fréchet differentiation is a linear operation: If $Q_1, Q_2 : D \to Y$ are both differentiable at $x_0 \in D$, then for any $a, b \in \mathcal{C}$

$$(aQ_1 + bQ_2)'(x_0) = aQ_1'(x_0) + bQ_2'(x_0) \tag{3.62}$$

2. Denote $\mathcal{R}(Q)$ as the range of the operator Q. Let $Q_1 : D \to Y$ and $Q_2 : \mathcal{R}(Q_1) \to Z$ be two differentiable operators at $x_0 \in D$ and at $Q_1(x_0) \in \mathcal{R}(Q_1)$, respectively. Then the composite operator $Q_2Q_1 : D \to Z$ defined by $Q_2Q_1(x) = Q_2(Q_1(x))$ is differentiable at $x_0 \in D$ with

$$(Q_2Q_1)'(x_0) = Q_2'[Q_1(x_0)]Q_1'(x_0) \tag{3.63}$$

By extending the above definition, it is possible to recursively define higher order derivatives of a nonlinear operator $Q : D \to Y$, which is omitted here.

Lemma 3.4 Let D be an open subset of X and suppose that the nth derivative $Q^{(n)}(x)$ of a nonlinear operator $Q : D \to Y$ is continuous on D. Then, for any $x, h \in X$ such that $x + h \in D$, it has

$$Q(x + h) = Q(x) + Q(x)'(h) + \cdots + \frac{1}{(n-1)!} Q^{(n-1)}(x)(h^{n-1}) + o(\|h\|_X^n) \tag{3.64}$$

Proof The proof can be completed by mathematical induction with repeated use of the definition. ■

Equation (3.64) is usually called the Taylor expansion of differentiable nonlinear operators.

Let $D \subseteq X$ be an open and convex subset. Let $\mathcal{A}(D, Y)$ be the family of operators $Q : D \to Y$ defined by

$$\mathcal{A} = \left\{ Q : D \to Y \middle| Q'(x) \text{ exists for each } x \in X \text{and} \sup_{x \in D} \|Q'(x)\|_Y < \infty \right\} \tag{3.65}$$

Lemma 3.5 Let $\mathrm{Lip}(D, Y)$ be the family of Lipschitz operators. Then $Q \in \mathcal{A}(D, Y)$ if and only if Q is differentiable on D and $Q \in \mathrm{Lip}(D, Y)$. Moreover, if $Q \in \mathcal{A}(D, Y)$ with the operator seminorm defined by

$$\|Q\| = \sup_{\substack{x_1, x_2 \in D \\ x_1 \neq x_2}} \frac{\|Q(x_1) - Q(x_2)\|_Y}{\|x_1 - x_2\|_X} \tag{3.66}$$

then we have

$$\|Q\| = \sup_{x \in D} \|Q'(x)\| \tag{3.67}$$

where $Q'(x)$ is the F-derivative of Q at $x \in D$. Finally, $\mathcal{A}(D, Y)$ is a closed subspace of the Banach space $\mathrm{Lip}(D, Y)$ under the Lipschitz norm $\|\cdot\|_{\mathrm{Lip}}$. Consequently, $\mathcal{A}(D, Y)$ is a Banach space under the Lipschitz norm [1].

Proof Let

$$L = \sup_{x \in D} \|Q'(x)\|$$

Then $Q \in \mathcal{A}(D, Y)$ implies $L < \infty$, so that by the F-differentiability of Q

$$\|Q(x_1) - Q(x_2)\|_Y \leq L\|x_1 - x_2\|_X$$

for all $x_1, x_2 \in D$. Hence, $Q \in \mathrm{Lip}(D, Y)$. Conversely, suppose that $Q \in \mathrm{Lip}(D, Y)$. Since Q is F-differentiable, $Q'(x)$ exists for each $x \in D$ so that

$$\begin{aligned}\|Q\| &= \sup_{\substack{x_1, x_2 \in D \\ x_1 \neq x_2}} \frac{\|Q(x_1) - Q(x_2)\|_Y}{\|x_1 - x_2\|_X} \\ &\leq L = \sup_{x \in D} \|Q'(x)\|\end{aligned}$$

For $\epsilon > 0$, let $x \in D$ be such that $\|Q'(x)\| \geq L - \epsilon$. For such an $x \in D$ and $x_0 \in X$ satisfying $\|x_0\|_X = 1$ and $\|Q'(x)(x_0)\| \geq (L - \epsilon) - \epsilon$, it has that

$$\begin{aligned}
\|Q\| &\geq \overline{\lim}_{\lambda \to 0+} \frac{\|Q(x + \lambda x_0) - Q(x)\|_Y}{\|\lambda x_0\|_X} \\
&\geq \overline{\lim}_{\lambda \to 0+} \frac{\|Q'(x)(\lambda x_0)\|_Y}{\|\lambda x_0\|_X} - \frac{\|Q(x + \lambda x_0) - Q(x) - Q'(x)(\lambda x_0)\|_Y}{\|\lambda x_0\|_X} \\
&= \frac{\|Q'(x)(x_0)\|_Y}{\|x_0\|_X} \\
&\geq (L - \epsilon) - \epsilon
\end{aligned}$$

which implies that $\|Q(x)\| = L$ and $Q \in \mathrm{Lip}(D, Y)$. The proof of the last assertion is similar to Lemma 2.1, which is omitted here. ■

Lemma 3.6 [1] Let X be a Banach space and $D \subseteq X$ be an open and convex subset. Suppose that Q is an invertible element in $\mathrm{Lip}(X)$ and for a fixed $x \in D$ the F-derivative $Q'(x)$ exists. Then, $Q'(x)$ is an invertible element in $\mathcal{L}(X)$, where $\mathcal{L}(X) = \mathcal{L}(X, X)$ denotes the family of bounded linear operators from X to itself.

Proof Let $C(x_0) = Q(x + x_0) - Q(x)$ for all $x_0 \in X$. Then it is easily seen that $C, C^{-1} \in \mathrm{Lip}(X)$, $C(0) = 0$, and (the bounded linear operator) $C'(0) = Q'(x)$. Consequently, we have that

$$\lim_{h \to 0} \frac{\|C(h)\|_X}{\|h\|_X} = \lim_{h \to 0} \frac{\|Q'(x)(h)\|_X}{\|h\|_X}$$

where it should be noted that $Q'(x)$ on the right-hand side is a bounded linear operator for the fixed $x \in X$, so that the limit is only a formal operation. Now, let $x_0 \in X$ with $\|x_0\|_X = 1$. Since $Q'(x)$ is linear, it follows from the above formal equality that

$$\|[Q'(x)]'(x_0)\|_X = \lim_{\lambda \to 0+} \frac{\|Q'(x)(\lambda x_0)\|_X}{\|\lambda x_0\|_X} = \lim_{\lambda \to 0+} \frac{\|C(\lambda x_0)\|_X}{\|\lambda x_0\|_X}$$

Then, observe that Q is invertible in $\mathrm{Lip}(X)$. It has that $\|C^{-1}\|_{\mathrm{Lip}} < \infty$, so that

$$\|\tilde{x}\|_X = \|C^{-1}[C(\tilde{x})]\|_X \leq \|C^{-1}\| \, \|C(\tilde{x})\|_X$$

or

$$\frac{\|C(\tilde{x})\|_X}{\|\tilde{x}\|_X} \geq \|C^{-1}\|^{-1}$$

for all $\tilde{x} \in X$. It follows that

$$\|[Q'(x)]'(x_0)\|_X \geq \|C^{-1}\|^{-1}$$

for all $x_0 \in X$ with $\|x_0\|_X = 1$. Consequently,

$$\|[Q'(x)]'(\tilde{x})\|_X \geq \|C^{-1}\|^{-1}\|\tilde{x}\|_X$$

for all $\tilde{x} \in X$. This implies that $Q'(x)$ is injective. Moreover, it can be shown that the range of $Q'(x)$ is closed. In fact, assume that $\{x_k\}$ and $\{y_k\}$ are two sequences in X such that $y_k = Q'(x)(x_k)$ and $\|y_k - y\| \to 0$ as $k \to \infty$. Then, since

$$\begin{aligned}\|x_n - x_m\|_X &\leq \|C^{-1}\|\|Q'(x)(x_n) - Q'(x)(x_m)\|_X \\ &\leq \|C^{-1}\|\|y_n - y_m\|_X \to 0\end{aligned}$$

as $n, m \to \infty$, and since X is complete, we have $x_k \to x^*$ for some $x^* \in X$. Hence,

$$Q'(x)(x^*) = \lim_{k\to\infty} Q'(x)(x_k) = \lim_{k\to\infty} y_k = y$$

which implies that the range of $Q'(x)$ is closed. Furthermore, it can be shown that the range of $Q'(x)$ is dense in X, which is omitted here. Thus, $Q'(x)$ is both injective and surjective and thereby invertible. Finally, since

$$\|Q'(x)(\tilde{x})\|_X \geq \|C^{-1}\|^{-1}\|\tilde{x}\|_X$$

for all $\tilde{x} \in X$, we have $[Q'(x)]^{-1} \in \mathcal{L}(X)$ with $\|[Q'(x)]^{-1}\| \leq \|C^{-1}\|$. ■

Definition 3.4 An unstable operator $Q : \mathcal{D}(Q) \subseteq X^e \to \mathcal{R}(Q) \subseteq Y^e$ is said to be stabilizable if there exists an operator $P : \mathcal{D}(Q) \to \mathcal{D}(Q)$ such that the composite operator QP is stable from $X_i \subseteq \mathcal{D}(Q)$ to $Y_0 = \mathcal{R}(Q)$.

3.4.3 Differentiable Operator-Based Nonlinear Robust Control for MIMO Nonlinear Systems Using Controller Factorization

To consider the operator-based control design of a MIMO nonlinear system, first a two-input, two-output system is described using Figure 3.32 based on the control design of a SISO nonlinear system.

Concerning the two-input, two-output nonlinear closed-loop system shown in Figure 3.32, let the input space, output space, and quasi-state space be U, V, W, respectively, and assume that the nominal plant $P_i : U \to V (i = 1, 2)$ has right factorization

$$P_i = N_i D_i^{-1} \qquad i = 1, 2 \tag{3.68}$$

such that $D_i : W \to U$ and $N_i : W \to V (i = 1, 2)$ are stable operators and D_i is invertible, x_i denotes the output of plant B_i^{-1}, and z_i is the control input of P_i satisfying

$$z_1(t) = x_1(t) + G_{12}(x_2)(t) \qquad z_2(t) = x_2(t) + G_{21}(x_1)(t)$$

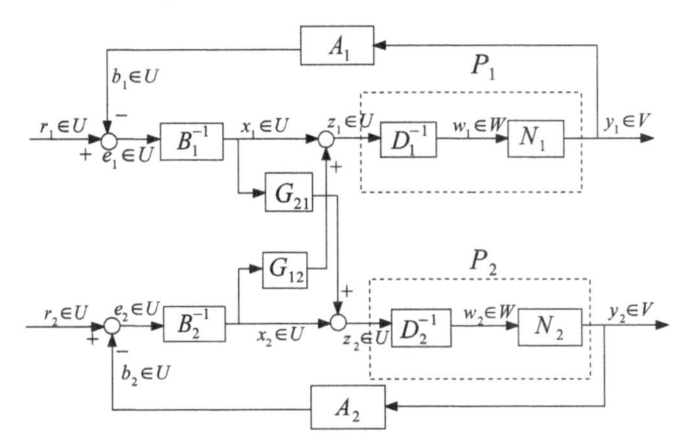

FIGURE 3.32 Two-input, two-output nonlinear feedback system.

where G_{12} is the operator of the effect from x_2 to x_1 and G_{21} is the operator of the effect from x_1 to x_2. If the two subsystems are independent, namely, $G_{12}(x_2)(t) = 0$, $G_{21}(x_1)(t) = 0$, the system is divided into two independent subsystems. Let

$$A_i N_i + B_i D_i = M_i \quad \text{for } M_i \in \mathcal{U}(W, U) \qquad i = 1, 2 \tag{3.69}$$

where $\mathcal{U}(W, U)$ is the set of unimodular operators, operator $A_i : V \rightarrow U$ is stable, and $B_i : U \rightarrow U$ is stable and invertible. Then the system is stable because the two subsystems are stable.

Let $C_{[0,\infty)}$ be the space of continuous functions and $C^1_{[0,\infty)}$ be the subspace of $C_{[0,\infty)}$ which consists of all the functions having a continuous first derivative and being stable. To consider the BIBO stability of the system input and output shown in Figure 3.32, assume that $B_i \in C^1_{[0,\infty)}$ and its derivative mapping is L_i. Then the following relations are derived according to Taylor expansion of B_i:

$$\begin{aligned} B_1(x_1)(t) &= B_1[z_1 - G_{12}(x_2)](t) \\ &= B_1 D_1(w_1)(t) - L_1[D_1(w_1)][G_{12}(x_2)](t) + o(D_1(w_1), G_{12}(x_2))(t) \end{aligned} \tag{3.70}$$

$$\begin{aligned} B_2(x_2)(t) &= B_2[z_2 - G_{21}(x_1)](t) \\ &= B_2 D_2(w_2)(t) - L_2[D_2(w_2)][G_{21}(x_1)](t) + o(D_2(w_2), G_{21}(x_1))(t) \end{aligned} \tag{3.71}$$

where operator $o(\cdot)$ approaches zero. In particular, the latter part of (3.70) and (3.71) can be considered to be the operator concerned with $D_i(w_i)$, which is denoted by ΔB_i and is assumed to be stable, that is, $B_i(x_i)(t) = B_i[D_i(w_i)](t) + \Delta B_i[D_i(w_i)](t)$. Then the stability can be guaranteed by the following theorem.

Theorem 3.2 Consider the nonlinear feedback control system shown in Figure 3.32, which is well-posed, and assume the plant P_i $(i = 1, 2)$ has right coprime factorization as (3.69). Let D^e be a linear subspace of the extended linear space U^e associated with a given Banach space U_B, and let $\Delta B_i D_i M_i^{-1} \in \mathrm{Lip}(D^e)$. The Bezout identities of the nominal plant and the exact plant are $A_i N_i + B_i D_i = M_i \in \mathcal{U}(W, U)$ and $A_i N_i + B_i D_i + \Delta B_i D_i = \tilde{M}_i (i = 1, 2)$, respectively. If

$$\|\Delta B_i D_i M_i^{-1}\| < 1 \qquad i = 1, 2 \tag{3.72}$$

then the nonlinear system is stable.

Proof For any $r_1, r_2 \in U_s$, the system has that

$$r_1(t) = A_1 N_1(w_1)(t) + B_1[D_1(w_1) - G_{12}(x_2)](t) \tag{3.73}$$

$$r_2(t) = A_2 N_2(w_2)(t) + B_2[D_2(w_2) - G_{21}(x_1)](t) \tag{3.74}$$

From the assumption, it follows that

$$r_i(t) = A_i N_i(w_i)(t) + B_i D_i(w_i)(t) + \Delta B_i D_i(w_i)(t) \qquad i = 1, 2 \tag{3.75}$$

Based on Lemma 2.7, $\tilde{M}_i \in \mathcal{U}(W, U)$ $(i = 1, 2)$. Since the system is well-posed, for any $r_i \in U_s$, $w_i \in W_s$. Further, $y_i = N_i w_i$, $b_i = A_i y_i$, and $e_i = r_i - b_i$. Thus the stability of A_i, N_i, B_i, D_i, and ΔB_i implies that $y_i \in V_s$, $b_i \in U_s$, and $e_i \in U_s$. Thus the system is stable. ∎

Based on Theorem 3.2, the BIBO stability of the two-input, two-output nonlinear system is guaranteed by designing the operators A_i and B_i. In the following, the plant output tracking problem for the stabilizing system is considered. A tracking system is given as Figure 3.33, where C_i is a stable filter so that plant output $y_i(t)$ tracks to the reference input $r_i^*(t)$, and $r_i^* \in V (i = 1, 2)$ is the reference input of the system.

For the stabilizing system,

$$y_i(t) = N_i \tilde{M}_i^{-1} r_i(t) \tag{3.76}$$

If $N_i \tilde{M}_i^{-1} C_i = I$, obviously, the output tracking performance is realized. When B_i is linear, the system output is

$$y_1(t) = N_1 M_1^{-1}[r_1 + B_1 G_{12}(x_2)](t) \tag{3.77}$$

$$y_2(t) = N_2 M_2^{-1}[r_2 + B_2 G_{21}(x_1)](t) \tag{3.78}$$

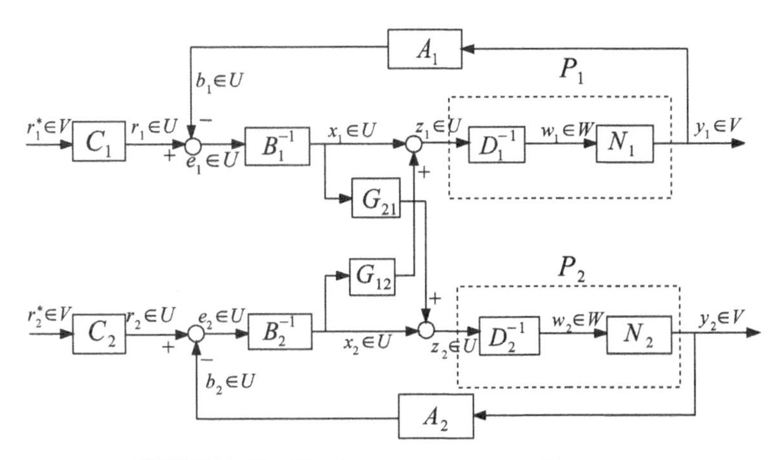

FIGURE 3.33 Two-input, two-output tracking system.

Then the plant output $y_i(t)$ tracks to the reference input $r_i(t)$ provided that

$$N_1 M_1^{-1}\{C_1(r_1^*) + B_1 G_{12} B_2^{-1}[C_2(r_2^*) - A_2 N_2(w_2)]\}(t) = I(r_1^*)(t) \quad (3.79)$$

$$N_2 M_2^{-1}\{C_2(r_2^*) + B_2 G_{21} B_1^{-1}[C_1(r_1^*) - A_1 N_1(w_1)]\}(t) = I(r_2^*)(t) \quad (3.80)$$

It is worth mentioning that the quasi-state $w_i(t)$ can be measured by designing the external observer and control operators A_i^*, B_i^* such that

$$A_i^* N_i + B_i^* D_i = I \qquad i = 1, 2 \quad (3.81)$$

where I is the identity operator.

3.4.3.1 Robust Control for MIMO Nonlinear Systems

Based on the control design for two-input, two-output nonlinear systems, the BIBO stability and output tracking analysis can be extended to MIMO systems using robust right coprime factorization. To demonstrate this, a MIMO nonlinear control system is described.

The operator theory MIMO nonlinear closed-loop feedback system is shown as Figure 3.34, where $\mathbf{P} = (P_1, P_2, \ldots, P_n) : U \to V$ is the nominal nonlinear plant. Let the input space, output space, and quasi-state space be U, V, W, respectively. Assume that plant $\mathbf{P}$ has right factorization $\mathbf{P} = \mathbf{N}\mathbf{D}^{-1}$, where $\mathbf{D} = (D_1, D_2, \ldots, D_n) : W \to U$ and $\mathbf{N} = (N_1, N_2, \ldots, N_n) : W \to V$ are stable operators, D_i is invertible such that $P_i = N_i D_i^{-1}$, and $\mathbf{A} = (A_1, A_2, \ldots, A_n) : V \to U$, $\mathbf{B} = (B_1, B_2, \ldots, B_n) : U \to U$ are stable control operators to be designed. Let the signals of reference input, error, control input, quasi-state, and plant output be $\mathbf{r} = (r_1, r_2, \ldots, r_n) \in U$, $\mathbf{e} = (e_1, e_2, \ldots, e_n) \in U$, $\mathbf{z} = (z_1, z_2, \ldots, z_n) \in U$, $\mathbf{w} = (w_1, w_2, \ldots, w_n) \in W$, and $\mathbf{y} = (y_1, y_2, \ldots, y_n) \in V$, respectively. Usually, the given plant $\mathbf{P}$ is unstable.

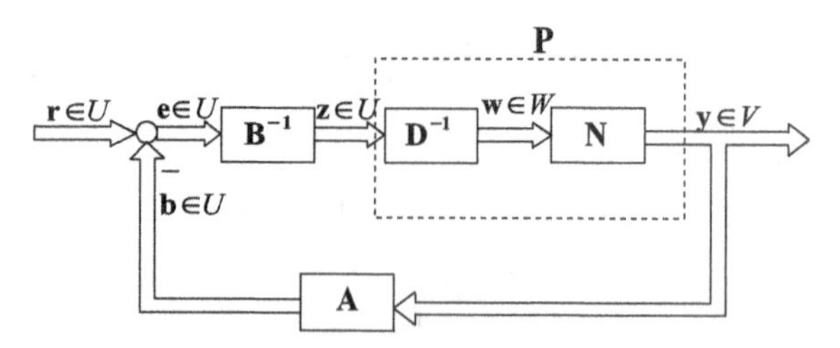

FIGURE 3.34 MIMO nonlinear feedback systems.

It should be mentioned that every input signal $r_i \in \mathbf{r}$ belongs to a subspace U_i of the input space U, namely, $r_i \in U_i \subseteq U$. Similarly, $e_i \in U_i \subseteq U$, $z_i \in U_i \subseteq U$, $w_i \in W_i \subseteq W$, and $y_i \in V_i \subseteq V$ $(i = 1, 2, \ldots, n)$.

To consider the MIMO nonlinear system, the nominal plant P_i $(i = 1, 2, \ldots, n)$ must have right coprime factorization, that is, the Bezout identity

$$A_i N_i + B_i D_i = M_i \quad \text{for } M_i \in \mathcal{U}(W, U) \qquad i = 1, 2, \ldots, n \tag{3.82}$$

is satisfied, where $\mathcal{U}(W, U)$ is the set of unimodular operators. Then the MIMO system is BIBO stable if there are no coupling effects in the system.

In fact, the coupling effects always exist between each two subsystems, which can be summarized as the internal operators concerned with the control inputs. To demonstrate in detail, assume that x_i is the output of plant B_i^{-1} and z_i is the control input of P_i such that

$$z_i(t) = x_i(t) + \sum_{\substack{j=1 \\ j \neq i}}^{n} G_{ij}(x_j)(t) \qquad i = 1, 2, \ldots, n \tag{3.83}$$

where the internal operator G_{ij} $(i, j = 1, 2, \ldots, n,\ j \neq i)$ is related to the effect from the input of the jth subsystem to the input of the ith subsystem.

The coupling effects also can be summarized as

$$z_i(t) = x_i(t) + \sum_{\substack{j=1 \\ j \neq i}}^{n} H_{ij}(y_j)(t) \qquad y_i(t) = P_i(z_i)(t) \tag{3.84}$$

$$y_i(t) = P_i(x_i)(t) + \sum_{\substack{j=1 \\ j \neq i}}^{n} J_{ij}(x_j)(t) \tag{3.85}$$

$$y_i(t) = P_i(x_i)(t) + \sum_{\substack{j=1 \\ j \neq i}}^{n} Q_{ij}(y_j)(t) \tag{3.86}$$

where H_{ij}, J_{ij}, and $Q_{ij}(i, j = 1, 2, \ldots, n,\ j \neq i)$ are internal operators in relation to most possible kinds of coupling effects. Without loss of generality, we only consider one case in this book.

Similar to the analysis of the two-input, two-output system, suppose B_i is differentiable and its derivative mapping is L_i. According to the Taylor expansion of B_i,

$$B_i \left(D_i(w_i) - \sum_{\substack{j=1 \\ j \neq i}}^{n} G_{ij}(x_j) \right)(t)$$

$$= B_i D_i(w_i)(t) - L_i[D_i(w_i)] \left(\sum_{\substack{j=1 \\ j \neq i}}^{n} G_{ij}(x_j) \right)(t) + o\left(D_i(w_i), \sum_{\substack{j=1 \\ j \neq i}}^{n} G_{ij}(x_j) \right)(t) \tag{3.87}$$

where the operator $o(\cdot)$ approaches zero. The latter part of (3.87) can be denoted by ΔB_i and is assumed to be stable, that is, $B_i(x_i)(t) = B_i[D_i(w_i)](t) + \Delta B_i[D_i(w_i)](t)$ and thus the following theorem is obtained.

Theorem 3.3 Consider the nonlinear feedback control system shown in Figure 3.34, which is well-posed, and assume the nominal plant P_i $(i = 1, 2, \ldots, n)$ has right coprime factorization as (3.82). Let D^e be a linear subspace of the extended linear space U^e associated with a given Banach Space U_B, and let $\Delta B_i D_i M_i^{-1} \in \mathrm{Lip}(D^e)$. The Bezout identities of the nominal plant and the exact plant are $A_i N_i + B_i D_i = M_i \in \mathcal{U}(W, U)$ and $A_i N_i + B_i D_i + \Delta B_i D_i = \tilde{M}_i$, respectively, where $B_i \in C^1_{[0,\infty)}$. If

$$\|\Delta B_i D_i M_i^{-1}\| < 1 \qquad i = 1, 2, \ldots, n \tag{3.88}$$

then the MIMO nonlinear system is stable.

Proof Similar to Theorem 3.2, the system is proved BIBO stable. ■

To consider the output tracking performance of the stabilizing system, the tracking system is given in Figure 3.35, where, $\mathbf{r}^* = (r_1^*, r_2^*, \ldots, r_n^*) \in V$ is the reference input of the system, and the stable tracking filter $\mathbf{C} = (C_1, C_2, \ldots, C_n) : V \to U$ is designed with a view to making the plant output $y_i(t)$ track to the reference input $r_i(t)$. The plant output of the system is obtained as

$$y_i(t) = N_i \tilde{M}_i^{-1} r_i(t) \tag{3.89}$$

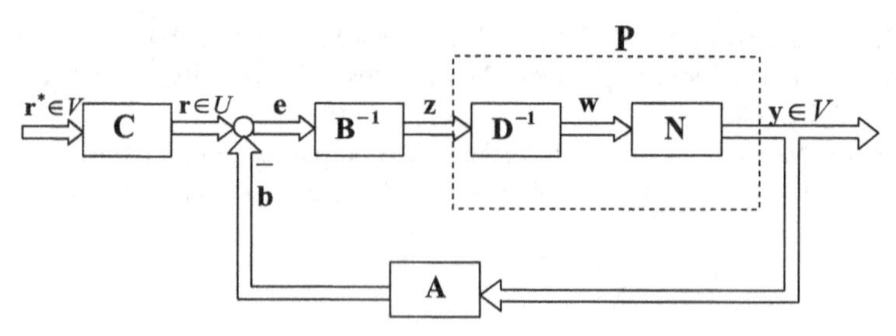

FIGURE 3.35 Tracking design of MIMO nonlinear systems.

Consequently, if $N_i \tilde{M}_i^{-1} C_i = I$, the output tracking performance can be realized. Also, for B_i linear, the system output is

$$y_i(t) = N_i(A_i N_i + B_i D_i)^{-1} \left[r_i + B_i \left(\sum_{\substack{j=1 \\ j \neq i}}^{n} G_{ij}(x_j) \right) \right](t) \tag{3.90}$$

Thus, the output tracking performance can be realized provided that

$$N_i M_i^{-1} \left(C_i(r_i^*) + B_i \sum_{\substack{j=1 \\ j \neq i}}^{n} G_{ij} B_j^{-1}[C_j(r_j^*) - A_j N_j(w_j)] \right)(t)$$
$$= I(r_i^*)(t) \qquad i = 1, 2, \ldots, n \tag{3.91}$$

where the quasi-state $w_i(t)$ $(i = 1, 2, \ldots, n)$ is measurable [45].

Similarly, operator theory–based robust control for MIMO nonlinear systems with uncertainties as shown in Figure 3.36 is considered, where $\mathbf{P} = (P_1, \ldots, P_n)$ is the nominal plant and $\Delta\mathbf{P} = (\Delta P_1, \ldots, \Delta P_n)$ is plant uncertainty, and the output tracking

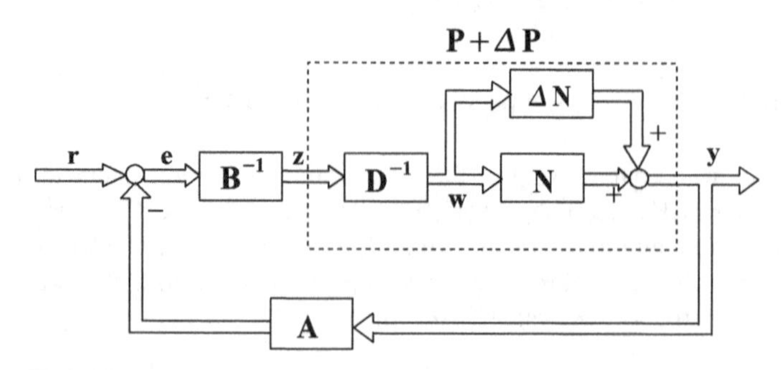

FIGURE 3.36 MIMO nonlinear feedback systems with uncertainties.

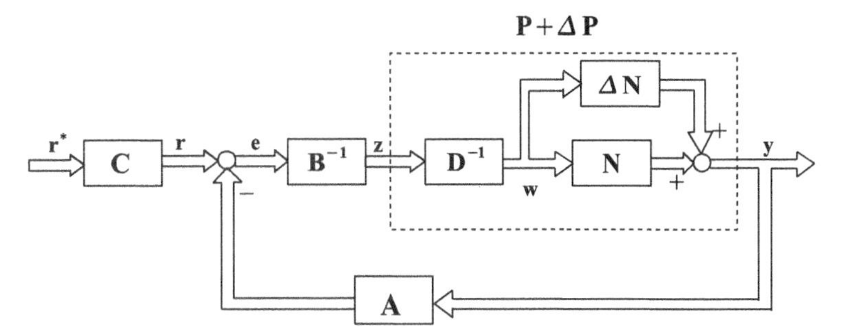

FIGURE 3.37 Tracking design of MIMO nonlinear systems with uncertainties.

system is given as Figure 3.37, where $\mathbf{C} = (C_1, C_2, \ldots, C_n)$ is the stable nonlinear tracking filter for making the output $\mathbf{y} = (y_1, y_2, \ldots, y_n)$ track to the reference input $\mathbf{r}^* = (r_1^*, r_2^*, \ldots, r_n^*)$. Suppose that the nominal plant $\mathbf{P}$ and real plant $\tilde{\mathbf{P}} = \mathbf{P} + \Delta\mathbf{P}$ have right factorization as $\mathbf{P} = ND^{-1}$ and $\tilde{\mathbf{P}} = \tilde{\mathbf{N}}\mathbf{D}^{-1} = (\mathbf{N} + \Delta\mathbf{N})\mathbf{D}^{-1}$, namely

$$P_i = N_i D_i^{-1} \tag{3.92}$$

$$P_i + \Delta P_i = (N_i + \Delta N_i) D_i^{-1} \tag{3.93}$$

where N_i, ΔN_i, and D_i are stable such that D_i^{-1} is invertible, ΔN_i is unknown but the upper and lower bounds are known.

As has been stated, if there exist no coupling effects, that is, $G_{ij}(x_j) = 0$, then based on the robust right coprime factorization approach (i.e., Lemma 2.7), the system is stable under the condition of (3.82) and

$$\left\| \left(A_i(N_i + \Delta N_i) - A_i N_i \right) M^{-1} \right\| < 1 \tag{3.94}$$

because of the fact that every subsystem is robustly BIBO stable under the two conditions. Consequently, for the MIMO nonlinear system with uncertainties and coupling effects, the Bezout identities of the nominal plant and the exact plant are $A_i N_i + B_i D_i = M_i \in \mathcal{U}(W, U)$ and $A_i(N_i + \Delta N_i) + (B_i + \Delta B_i) D_i = \hat{M}_i$, respectively. Then robust BIBO stability can be guaranteed provided that

$$\left\| \left[A_i(N_i + \Delta N_i) - A_i N_i + \Delta B_i D_i \right] M_i^{-1} \right\| < 1 \qquad i = 1, 2, \ldots, n \tag{3.95}$$

and the output tracking performance can be realized under the condition

$$(N_i + \Delta N_i) \hat{M}_i^{-1} C_i = I \tag{3.96}$$

or

$$(N_i + \Delta N_i)M_i^{-1}\left(C_i(r_i^*) + B_i \sum_{\substack{j=1 \\ j \neq i}}^{n} G_{ij}B_j^{-1}\left[C_j(r_j^*) - A_j(N_j + \Delta N_j)(w_j)\right]\right)(t)$$

$$= I(r_i^*)(t) \qquad i = 1, 2, \ldots, n \tag{3.97}$$

with B_i being linear.

In the following example, the system control design for two-input, two-output unstable plant is considered, where the two plants are the same and cited from [3] as follows:

$$P_i(z_i)(t) = \int_0^t z_i^{1/3}(\tau)\,d\tau + e^{t/3}z_i^{1/3}(t) \qquad i = 1, 2 \tag{3.98}$$

where the input space is $U = C_{[0,\infty)}$ and the output space is $V = \{\mu + e^{t/3}\mu' | \mu \in C^1_{[0,\infty)}\} \subset U$, $z_i \in U$ with $P_i(z_i) \in V$, namely, it has that $\mathcal{D}(P_i) = U$, $\mathcal{R}(P_i) \subseteq V$. In this example, $W = U$, $W_s = U_s$ are selected, and the norm is defined as the sup-norm

$$\|u\|_\infty = \sup_{t\in[0,\infty)} |u(t)| \tag{3.99}$$

with $U_s = \{u(t) : u \in U, \ \|u\|_\infty < \infty\}$, $U^e = \{u(t) : u \in U, \ \|u_T\|_\infty < \infty$ for all $T < \infty\}$.

To start with, it should be pointed out that the given plant P_i is unstable. Indeed, there is a $\tilde{z}_i \in U_s$ (e.g., $\tilde{z}_i = 1$) such that

$$y_i = P_i(\tilde{z}_i)(t) = \int_0^t \tilde{z}_i^{1/3}(\tau)\,d\tau + e^{t/3}\tilde{z}_i^{1/3}(t) \to \infty \tag{3.100}$$

which is not in V_s.

The right factorization of the plant is chosen as

$$N(w_i)(t) = \int_0^t e^{-\tau/3}w_i^{1/3}(\tau)\,d\tau + w_i^{1/3}(t)$$

$$D_i(w_i)(t) = e^{-t}w_i(t) \qquad i = 1, 2 \tag{3.101}$$

it can be proved that $N_i : W \to V$ and $D_i : W \to U$ are stable such that $P_i = N_i D_i^{-1}$.

The internal operator G_{ij} in relation to the coupling effects is assumed to be

$$G_{12}(x_2)(t) = \tfrac{1}{3}I(x_2)(t) \tag{3.102}$$

$$G_{21}(x_1)(t) = \tfrac{1}{2}I(x_1)(t) \tag{3.103}$$

where x_i is the controller output of every subsystem. Based on the design scheme, the controllers are designed as

$$A_i(y_i)(t) = \begin{cases} (e^t - 1)(\mu_i')^3(t) & \text{if } y_i = \mu_i + e^{t/3}\mu_i' \\ 0 & \text{otherwise} \end{cases}$$
$$B_i(x_i)(t) = I(x_i)(t) \tag{3.104}$$

from which it follows that

$$A_i N_i(w_i)(t) = (1 - e^{-t})w_i(t)$$
$$B_i D_i(w_i)(t) = e^{-t}(w_i)(t) \tag{3.105}$$

that is, the Bezout identity

$$A_i N_i + B_i D_i = I \tag{3.106}$$

is satisfied. Also, for a given reference input

$$r_i^*(t) = \int_0^t e^{-\tau/3} s_i^{1/3}(\tau)\, d\tau + s_i^{1/3}(t) \in V \tag{3.107}$$

where $s_1 = 0.5s_2$, $s_i(t)$ is arbitrarily chosen to make the bounded input. According to the design scheme,

$$\Delta B_1 = -\tfrac{3}{5}I \qquad \Delta B_2 = -\tfrac{1}{10}I \tag{3.108}$$

which implies that the inequality (3.72) is satisfied. Also, to achieve good output tracking performance, the tracking filter is designed as

$$r_1(t) = C_1(r_1^*)(t) = (1 - \tfrac{3}{5}e^{-t})s_1(t)$$
$$r_2(t) = C_2(r_2^*)(t) = (1 - \tfrac{1}{10}e^{-t})s_2(t) \tag{3.109}$$

from which it follows that $y_i(t) = r_i(t)$.

In the simulation, the variables in the system are selected as $s_1 = 1/2e^{-10t}$, $s_2 = e^{-10t}$, and the simulation results are shown as Figures 3.38–3.41. The reference input r_i^* and plant output y_i $(i = 1, 2)$ of the system are shown in Figures 3.38 and 3.39 (the reference inputs and plant outputs coincide), the filter output r_i $(i = 1, 2)$ is shown in Figure 3.40, and the control input z_i $(i = 1, 2)$ is given in Figure 3.41. The simulation results confirm the effectiveness of the design scheme.

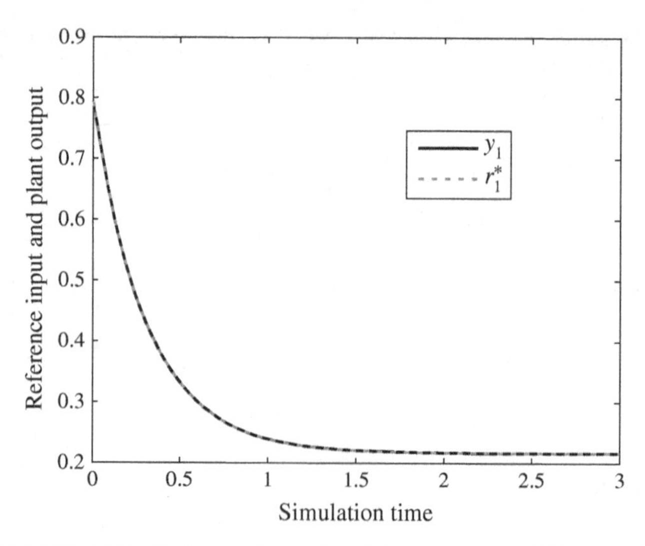

FIGURE 3.38 Reference input r_1^* and plant output y_1 of Example 1.

3.4.4 Nonlinear Robust Control for MIMO Nonlinear Systems by Considering Coupling Effects as Uncertainties of Plants

Similar to [7], one way for considering the coupling effects is to regard them as the effects from uncertainties of the plant. Suppose that G_{ij} is a stable operator and the effects from the coupling effects are considered as the uncertainties of D_i, denoted

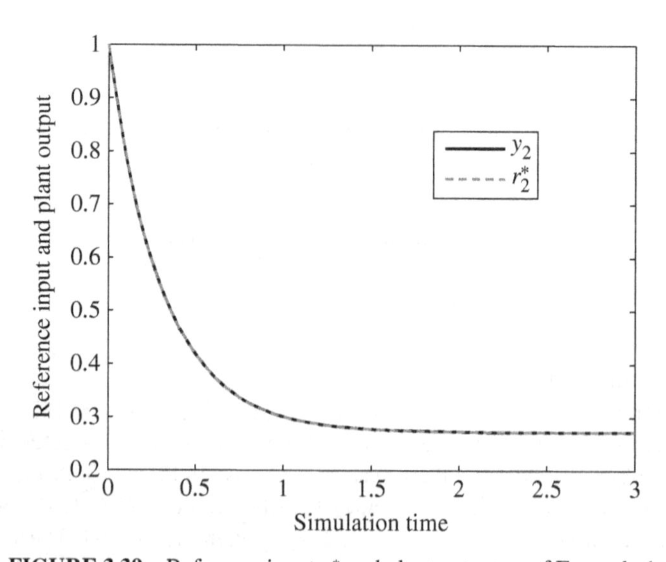

FIGURE 3.39 Reference input r_2^* and plant output y_2 of Example 1.

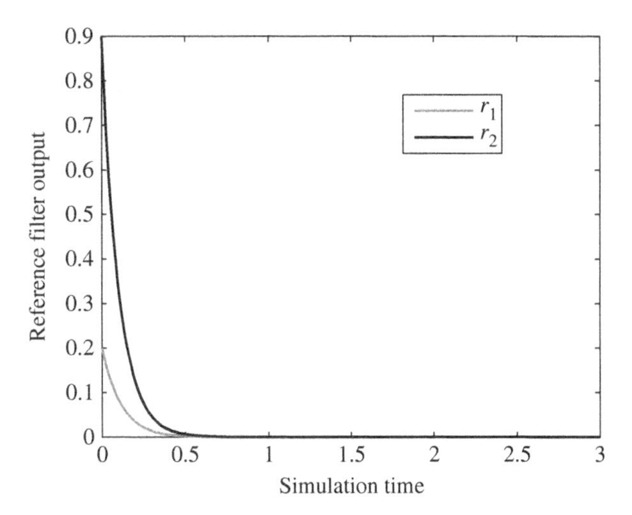

FIGURE 3.40 Tracking filter outputs r_1 and r_2 of Example 1.

by ΔD_i, such that ΔD_i is stable and $D_i + \Delta D_i$ is invertible. That is,

$$x_i(t) + \sum_{\substack{j=1 \\ j \neq i}}^{n} G_{ij}(x_j)(t) = (D_i + \Delta D_i)(w_i)(t) \qquad i = 1, 2, \ldots, n \quad (3.110)$$

Then, similar to Lemma 2.7, the following condition for guaranteeing the robust BIBO stability is derived.

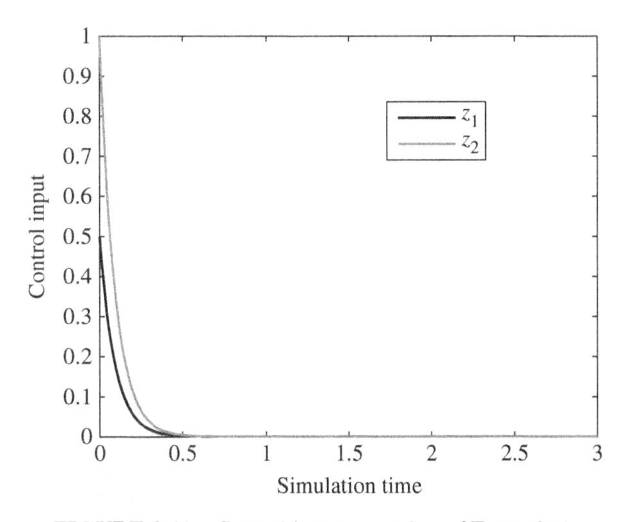

FIGURE 3.41 Control inputs z_1 and z_2 of Example 1.

Theorem 3.4 Let D^e be a linear subspace of the extended linear space U^e, and let $\left[A_i(N_i + \Delta N_i) - A_i N_i + B_i(D_i + \Delta D_i) - B_i D_i\right] M_i^{-1} \in \mathrm{Lip}(D^e)$. Assume that the MIMO nonlinear feedback control system with uncertainties shown in Figure 3.36 is well-posed. The Bezout identities of the nominal plant and the perturbed plant are $A_i N_i + B_i D_i = M_i \in \mathcal{U}(W, U)$ and $A_i(N_i + \Delta N_i) + B_i(D_i + \Delta D_i) = \hat{M}_i$, respectively. Then the system is stable under the condition of controllers A_i, B_i to satisfy (3.82) and

$$\left\| \left[A_i(N_i + \Delta N_i) - A_i N_i + B_i(D_i + \Delta D_i) - B_i D_i\right] M_i^{-1} \right\| < 1 \quad (3.111)$$

Proof The proof is similar to Lemma 2.7 and is omitted here. ■

It is also conceivable for the system without perturbation, that is, the system shown in Figure 3.34, to be BIBO stable under the condition of (3.82) and

$$\left\| \left(B_i(D_i + \Delta D_i) - B_i D_i\right) M_i^{-1} \right\| < 1 \quad (3.112)$$

In the following, the output tracking performance is considered for the stabilizing system. Based on the result of [21], the plant output of the tracking system with uncertainties shown in Figure 3.37 is

$$y_i(t) = (N_i + \Delta N_i)\hat{M}_i^{-1} C_i(r_i^*)(t) \quad (3.113)$$

and the output of the tracking system without uncertainties shown in Figure 3.35 is

$$y_i(t) = N_i \tilde{M}_i^{-1} C_i(r_i^*)(t) \quad (3.114)$$

where $\hat{M}_i = A(N_i + \Delta N_i) + B_i(D_i + \Delta D_i)$ and $\tilde{M}_i = AN_i + B_i(D_i + \Delta D_i)$. Then the output tracking performance of the MIMO nonlinear system with and without uncertainties can be realized respectively provided that

$$(N_i + \Delta N_i)\hat{M}_i^{-1} C_i = I \quad (3.115)$$

$$N_i \tilde{M}_i^{-1} C_i = I \quad (3.116)$$

From the analysis, it can be seen that, by using the robust right coprime factorization approach, the effects caused by the interactions cannot be transmitted back to the error signal. This is one of the advantages of robust right coprime factorization.

In the example given in Section 3.4.3, the considered plants of the two-input, two-output system are the same. In this example, a two-input, two-output system with different plants is considered, where one plant and its right factorization are given as

$$\begin{aligned} P_1(z_1)(t) &= \int_0^t z_1^{1/3}(\tau)\,d\tau + e^{t/3}z_1^{1/3}(t) \\ N_1(w_1)(t) &= \int_0^t e^{-\tau/3}w_1^{1/3}(\tau)\,d\tau + w_1^{1/3}(t) \\ D_1(w_1)(t) &= e^{-t}w_1(t) \end{aligned} \tag{3.117}$$

and the other plant is given as

$$\begin{aligned} P_2(z_2)(t) &= \int_0^t z_2(\tau)\,d\tau \\ N_2(w_2)(t) &= \int_0^t (1+e^{\tau})^{-1}w_2(\tau)\,d\tau \\ D_2(w_2)(t) &= (1+e^{t})^{-1}w_2(t) \end{aligned} \tag{3.118}$$

where $z_i \in U$ is the control input with $P_i(z_i) \in V$, $U_1 = U_2 = U = C_{[0,\infty)}$, $V_1 = \{\mu + e^{t/3}\mu' | \mu \in C^1_{[0,\infty)}\} \subseteq U$, $V_2 = U$, $V = V_1 \cup V_2$, $\mathcal{D}(P_i) = U$, and $\mathcal{R}(P_i) \subseteq V$. Also, P_i is unstable and has right factorization as $P_i = N_i D_i^{-1}$, where $N_i : W \to V$ and $D_i : W \to U$, N_i, D_i are stable, and D_i is invertible. In this example, $W = U$, $W_s = U_s$. The internal operators relating to the coupling effects between the two plants are assumed to be known as

$$\begin{aligned} G_{12}(x_2)(t) &= I(x_2)(t) \\ G_{21}(x_1)(t) &= (1+e^{-t})^{-1}(x_1)(t) \end{aligned} \tag{3.119}$$

where x_i is the controller output of every subsystem.

Based on the design scheme, the stable controllers are designed as

$$\begin{aligned} A_1(y_1)(t) &= \begin{cases} (e^t - 1)(\mu_1')^3(t) & \text{if } y_1 = \mu_1 + e^{t/3}\mu_1' \\ 0 & \text{otherwise} \end{cases} \\ B_1(x_1)(t) &= I(x_1)(t) \\ A_2(y_2)(t) &= (1+e^{-t})^{-1}y_2'(t) \\ B_2(x_2)(t) &= I(x_2)(t) \end{aligned} \tag{3.120}$$

Similar to Example 1,

$$\begin{aligned} A_1N_1(w_1)(t) &= (1-e^{-t})w_1(t) \\ A_2N_2(w_2)(t) &= (1+e^{-t})^{-1}w_2'(t) \end{aligned} \tag{3.121}$$

that is, the Bezout identity

$$A_i N_i(w_i)(t) + B_i D_i(w_i)(t) = I(w_i)(t) \qquad i = 1, 2 \tag{3.122}$$

is implied.

The reference input of the system is given as

$$\begin{aligned} r_1^*(t) &= \int_0^t e^{-\tau/3} s_1^{1/3}(\tau)\, d\tau + s_1^{1/3}(t) := u_1^* + e^{t/3} u_1^{*\prime} \in V \\ r_2^*(t) &= \int_0^t (1 + e^{\tau})^{-1} s_2(\tau)\, d\tau \in V \end{aligned} \tag{3.123}$$

where $s_i(t)$ is an arbitrary function which can ensure the boundedness of $r_i(t)$, without loss of generality, and $s_1 = s_2 = s(t)$. Thus

$$\Delta D_i = \frac{1}{1 + e^t} I$$

and I is the identity operator, which implies the stability condition. The following relations are then derived:

$$\begin{aligned} w_1(t) &= \frac{2 + e^{-t}}{3 + e^{-t}} r_1(t) + \frac{1 + e^{-t}}{3 + e^{-t}} r_2(t) \\ w_2(t) &= \frac{1 + e^{-t}}{3 + e^{-t}} r_1(t) + \frac{2 + 2e^{-t}}{3 + e^{-t}} r_2(t) \end{aligned} \tag{3.124}$$

In other words, for any given $r_1, r_2 \in U_s$, it follows that $w_1, w_2 \in W_s$. Also, to consider the output tracking performance, based on the design scheme, the tracking filters are designed as

$$r_1(t) = C_1(r_1^*)(t) = s_1(t) \tag{3.125}$$

$$r_2(t) = C_2(r_2^*)(t) = \frac{e^t}{e^t + 1} s_2(t) \tag{3.126}$$

It follows that $y_i = r_i^*$. That is, the output tracking performance is realized.

In the simulation, if $s_1(t) = s_2(t) = e^{-t}$, then the simulation results are shown as Figures 3.42–3.45. The reference input r_i^* and plant output y_i $(i = 1, 2)$ are drawn as Figures 3.42 and 3.43, where the reference inputs r_i^* and plant outputs y_i coincide. Reference filter output r_i and control input z_i $(i = 1, 2)$ are shown in Figures 3.44 and 3.45. In summary, for bounded reference input r_i^* $(i = 1, 2)$, the filter output r_i, control input z_i, and plant output y_i $(i = 1, 2)$ are all bounded correspondingly, that is, the BIBO stability of the designed feedback system is guaranteed. Moreover, the

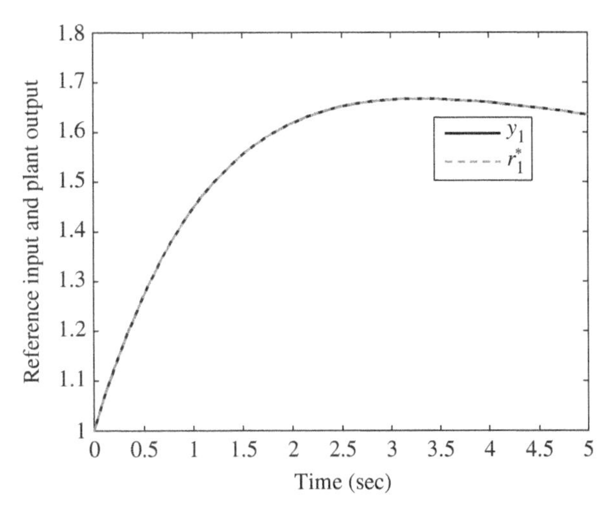

FIGURE 3.42 Reference input r_1^* and plant output y_1 of Example 2.

output tracking performance is realized for the fact that the reference inputs and plant outputs coincide.

3.4.5 Nonlinear Robust Control for MIMO Nonlinear Systems by Right Factorizing Coupling Operators

In Section 3.4.4, the coupling effects are regarded as the uncertainties of the plants. However, in some cases, $D_i + \Delta D_i$ is not invertible, namely, (3.110) is difficult to be satisfied. Thus, further discussion is given by factorizing the internal operators.

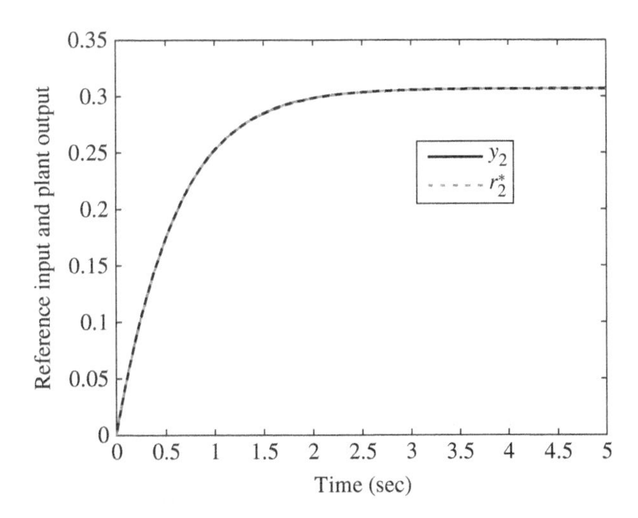

FIGURE 3.43 Reference input r_2^* and plant output y_2 of Example 2.

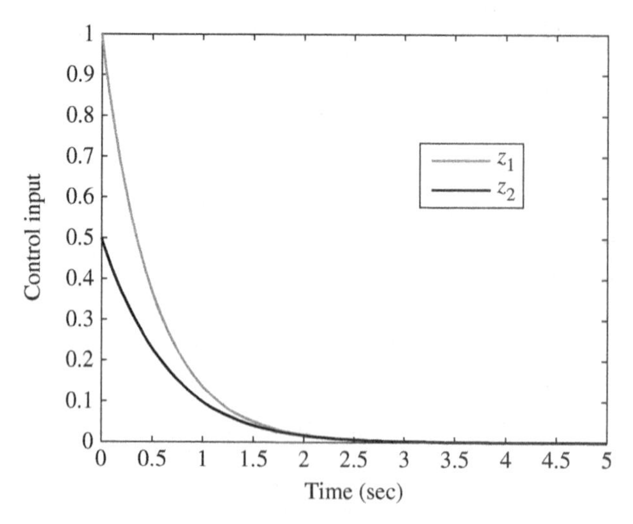

FIGURE 3.44 Control inputs z_1 and z_2 of Example 2.

For brevity, the two-input, two-output nonlinear plant is considered first. Concerning the two-input, two-output nonlinear feedback control system shown in Figure 3.32, suppose that the system is well-posed and the internal operators G_{12}, G_{21} in relation to the coupling effects are linear and can be factorized as

$$G_{12} = T_{12}D_2^{-1} \tag{3.127}$$

$$G_{21} = T_{21}D_1^{-1} \tag{3.128}$$

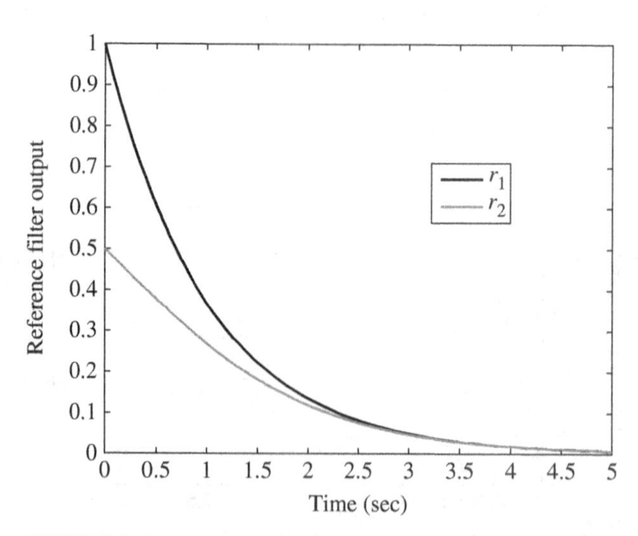

FIGURE 3.45 Tracking filter outputs r_1 and r_2 of Example 2.

where $T_{12} : W_2 \to U_1$ and $T_{21} : W_1 \to U_2$ are stable, that is, G_{12} and G_{21} have right factorization. In most cases the effect from one system to another is equal to or relatively smaller than itself, that is, $\|G_{12}(x_2)\| \leq \|x_2\|$ and $\|G_{21}(x_1)\| \leq \|x_1\|$. Assume that both effects are smaller than itself, it implies that $I - G_{12}G_{21}$ and $I - G_{21}G_{12}$ are invertible according to Lemma 2.6. Thus, the signals of the system have the relations

$$D_1(w_1)(t) - x_1(t) = T_{12}(w_2)(t) - G_{12}G_{21}(x_1)(t) \tag{3.129}$$

$$D_2(w_2)(t) - x_2(t) = T_{21}(w_1)(t) - G_{21}G_{12}(x_2)(t) \tag{3.130}$$

It follows that

$$(I - G_{12}G_{21})x_1(t) = D_1(w_1)(t) - T_{12}(w_2)(t) \tag{3.131}$$

$$(I - G_{21}G_{12})x_2(t) = D_2(w_2)(t) - T_{21}(w_1)(t) \tag{3.132}$$

Assume that $B_1^* = B_1(I - G_{12}G_{21})^{-1}$ and $B_2^* = B_2(I - G_{21}G_{12})^{-1}$ satisfy the addition rule and can be stabilized by D_1, T_{12} and D_2, T_{21}, respectively. If operators A_i, B_i^* satisfy the Bezout identity,

$$A_i N_i + B_i^* D_i = \tilde{M}_i \qquad \tilde{M}_i \in \mathcal{U}(W, U) \qquad i = 1, 2 \tag{3.133}$$

then

$$r_1(t) = \tilde{M}_1(w_1)(t) - B_1^* T_{12}(w_2)(t) \tag{3.134}$$

$$r_2(t) = \tilde{M}_2(w_2)(t) - B_2^* T_{21}(w_1)(t) \tag{3.135}$$

Since $B_1^* T_{12}$ and $B_2^* T_{21}$ are stable, $\tilde{M}_i$ is a unimodular operator and the system is well-posed, then for any $(r_1, r_2) \in U_s$, it has that $(w_1, w_2) \in W_s$. Further, $y_i = N_i w_i$, $b_i = A_i y_i$, and $e_i = u_i - b_i$, the stability of A_i, N_i implies that $y_i \in V_s$, $b_i \in U_s$, and $e_i \in U_s$ $(i = 1, 2)$. Thus, the system is BIBO stable.

As a consequence, n-input, n-output nonlinear plants with coupling effects can be considered similarly.

Theorem 3.5 Assume that the MIMO nonlinear feedback control system shown in Figure 3.34 is well-posed, and the nominal plant P_i $(i = 1, 2, \ldots, n)$ has right coprime factorization as (3.82). The internal operator G_{ij} $(i \neq j, i, j = 1, 2, \ldots, n)$

is linear and has right factorization as $G_{ij} = T_{ij}D_j^{-1}$ such that $T_{ij} : W_j \to U_i$ is stable. Suppose that

$$B_i^* = B_i \left(I - \sum_{\substack{j=1 \\ j \neq i}}^{n} G_{ij}G_{ji} \right)^{-1}$$

satisfies the addition rule and can be stabilized by D_i and T_{ij}, respectively. If operators A_i, B_i^* satisfy the Bezout identity

$$A_i N_i + B_i^* D_i = \tilde{M}_i \qquad \tilde{M}_i \in \mathcal{U}(W, U) \qquad i = 1, 2, \ldots, n \tag{3.136}$$

then the system is BIBO stable.

Proof Based on the assumptions, the feedback system has that

$$r_i(t) = \tilde{M}_i(w_i)(t) - \sum_{\substack{j=1 \\ j \neq i}}^{n} B_i^* T_{ij}(w_j)(t) \qquad i = 1, 2, \ldots, n \tag{3.137}$$

Thus, similar to the analysis of the two-input, two-output system, the BIBO stability of the system is guaranteed by the assumption. ■

Then, for the MIMO nonlinear system with uncertainties shown in Figure 3.36, the following corollary is obtained for guaranteeing robust BIBO stability.

Corollary 3.1 Let D^e be a linear subspace of the extended linear space U^e, and let $(A_i(N_i + \Delta N_i) - A_i N_i)\tilde{M}_i^{-1} \in \text{Lip}(D^e)$. Considering the MIMO nonlinear control system with uncertainties shown in Figure 3.36, which has the same assumptions as Theorem 3.5, the Bezout identities of the nominal plant and the perturbed plant are $A_i N_i + B_i^* D_i = \tilde{M}_i \in \mathcal{U}(W, U)$ and $A_i(N_i + \Delta N_i) + B_i^* D_i = \hat{M}_i$, respectively. Under the condition of controller A_i to satisfy

$$\|(A_i(N_i + \Delta N_i) - A_i N_i)\tilde{M}_i^{-1}\| < 1 \qquad i = 1, 2, \ldots, n \tag{3.138}$$

the system is robustly BIBO stable.

Proof By using the robust right coprime factorization condition of Lemma 2.7 and Theorem 3.5, the robust BIBO stability of the system is guaranteed by the assumption. ■

Thus, for the stabilizing system, the output tracking performance can be realized by designing tracking filter C_i shown in Figure 3.37. When the quasi-state w_i of the MIMO nonlinear systems is observable, and if $C_i = C_{i1} + C_{i2}$ $(i = 1, 2, \ldots, n)$ satisfies

$$(N_i + \Delta N_i)\hat{M}_i^{-1} C_{i1} = I \tag{3.139}$$

$$C_{i2}(r_i^*)(t) + \sum_{\substack{j=1 \\ j \neq i}}^{n} B_i^* T_{ij}(w_j)(t) = 0 \qquad i = 1, 2, \ldots, n \tag{3.140}$$

then the plant output tracks to the reference input because of the fact that

$$y_i(t) = (N_i + \Delta N_i)\hat{M}_i^{-1} \left(C_i(r_i^*) + \sum_{\substack{j=1 \\ j \neq i}}^{n} B_i^* T_{ij}(w_j) \right)(t) \tag{3.141}$$

Moreover, for the system without uncertainties, the output tracks to the reference input provided that

$$N_i \tilde{M}_i^{-1} C_{i1} = I \tag{3.142}$$

$$C_{i2}(r_i^*)(t) + \sum_{\substack{j=1 \\ j \neq i}}^{n} B_i^* T_{ij}(w_j)(t) = 0, \qquad i = 1, 2, \ldots, n. \tag{3.143}$$

The case that the internal operator G_{ij} $(i \neq j, i, j = 1, 2 \ldots, n)$ has right factorization equals to a certain degree that the quasi-state signal w_i of plant P_i affects the control input signal z_j of plant P_j directly.

According to Sections 3.4.4 and 3.4.5, the starting points are both from the coupling effects. Thus, in this example, the same plants given in Section 3.4.4 are considered for demonstrating the theoretical analysis of this chapter, that is, the plants are given as

$$\begin{aligned}
P_1(z_1)(t) &= \int_0^t z_1^{1/3}(\tau)\,d\tau + e^{t/3} z_1^{1/3}(t) \\
N_1(w_1)(t) &= \int_0^t e^{-\tau/3} w_1^{1/3}(\tau)\,d\tau + w_1^{1/3}(t) \\
D_1(w_1)(t) &= e^{-t} w_1(t) \\
P_2(z_2)(t) &= \int_0^t z_2(\tau)\,d\tau \\
N_2(w_2)(t) &= \int_0^t (1 + e^{\tau})^{-1} w_2(\tau)\,d\tau \\
D_2(w_2)(t) &= (1 + e^{t})^{-1} w_2(t)
\end{aligned} \tag{3.144}$$

and the internal operators relating to the coupling effects between the two plants are

$$\begin{aligned} G_{12}(x_2)(t) &= I(x_2)(t) \\ G_{21}(x_1)(t) &= (1+e^{-t})^{-1}(x_1)(t) \end{aligned} \tag{3.145}$$

where x_i is the controller ouptut. According to the design scheme, operators G_{12} and G_{21} are divided into two parts as

$$\begin{aligned} T_{12}(w_2)(t) &= (1+e^{t})^{-1}(w_2)(t) \\ D_2(w_2)(t) &= (1+e^{t})^{-1}(w_2)(t) \\ T_{21}(w_1)(t) &= (1+e^{t})^{-1}(w_1)(t) \\ D_1(w_1)(t) &= e^{-t}(w_1)(t) \end{aligned} \tag{3.146}$$

and the controllers are the same with Example 2, that is,

$$\begin{aligned} A_1N_1(w_1)(t) &= (1-e^{-t})w_1(t) \\ B_1D_1(w_1)(t) &= e^{-t}w_1(t) \\ A_2N_2(w_2)(t) &= (1+e^{-t})^{-1}w_2(t) \\ B_2D_2(w_2)(t) &= (1+e^{t})^{-1}w_2(t)(x_2)(t) \end{aligned} \tag{3.147}$$

Thus the Bezout identities of the subsystems shown in (3.136) are obtained as

$$\begin{aligned} \tilde{M}_1(w_1)(t) &= 2I(w_1)(t) \\ \tilde{M}_2(w_2)(t) &= (2I-(1+e^{t})^{-1})(w_2)(t) \end{aligned} \tag{3.148}$$

which are unimodular operators. Based on Theorem 3.4, the system is BIBO stable. For considering the output tracking performance, the same reference input as Section 3.4.4 is chosen, that is,

$$\begin{aligned} r_1^*(t) &= \int_0^t e^{-\tau/3}s_1^{1/3}(\tau)\,d\tau + s_1^{1/3}(t) := u_1^* + e^{t/3}u_1^{*\prime} \in V \\ r_2^*(t) &= \int_0^t (1+e^{\tau})^{-1}s_2(\tau)\,d\tau \in V \end{aligned} \tag{3.149}$$

where $s_i(t)$ is arbitrarily selected, making the input $r_i(t)$ bounded, that is, $e^{t/3}u_1^{*\prime}$ is bounded. Thus the tracking filters are designed as

$$C_{11}(r_1^*)(t) = \begin{cases} 2e^t (u_1^{*\prime})^3(t) & \text{if } r_1^* = u_1^* + e^{t/3} u_1^{*\prime} \\ 0 & \text{otherwise} \end{cases}$$

$$C_{12}(r_1^*)(t) = \begin{cases} -e^t (u_1^{*\prime})^3(t) & \text{if } r_1^* = u_1^* + e^{t/3} u_1^{*\prime} \\ 0 & \text{otherwise} \end{cases}$$

$$C_{21}(r_2^*)(t) = (1 + 2e^t)(r_2^*)'$$

$$C_{22}(r_2^*)(t) = -(1 + e^t)(r_2^*)'$$

it follows that

$$r_1(t) = C_1(r_1^*)(t) = s_1(t)$$
$$r_2(t) = C_2(r_2^*)(t) = (1 + e^{-t})^{-1} s_2(t).$$

That is, the tracking filters are the same as in Example 2. It can be seen that the same controllers and tracking filters are derived even using different control design schemes. Thus, for further comparison with the two design schemes given in Sections 3.4.4 and 3.4.5, the robustness is considered. Suppose the plant uncertainties are

$$\begin{aligned} \Delta P_1(z_1)(t) &= \Delta_1 \int_0^t z_1^{1/3}(\tau)\, d\tau + \Delta_2 e^{t/3} z_1^{1/3}(t) \\ \Delta N_1(w_1)(t) &= \Delta_1 \int_0^t e^{-\tau/3} w_1^{1/3}(\tau)\, d\tau + \Delta_2 w_1^{1/3}(t) \\ D_1(w_1)(t) &= e^{-t} w_1(t) \\ \Delta P_2(z_2)(t) &= \Delta_3 \int_0^t z_2(\tau)\, d\tau \\ \Delta N_2(w_2)(t) &= \Delta_3 \int_0^t (1 + e^{\tau})^{-1} w_2(\tau)\, d\tau \\ D_2(w_2)(t) &= (1 + e^t)^{-1} w_2(t) \end{aligned} \tag{3.150}$$

where Δ_i $(i = 1, 2, 3)$ is bounded. The controllers are designed as

$$\begin{aligned} A_1(y_1)(t) &= \begin{cases} \left[(e^t (1 + \Delta_2)^3 - (1 + \Delta_1)^3\right](\mu_1')^3(t), \\ \quad \text{if } y_1 = (1 + \Delta_1)\mu_1 + e^{t/3}(1 + \Delta_2)\mu_1' \\ 0 \quad \text{otherwise} \end{cases} \\ B_1(x_1)(t) &= I(x_1)(t) \\ A_2(y_2)(t) &= (1 + \Delta_3)(1 + e^{-t})^{-1} z_2(t) \quad \text{if } y_2 = (1 + \Delta_3) \int_0^t z_2(\tau)\, d\tau \\ B_2(x_2)(t) &= I(x_2)(t) \end{aligned} \tag{3.151}$$

from which it follows that

$$A_1(N_1 + \Delta N_1)(w_1)(t) = \left[(\Delta_2 + 1)^3 - (\Delta_1 + 1)^3 e^{-t}\right] w_1(t)$$
$$A_2(N_2 + \Delta N_2)(w_2)(t) = \frac{\Delta_3}{1 + e^{-t}} w_2(t)$$

Thus,

$$A_1(N_1 + \Delta N_1)(w_1)(t) - A_1 N_1(w_1)(t)$$
$$= \left[\Delta_2^3 + 3\Delta_2^2 + 3\Delta_2 - (\Delta_1^3 + 3\Delta_1^2 + 3\Delta_1)e^{-t}\right] w_1(t) \tag{3.152}$$

$$A_2(N_2 + \Delta N_2)(w_2)(t) - A_2 N_2(w_2)(t) = \frac{\Delta_3}{1 + e^{-t}} w_2(t) \tag{3.153}$$

Based on Section 3.4.4, the robustness can be guaranteed under the conditions

$$\left\| A_1(N_1 + \Delta N_1) - A_1 N_1 \right\|$$
$$= \left\| \Delta_2^3 + 3\Delta_2^2 + 3\Delta_2 - (\Delta_1^3 + 3\Delta_1^2 + 3\Delta_1)e^{-t} \right\| < 1 \tag{3.154}$$

and

$$\left\| A_2(N_2 + \Delta N_2) - A_2 N_2 \right\| = \left\| \frac{\Delta_3}{1 + e^{-t}} \right\| < 1 \tag{3.155}$$

However, based on Section 3.4.5, the robust conditions are

$$\left\| [A_1(N_1 + \Delta N_1) - A_1 N_1]\tilde{M}_1^{-1} \right\|$$
$$= \left\| \Delta_2^3 + 3\Delta_2^2 + 3\Delta_2 - (\Delta_1^3 + 3\Delta_1^2 + 3\Delta_1)e^{-t} \right\| < 1/2 \tag{3.156}$$
$$\left\| [A_2(N_2 + \Delta N_2) - A_2 N_2]\tilde{M}_2^{-1} \right\| = \left\| \frac{\Delta_3}{1 + 2e^{-t}} \right\| < 1$$

By comparing the two cases, it can be known that if the effect from controller A_i is in accordance with the whole system, that is, identical with M_i, then Section 3.4.5 provides a more precise range for the uncertainties than Section 3.4.4.

In the design scheme, the sufficient condition for realizing perfect output tracking performance is given. However, the fact that ΔN_i is unknown generates difficulties in designing controllers to satisfy the condition. The discussion in this chapter on the output tracking performance of nonlinear systems with uncertainty details a way to solve this problem. In brief, one SISO nonlinear system shown with uncertainty in Figure 3.46 is taken into consideration.

Concerning the nonlinear feedback system with uncertainty shown in Figure 3.46, assume that the input and output spaces are the same. The uncertainty ΔP is unknown, which causes difficulties in designing controllers to obtain the desired performance.

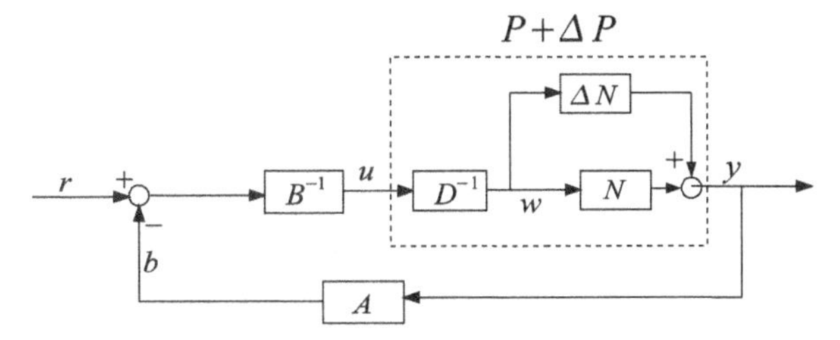

FIGURE 3.46 Nonlinear system with uncertainty.

Thus, how to get the mismatch between the exact and nominal plants is an important issue. A mismatch prediction structure is designed in Figure 3.47, where ΔP^* is the desired mismatch operator which equals the plant perturbation ΔP.

Figure 3.47 implies that

$$\alpha = (\tilde{P} - P)\beta \tag{3.157}$$

that is, the feedback structure of I, $\tilde{P}$, and P in the dashed line constructs an equivalent inversion of ΔP. In this structure, ΔP^* is designed to make $\delta = \beta$. Also, ΔP^* is assumed to have right factorization $\Delta P^* = \Delta N^* D^{-1}$, where, ΔN^* is assumed to be stable and bounded.

Then, the system is robustly BIBO stable and output tracks to the reference input provided that

$$AN + BD = M \tag{3.158}$$

$$A(\Delta N^* + N) + BD = \tilde{M} \tag{3.159}$$

$$\|(A(N + \Delta N^*) - AN)M^{-1}\| < 1 \tag{3.160}$$

$$(N + \Delta N^*)\tilde{M}^{-1} = I \tag{3.161}$$

If $N + \Delta N^*$ is invertible and $(N + \Delta N^*)^{-1}$ is stable, that is, $N + \Delta N^*$ is a unimodular operator, these conditions can be summarized in one equation which

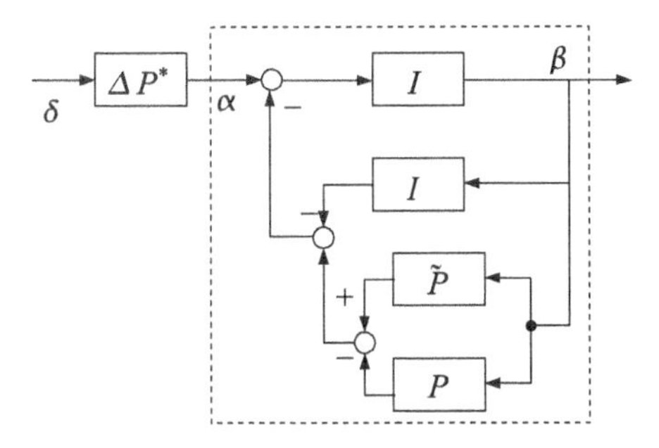

FIGURE 3.47 Prediction structure of plant uncertainty.

also guarantees the robust BIBO stability and is able to realize the output tracking performance.

Theorem 3.6 Assume that the nonlinear system with uncertainty shown in Figure 3.46 is well-posed. If

$$A(N + \Delta N^*) + BD = N + \Delta N^* \tag{3.162}$$

then the output tracks to the reference input.

Proof Based on Lemma 2.7, the system with uncertainty is BIBO stable under the condition of (3.162).

Further, on the stabilizing system, (3.162) implies that

$$y(t) = (N + \Delta N^*)[A(N + \Delta N^*) + BD]^{-1} r(t) = r(t)$$

that is, output tracking is realized. ■

By Theorem 3.1 the perfect tracking can be realized under the condition that uncertainty can be predicted accurately. Also, this result can be used in the MIMO nonlinear control process. However, in practice, there always exists error in the prediction process. To avoid this problem, a control design for a nonlinear feedback system is considered, shown in Figure 3.48, in which controller B is constructed by A and a feedback loop. Then

$$e(t) = (I - A)(x)(t) \qquad x(t) = \tilde{P}(u)(t) \tag{3.163}$$

which implies that

$$A(N + \Delta N) + BD = N + \Delta N \tag{3.164}$$

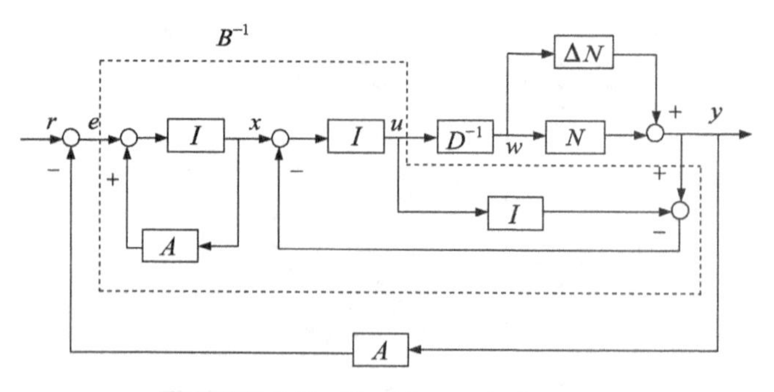

FIGURE 3.48 Tracking control system.

Thus, output tracking is realized. In detail, if the stable operator A satisfies that $I - A$ can stabilize the plant with uncertainty, then the plant output tracks to the reference input.

In the control design scheme shown in Figure 3.48, mismatch between the nominal plant and the real plant does not need to be known. Also, the controller B is constructed by controller A, that is, the design process is simplified to a great degree. However, the control problem becomes complex if it is extended to the MIMO control process directly. Therefore, extension of the generalized quantitative design for MIMO system output tracking is not considered here.

3.4.6 Operator-Based Nonlinear Robust Control for MIMO Nonlinear Systems with Unknown Coupling Effects

Operator-based robust right coprime factorization is used to study MIMO nonlinear systems with uncertainties and known coupling effects, where robust BIBO stability and the desired output tracking are considered using the perturbed Bezout identities which consist of the controlled plant operators and the operators in relation to the known coupling effects. However, in many cases, the coupling effects are unknown, which causes difficulties in the control design process. Further consideration is given in Section 3.4.6 on solving the problem and improving the control performance.

Generally speaking, the coupling effects in plants have often been considered in the control design of MIMO nonlinear systems. However, the mutual interference caused by the effects of the controllers and plant outputs is always ignored but important for MIMO nonlinear system design because this kind of coupling effect has always existed in practice, influencing the performance of the controlled system. In this chapter, both kinds of the coupling effects are considered. For MIMO nonlinear systems with uncertainties and unknown coupling effects, BIBO stability can be guaranteed using the generalized Lipschitz operator and contraction mapping theorem. Moreover, a control structure is introduced which ensures that the desired output tracking performance can be realized. In the consideration of the MIMO system design, the system is usually assumed to be well-posed. To satisfy the assumption, a discussion on operator-based well-posedness for MIMO nonlinear systems is also given. First, a nonlinear control and tracking design for MIMO nonlinear systems with unknown coupling effects is considered, where the coupling effects include the existing one in the plants and the one caused by controllers and plant outputs. To consider the BIBO stability of the MIMO nonlinear system, operator-based decoupling control realized by feedback property is introduced, and a sufficient condition is obtained using the operator-based approach and contraction mapping theorem. Moreover, for realizing the desired output tracking performance, a control structure is shown for the stabilized system. Second, an example of the application to temperature control of a three-input, three-output aluminum plate thermal process is presented, where the simulation and experimental results confirm the effectiveness of the design scheme. Finally, the well-posedness property of MIMO systems is discussed using an operator-based implicit function theorem.

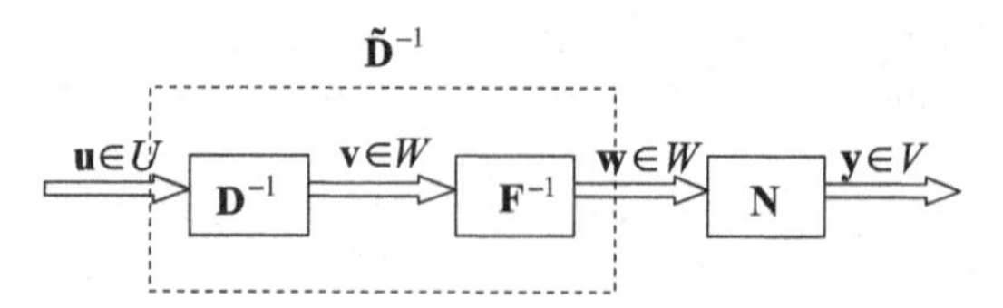

FIGURE 3.49 MIMO nonlinear plants with coupling effects.

The contraction mapping theorem in an extended form for Lipschitz operator is introduced.

Contraction: Let X be a normed linear space and let D be a subset of X. A Lipschitz operator $Q : D \to X$ is called a contraction on D if its norm $\|Q\| < 1$ on D.

Lemma 3.7 [1] Let X be a Banach space and let $Q : X \to X$ be an operator (not necessarily Lipschitz). If there exists an $n \geq 1$ such that Q^n is Lipschitz on X with $L = \|Q^n\| < 1$, then there exists a unique element $x^* \in X$ such that $Q(x^*) = x^*$.

For the coupling effects existing in a plant, one general method is to decouple the effects. In the following, operator-based decoupling control by feedback is considered.

A MIMO nonlinear plant with coupling effects is described in Figure 3.49, where U and V denote the input space and output space, $\mathbf{u} = (\mathbf{u}_1, \mathbf{u}_2, \ldots, \mathbf{u}_n)$ and $\mathbf{y} = (\mathbf{y}_1, \mathbf{y}_2, \ldots, \mathbf{y}_n)$ are the input and output of a nominal plant $\mathbf{P} = (\mathbf{P}_1, \mathbf{P}_2, \ldots, \mathbf{P}_n) : \mathbf{U} \to \mathbf{V}$, and the plant has right factorization as $\mathbf{P} = \mathbf{N}\tilde{\mathbf{D}}^{-1}$, where $\mathbf{N} = (\mathbf{N}_1, \mathbf{N}_2, \ldots, \mathbf{N}_n) : \mathbf{W} \to \mathbf{V}$ is nonlinear and stable, $\tilde{\mathbf{D}} = \mathbf{DF} : \mathbf{W} \to \mathbf{U}$ such that $\mathbf{D} = (\mathbf{D}_1, \mathbf{D}_2, \ldots, \mathbf{D}_n)$ is diagonalizable. Suppose that D_i is linear, stable, and invertible (that D_i is linear does not mean that D_i^{-1} is also linear) and $\mathbf{F}$ is an affine operator in relation to the coupling effects such that

$$\mathbf{v(t)} = \mathbf{F(w)(t)} = \mathbf{F} \cdot \mathbf{w(t)} + \mathbf{f(t)} \tag{3.165}$$

where, $\mathbf{w} = (\mathbf{w}_1, \mathbf{w}_2, \ldots, \mathbf{w}_n) \in \mathbf{W}$ is quasi-state, W is quasi-state space, $\mathbf{v} = (\mathbf{v}_1, \mathbf{v}_2, \ldots, \mathbf{v}_n) \in \mathbf{W}$, v_i is the output of D_i^{-1}, $F = (F_{ij})_{n \times n}$ is a constant representing the coupling relations between $\mathbf{w}$ and $\mathbf{v}$, and $\mathbf{f} = (\mathbf{f}_1, \mathbf{f}_2, \ldots, \mathbf{f}_n) \in \mathbf{W}$ is unknown but bounded.

Concerning the nonlinear control system design of the given MIMO nonlinear plant, a feedback control design is shown in Figure 3.50, where $\mathbf{S}$ is the feedback operator to be designed, $\mathbf{b} = (\mathbf{b}_1, \mathbf{b}_2, \ldots, \mathbf{b}_n)$ is the feedback signal with $\mathbf{b} = \mathbf{S(y)}$, namely, $b_i = \sum_{j=1}^{n} S_{ij}(y_j)(t)$, and S_{ij} is a stable nonlinear operator. Thus, the decoupling problem is solved by the following conclusion.

Theorem 3.7 Consider the MIMO nonlinear plants shown in Figure 3.50. If

$$S_{ij}N_j + F_{ij}D_i = \begin{cases} 0 & j \neq i \\ R_i & j = i \end{cases} \tag{3.166}$$

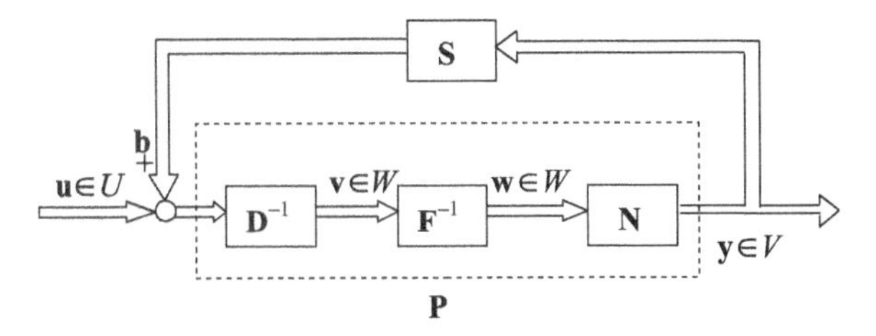

FIGURE 3.50 Decoupling structure of MIMO nonlinear systems.

then the plant is decoupled provided that R_i is invertible and stable, where 0 and R_i are operators.

Proof The feedback structure of Figure 3.50 indicates that

$$\begin{aligned}
u_i(t) &= D_i(v_i)(t) + \sum_{j=1}^{n} S_{ij}(y_j)(t) \\
&= D_i(f_i)(t) + \sum_{j=1}^{n} (S_{ij}N_j + D_i F_{ij})(w_j)(t) \\
&= D_i(f_i)(t) + R_i(w_i)(t)
\end{aligned} \tag{3.167}$$

This completes the proof. ■

As a result, the internal coupling effects are decoupled, and the feedback design reconstructs one right factorization as $\tilde{P}_i = N_i R_i^{-1}$, which implies that the system shown in Figure 3.50 is the same as the one in Figure 3.51, where the input of the feedback structure is equivalent to $u_i + \tilde{d}_i = u_i - D_i(f_i)$, $\tilde{\mathbf{d}} = (\tilde{d}_1, \tilde{d}_2, \ldots, \tilde{d}_n)$. Regard the internal signal $D_i[f_i(t)]$ as the disturbance to the ith feedback subsystem. Then the MIMO system is divided into n independent subsystems. If R_i^{-1} is stable, without any doubt, the MIMO system is BIBO stable. If R_i^{-1} is unstable, other controllers should be designed to guarantee the stability of the decoupled plant.

The operator-based feedback control system for the decoupled system is given in Figure 3.52, where $\mathbf{r} = (\mathbf{r}_1, \mathbf{r}_2, \ldots, \mathbf{r}_n) \in \mathbf{U}$, $\mathbf{e} = (\mathbf{e}_1, \mathbf{e}_2, \ldots, \mathbf{e}_n) \in \mathbf{U}$, $\mathbf{z} = (\mathbf{z}_1, \mathbf{z}_2, \ldots, \mathbf{z}_n) \in \mathbf{U}$, and $\mathbf{y} = (\mathbf{y}_1, \mathbf{y}_2, \ldots, \mathbf{y}_n) \in \mathbf{V}$ are reference input, error, control input, and plant output, respectively. Assume $\mathbf{A} = (\mathbf{A}_1, \mathbf{A}_2, \ldots, \mathbf{A}_n) : \mathbf{V} \to \mathbf{U}$ and $\mathbf{B} = (\mathbf{B}_1, \mathbf{B}_2, \ldots, \mathbf{B}_n) : \mathbf{U} \to \mathbf{U}$ are controllers to be designed (the so-called system design problem) with B_i being invertible. By this structure, the coupling effects may

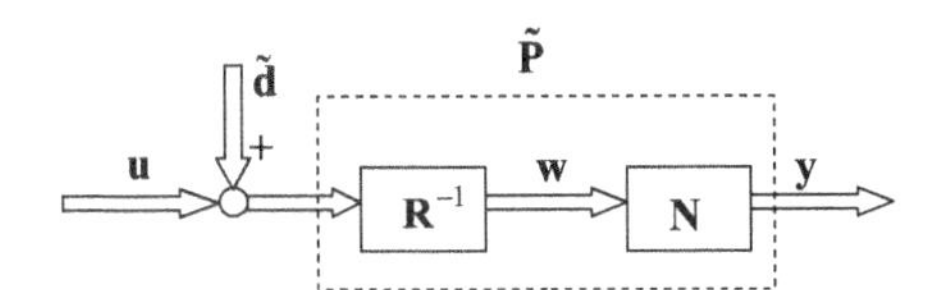

FIGURE 3.51 Equivalent system of Figure 3.50.

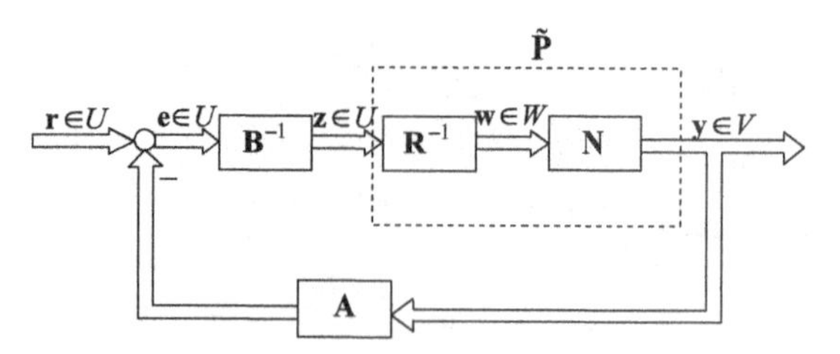

FIGURE 3.52 MIMO nonlinear feedback systems.

happen because of the effects generated by the controllers and plant outputs of the controlled systems, which are always ignored in many researches. In this work, this kind of coupling effect is also taken into consideration. To consider this kind of coupling effect in detail, assume that u_i is the output of the controller B_i and y_i is the plant output of the ith subsystem. As stated in Section 3.4, the coupling effects are demonstrated as follows:

$$z_i(t) = u_i(t) + \sum_{\substack{j=1 \\ j \neq i}}^{n} G_{ij}(u_j)(t) \qquad y_i = P_i(z_i)(t) \tag{3.168}$$

$$z_i(t) = u_i(t) + \sum_{\substack{j=1 \\ j \neq i}}^{n} H_{ij}(y_j)(t) \qquad y_i = P_i(z_i)(t) \tag{3.169}$$

$$y_i(t) = P_i(u_i)(t) + \sum_{\substack{j=1 \\ j \neq i}}^{n} J_{ij}(u_j)(t) \tag{3.170}$$

$$y_i(t) = P_i(u_i)(t) + \sum_{\substack{j=1 \\ j \neq i}}^{n} Q_{ij}(y_j)(t) \tag{3.171}$$

where G_{ij}, H_{ij}, J_{ij}, and Q_{ij} $(i, j = 1, 2, \ldots, n,\ j \neq i)$ are internal operators, denoting the sum of all coupling effects. For example, in (3.168) all of the coupling effects caused by the controller from the jth plant to the ith plant are denoted by the operator G_{ij}. Without loss of generality, for the coupling effects mentioned above, only the first kind is considered in this book. Thus, for the decoupled plant shown in Figure 3.50, the real control input is

$$z_i(t) = u_i(t) + \sum_{\substack{j=1 \\ j \neq i}}^{n} G_{ij}(u_j)(t) - D_i(f_i)(t) \tag{3.172}$$

$$:= u_i(t) + d_i(t) \tag{3.173}$$

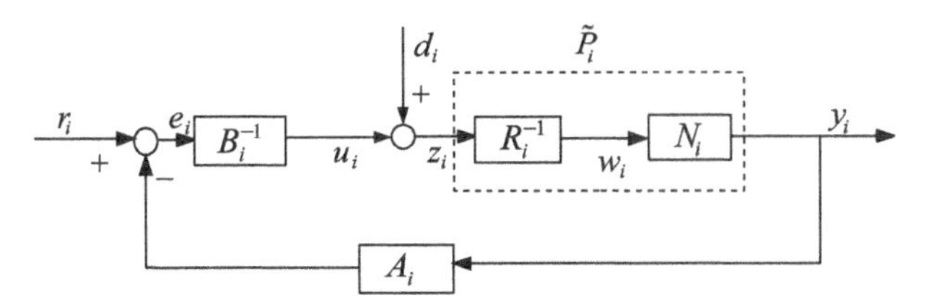

FIGURE 3.53 SISO feedback control system with disturbance.

Thus, to consider the MIMO nonlinear system shown in Figure 3.52, we first need to consider the SISO system with disturbance shown in Figure 3.53.

Theorem 3.8 Considering the system shown in Figure 3.53, the controller B_i is designed to have right factorization as $B_i = C_i R_i^{-1}$, where $C_i : W \rightarrow U$ is a unimodular operator. If C_i^{-1} and $A_i N_i$ are generalized Lipschitz operators with Lipschitz constants L_i and K_i such that $L_i K_i < 1$, then the nonlinear system is BIBO stable.

Proof Assume that there exists $d_i'(t) \in W$ such that

$$R_i^{-1}(u_i + d_i)(t) = R_i^{-1}(u_i)(t) + d_i'(t) = w_i(t) \tag{3.174}$$

where $d_i'(t)$ is unknown. Thus the system has

$$\begin{aligned} r_i(t) &= A_i N_i(w_i)(t) + B_i R_i(w_i - d_i')(t) \\ &= A_i N_i(w_i)(t) + C_i(w_i - d_i')(t) \end{aligned} \tag{3.175}$$

It follows that

$$w_i(t) = C_i^{-1}(r_i - A_i N_i w_i)(t) + d_i'(t) := f_i(w_i)(t) \tag{3.176}$$

Then, for any given $r_i \in U_s$

$$\begin{aligned} \| f_i(w_i) - f_i(\tilde{w}_i)\| &= \|C_i^{-1}(r_i - A_i N_i w_i)(t) - C_i^{-1}(r_i - A_i N_i \tilde{w}_i)(t)\| \\ &\leq L_i K_i \|w_i - \tilde{w}_i\| \end{aligned} \tag{3.177}$$

It can be known that $f(\cdot)$ is a contraction mapping under the condition of $L_i K_i < 1$. According to the concept of the Lipschitz operator and the contraction mapping theorem, w_i is uniquely determined by r_i and $w_i \in W_s$ for any $r_i \in U_s$. Also, the stability of A_i, N_i, R_i, C_i^{-1} implies that $y_i \in V_s$, $e_i \in U_s$, and $z_i \in U_s$, namely, the system is stable. Moreover, that C_i is unimodular indicates that $C_i^{-1} e_i \in W_s$, and d_i' is bounded because $w_i = C_i^{-1} e_i + d_i'$. ■

Thus, based on the above analysis, the stability of the MIMO system is equivalent to the stability of the SISO system with disturbance.

Theorem 3.9 Assume that the system shown in Figure 3.52 is well-posed and the controller B_i has right factorization as $B_i = C_i R_i^{-1}$, where $C_i : W \to U$ is a unimodular operator. Under the condition that the operators C_i^{-1} and $A_i N_i$ are generalized Lipschitz operators with Lipschitz constants L_i and K_i such that $L_i K_i < 1$, the MIMO nonlinear system is stable.

Proof Similar to the proof of Theorem 3.8, the system is stable under the assumption. ■

Also, if there are uncertainties in the system, they can be regarded as one part of the disturbance, and thus the system also can be stabilized using the analysis of Theorem 3.9. In the following, for the stabilized system, the output tracking performance is considered.

The output tracking system is shown in Figure 3.54, which is designed based on internal model control, where, $\mathbf{T} = (\mathbf{T}_1, \mathbf{T}_2, \ldots, \mathbf{T}_n)$ is the tracking operator, $\mathbf{r}^* = (\mathbf{r}_1^*, \mathbf{r}_2^*, \ldots, \mathbf{r}_n^*)$ is the reference input of the system, and $\tilde{\mathbf{P}} = (\tilde{\mathbf{P}}_1, \tilde{\mathbf{P}}_2, \ldots, \tilde{\mathbf{P}}_n)$ has $\tilde{P}_i = N_i R_i^{-1}$. Assume $\tilde{\mathbf{P}}_0 = (\tilde{\mathbf{P}}_{01}, \tilde{\mathbf{P}}_{02}, \ldots, \tilde{\mathbf{P}}_{0n})$, is the model of $\tilde{\mathbf{P}}$ and has right factorization as $\tilde{P}_{0i} = N_{0i} R_i^{-1}$ with $N_{0i} \to N_i$, $\phi = (\phi_1, \phi_2, \ldots, \phi_n)$ is a linear operator such that $\phi_i[\alpha(t)] \to 0$ for any bounded $\alpha(t)$, and ϕ_i^{-1} is linear.

Theorem 3.10 Consider the MIMO nonlinear feedback control system shown in Figure 3.54. Assume that

$$B_i R_i + A_i N_{0i} = \tilde{M}_i \tag{3.178}$$

where $\tilde{M}_i$ is a unimodular operator. If

$$N_i \tilde{M}_i^{-1} T_i = I \tag{3.179}$$

then the plant output tracks to the reference input.

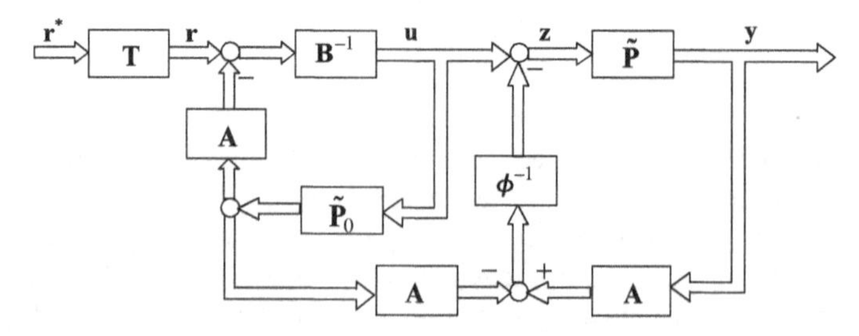

FIGURE 3.54 Tracking systems.

Proof Based on the system design scheme, the following relation is satisfied:

$$z_i(t) = u_i(t) + d_i(t) + \phi_i^{-1} A_i \tilde{P}_{0i}(u_i)(t) - \phi_i^{-1} A_i \tilde{P}_i(z_i)(t)$$

that is,

$$\phi_i^{-1}(\phi_i + A_i \tilde{P}_i)(z_i)(t) = d_i(t) + \phi_i^{-1}(\phi_i + A_i \tilde{P}_{0i})(u_i)(t)$$

Thus, when the model $\tilde{P}_{0i}$ arbitrarily approaches to the true plant $\tilde{P}_i$, the linearity of ϕ_i^{-1} and the property of ϕ_i imply that

$$z_i(t) \to u_i(t)$$

The Bezout identity (3.178) indicates that

$$u_i(t) = R_i \tilde{M}_i^{-1} T_i(r_i^*)(t)$$

and

$$y_i(t) = N_i R_i^{-1}(z_i)(t)$$

Then $y_i(t)$ tracks to $r_i^*(t)$ if (3.179) is satisfied. ■

In this section, the operator-based control design of a MIMO nonlinear plant with two kinds of coupling effects is considered. The system stability is guaranteed using the contraction mapping theorem and robust right coprime factorization. And output tracking is realized by the tracking design.

To illustrate the effectiveness of the operator-based robust control design scheme, an example of the temperature control of the three-input, three-output aluminum plate thermal process is given. Figure 3.55 shows a three-input, three-output aluminum plate thermal process device, and the photo of the process is given in Figure 3.56. The device includes three parts: aluminum part, interface part, and computer part.

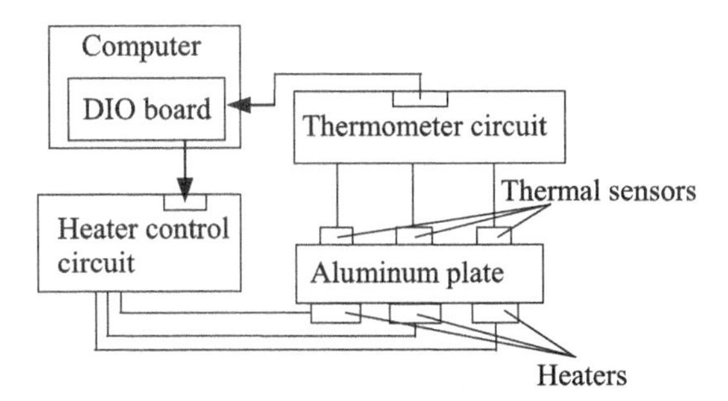

FIGURE 3.55 Three-input, three-output aluminum plate thermal process.

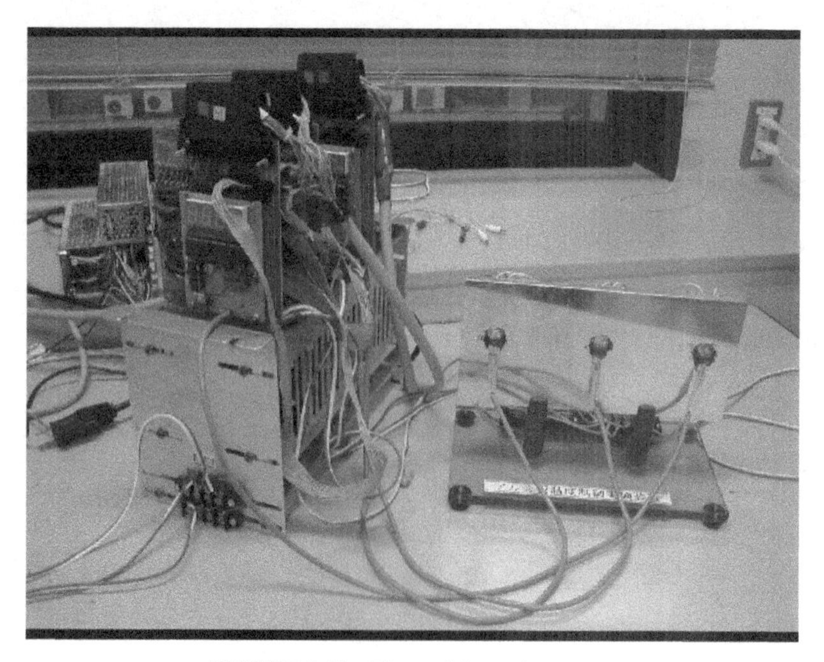

FIGURE 3.56 Photo of thermal process.

The first part consists of an aluminum board including three thermal sensors and three heaters.

The interface is composed of a pulse width modulation (PWM) circuit for heater control that has an oscillator, comparator, counter and solid-state relay (SSR), three thermometer circuits, and two digital input–output (DIO) boards. The PWM circuit creates clock pulses for controlling the output of the heater. Before this, 8-bit binary scale data are sent to the comparator from a computer which keeps the SSR on. If the data toward the comparator are greater than the original value from the computer, the SSR is turned off. The length of time that the SSR is switched on determines the power of the heater. To measure the temperature of an aluminum plate, a thermometer kit by Akizuki Denshi Tsusho Co. is used. Thermal sensors in the kit can measure between −40°C and 100°C, and the measured temperatures are outputted as a voltage which the analog-to-digital (A/D) converter (ICL7137) displays on a segment so that the temperature can be realized easily. The binary data are converted to base 10 at the computer. The DIO boards (PCI-2752C) are released by interface corporation, and each one has 32 digital input–output pins. They are connected to the A/D converter (ICL7137) on the thermometer circuits, the comparator (74LS85) on the PWM circuits, and the ground. There is a computer installed in the program that enables it to transfer the input or output digital signals to the DIO board in the computer.

The model in Figure 3.57 shows the three-input, three-output aluminum plate thermal process, where the plate board is divided into five parts corresponding to the

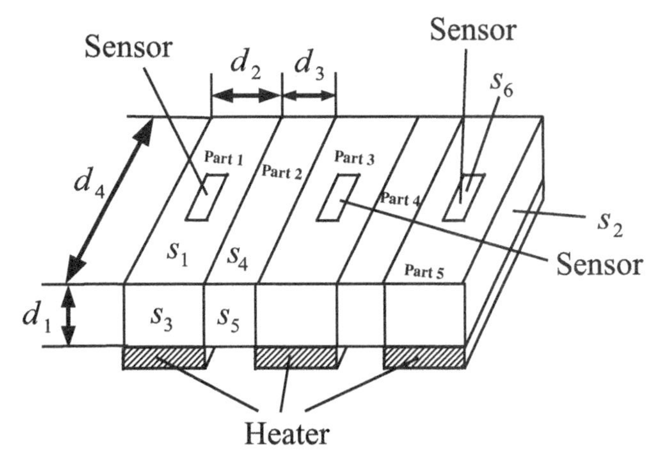

FIGURE 3.57 Model of thermal process.

distribution of thermal sensors and heaters. To get the mathematical model of the thermal process, three laws are used:

- Fourier's law of heat conduction:

$$q = -\lambda(d\theta/dn) \tag{3.180}$$

where $d\theta/dn$ (K/m) is the temperature gradient and q (W/m^2) is the heat flux.

- Newton's law of cooling:

$$q = \alpha(\theta_s - \theta_f) \tag{3.181}$$

where α is the heat transfer coefficient, θ_f (K) is the flowing air temperature, and θ_s (K) is the temperature of the object.

- Equation of the heat capacity and objects and their specific heat:

$$mcd\theta = dQ \tag{3.182}$$

where dQ[J] is the variation in heat capacities.

Based on these laws, the following equations are derived [45]:

Part 1:

$$\frac{d(\theta_1 - \theta_f)}{dt} m_1 c = u_1 - \left\{ \alpha(\theta_1 - \theta_f)(s_1 + s_2 + 2s_3 - s_6) + \lambda \frac{\theta_1 - \theta_2}{d_2} s_2 \right\} \tag{3.183}$$

TABLE 3.2 Parameters of Aluminum Plate

$d_1 = 0.01$ m	$d_2 = 0.06$ m	$d_3 = 0.035$ m	$d_4 = 0.12$ m
$s_1 = d_2 d_4$	$s_2 = d_1 d_4$	$s_3 = d_1 d_2$	
$s_4 = d_4 d_3$	$s_5 = d_1 d_3$	$s_6 = 2.7 * 10^{-8}$	

Part 2:

$$\frac{d(\theta_2 - \theta_f)}{dt} m_2 c = -\left\{\alpha(\theta_2 - \theta_f)(2s_4 + 2s_5) + \lambda \frac{\theta_2 - \theta_1}{d_3} s_2 + \lambda \frac{\theta_2 - \theta_3}{d_3} s_2\right\} \tag{3.184}$$

Part 3:

$$\frac{d(\theta_3 - \theta_f)}{dt} m_3 c = u_3 - \left\{\alpha(\theta_3 - \theta_f)(s_1 + 2s_3 - s_6) + \lambda \frac{\theta_3 - \theta_2}{d_2} s_2 + \lambda \frac{\theta_3 - \theta_4}{d_2} s_2\right\} \tag{3.185}$$

Part 4:

$$\frac{d(\theta_4 - \theta_f)}{dt} m_4 c = -\left\{\alpha(\theta_4 - \theta_f)(2s_4 + 2s_5) + \lambda \frac{\theta_4 - \theta_3}{d_3} s_2 + \lambda \frac{\theta_4 - \theta_5}{d_3} s_2\right\} \tag{3.186}$$

Part 5:

$$\frac{d(\theta_5 - \theta_f)}{dt} m_5 c = u_5 - \left\{\alpha(\theta_5 - \theta_f)(s_1 + s_2 + 2s_3 - s_6) + \lambda \frac{\theta_5 - \theta_4}{d_2} s_2\right\} \tag{3.187}$$

where $\theta_1, \ldots, \theta_5$ are the temperatures for parts 1–5, respectively. The utilized parameters are shown in Tables 3.2 and 3.3. Define $\theta_i - \theta_f = y_i(t)$ $(i = 1, \ldots, 5)$. Then the equations are transformed to the following differential equations:

TABLE 3.3 Parameters of Model

Specific heat of aluminum	$c = 900$ J/kg K
Thermal conductivity	$\lambda = 200$ W/m K
Heat transfer coefficient	$\alpha = 10$ W/m^2 K
Density of aluminum	$d = 2700$ kg/m^3
Mass of aluminum plate	$m_i = d_1 d_2 d_4 d$ kg, $i = 1,3,5$

Part 1:

$$\frac{dy_1(t)}{dt} = -E_1 y_1(t) + \frac{1}{m_1 c} u_1(t) + \frac{\lambda s_2}{m_1 c d_2} y_2(t) \tag{3.188}$$

Part 2:

$$\frac{dy_2(t)}{dt} = -E_2 y_2(t) + \frac{\lambda s_2}{m_2 c d_3} \left\{ y_1(t) + y_3(t) \right\} \tag{3.189}$$

Part 3:

$$\frac{dy_3(t)}{dt} = -E_3 y_3(t) + \frac{1}{m_3 c} u_3(t) + \frac{\lambda s_2}{m_3 c d_2} \left\{ y_2(t) + y_4(t) \right\} \tag{3.190}$$

Part 4:

$$\frac{dy_4(t)}{dt} = -E_4 y_4(t) + \frac{\lambda s_2}{m_4 c d_3} \left\{ y_3(t) + y_5(t) \right\} \tag{3.191}$$

Part 5:

$$\frac{dy_5(t)}{dt} = -E_5 y_5(t) + \frac{1}{m_5 c} u_5(t) + \frac{\lambda s_2}{m_5 c d_2} y_4(t) \tag{3.192}$$

where u_i and y_i $(i = 1, 3, 5)$ are process input and process output, y_2 and y_4 are the hypothetical process outputs of parts 2 and 4, respectively, and E_i is given below:

$$E_1 = \frac{1}{m_1 c} \left(\alpha(s_1 + s_2 + 2s_3 - s_6) + \frac{\lambda s_2}{d_2} \right) \tag{3.193}$$

$$E_2 = \frac{1}{m_2 c} \left(\alpha(2s_4 + 2s_5) + \frac{2\lambda s_2}{d_3} \right) \tag{3.194}$$

$$E_3 = \frac{1}{m_3 c} \left(\alpha(s_1 + 2s_3 - s_6) + \frac{2\lambda s_2}{d_2} \right) \tag{3.195}$$

$$E_4 = \frac{1}{m_4 c} \left(\alpha(2s_4 + 2s_5) + \frac{2\lambda s_2}{d_3} \right) \tag{3.196}$$

$$E_5 = \frac{1}{m_5 c} \left(\alpha(s_1 + s_2 + 2s_3 - s_6) + \frac{\lambda s_2}{d_2} \right) \tag{3.197}$$

Based on the control design scheme for MIMO nonlinear systems, the nominal plant is selected as

$$\dot{y}_i = \frac{1}{m_i c} u_i - E_i y_i \qquad i = 1, 3, 5 \tag{3.198}$$

with right factorization as

$$N_i: \ \dot{y}_i = w_i - E_i y_i$$
$$D_i^{-1}: \ w_i = \frac{1}{m_i c} u_i \tag{3.199}$$

The other parts of the system are considered as operators related to the coupling effects from the outputs to the outputs. That is,

$$\begin{array}{llll} H_{12} = H y_2(t) & H_{13} = 0 & H_{14} = 0 & H_{15} = 0 \\ H_{31} = 0 & H_{32} = H y_2(t) & H_{34} = H y_4(t) & H_{35} = 0 \\ H_{51} = 0 & H_{52} = 0 & H_{53} = 0 & H_{54} = H y_4(t) \end{array} \tag{3.200}$$

where $H = \lambda s_2/(mcd_2)$, $m = m_1 = m_3 = m_5$. Then, the control operators are designed as

$$B_i(x_i)(t) = \frac{b_i}{mc} x_i(t) \tag{3.201}$$
$$A_i(y_i)(t) = (1 - b_i)(\dot{y}_i + E_i y_i)(t) \tag{3.202}$$

such that $A_i N_i + B_i D_i = I$. In the real control process, the exothermic effect is nonlinear and unknown but bounded in practice. Thus, the effects caused in the exothermic process are considered as unknown but bounded disturbance $\tilde{d}_i$ ($i = 1, 3, 5$). For rejecting the effects, $\phi_i(\alpha_i(t)) = (1/n) \cdot \alpha_i(t)$ is selected for arbitrary bounded $\alpha_i(t)$, where $\alpha_i(t)/n$ can be made arbitrarily small by selecting n large enough. Then the tracking operator is designed as

$$T_i(r_i^*)(t) = E_i(r_i^*)(t) + (1 - E_i)e^{-t}(r_i^*)(t) \qquad i = 1, 3, 5 \tag{3.203}$$

In the simulation, a bounded uncertainty ΔP_i is included which refers to the unmodeled uncertainty and is always bounded. Without loss of generality, suppose that N_i has a bounded uncertainty ΔN_i such that the operators and $N_i + \Delta N_i$ are stable and

$$P_i + \Delta P_i = (N_i + \Delta N_i) D_i^{-1}$$
$$(N_i + \Delta N_i)(w_i)(t) = (e^{-E_i t} + \Delta_i) \int e^{E_i \tau} w_i(\tau) d\tau \qquad i = 1, 3, 5 \tag{3.204}$$

where Δ_i is assumed to be variable, making ΔN_i stable with known bound. Then

$$\Delta_i = 0.001 \text{ rand } \frac{e^{-E_i t}}{(1 - b_i)[u_i(t)/(m_i c) + E_i y_i(t)]} \qquad i = 1, 3, 5$$

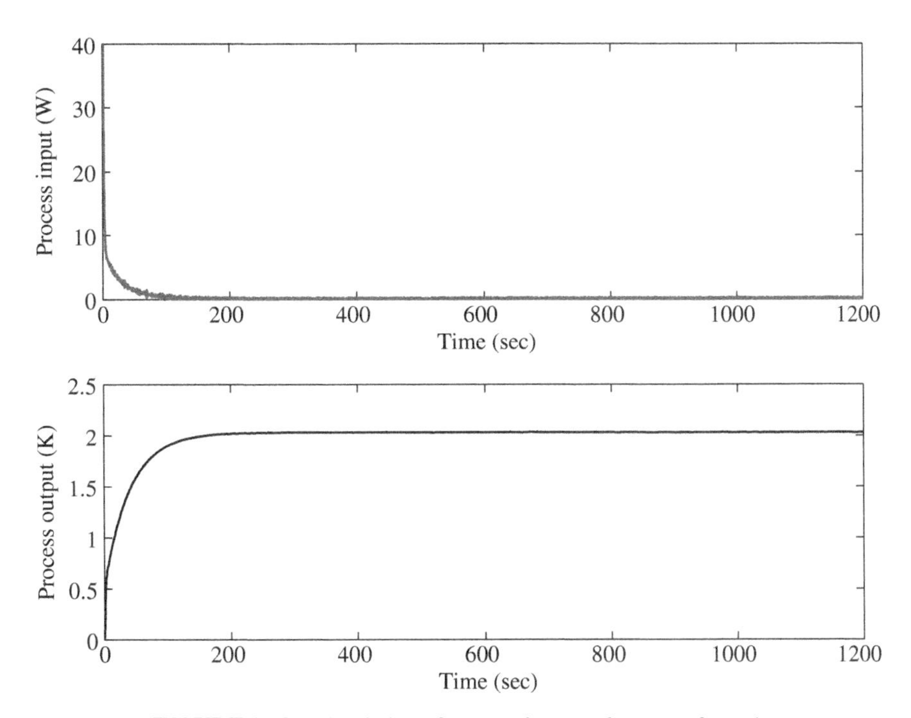

FIGURE 3.58 Simulation of process input and output of part 1.

In the simulation, the reference temperature is chosen as $r_i^* = 2k$ $(i = 1, 3, 5)$, that is, the objective is to heat 2°C for every heater by using the tracking structure. The parameter is selected as $b_i = 0.55$ and $n = 1000$ in the simulation. Then the simulation results about the process input and output of parts 1, 3, and 5 are shown in Figures 3.58–3.60, respectively. Also, the process outputs of the three parts are given in Figure 3.61. From the simulation results we can see that the input of part 3 is different from parts 1 and 5 because part 3 is influenced by two parts: 2 and 4. The output of part 3 is the best one because the heater of part 3 conducts more energy considering the neighboring two parts, while parts 1 and 3 are only influenced by one neighboring part. The results confirm the effectiveness of the design scheme.

An experiment on the temperature tracking of the MIMO system is conducted. Table 3.4 describes the parameters and conditions in this experiment. Figures 3.62–3.66 are the experimental results. Figure 3.62 shows process outputs of parts 1, 3, and 5 (temperature in parts 1, 3, and 5), where the initial temperatures of the three parts are 26.2°C, 27.8°C, and 27.2°C, respectively. Other parts (parts 2 and 4) calculated by differential equations (3.189) and (3.191) are in Figure 3.63. Figures 3.64–3.66 show the process input and output of parts 1, 3, and 5, respectively.

3.4.6.1 Discussion In the design scheme, the system is always assumed to be well-posed. The notion of well-posedness is of great importance in many researches, which ensures the existence and uniqueness for the signals of a system.

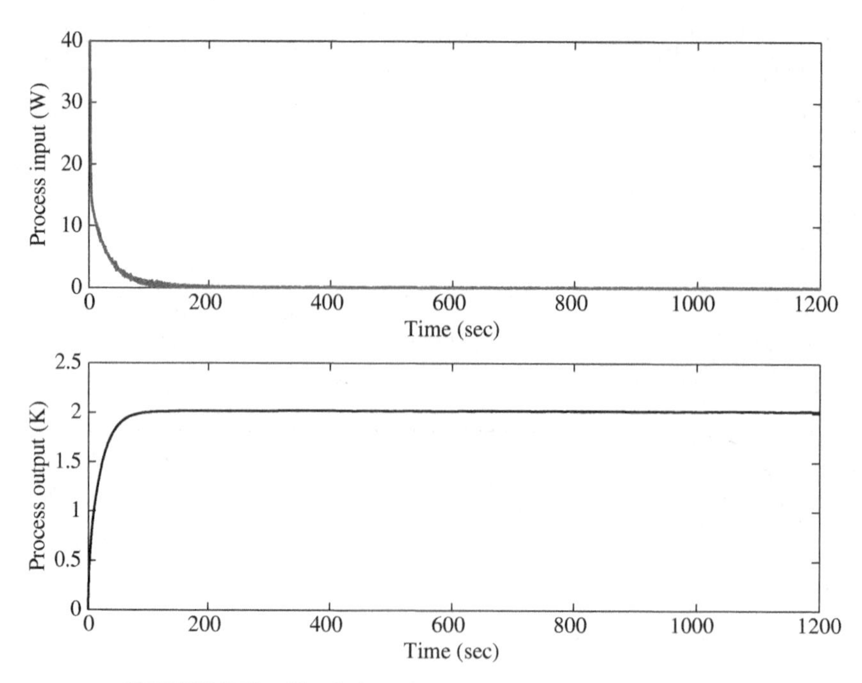

FIGURE 3.59 Simulation of process input and output of part 3.

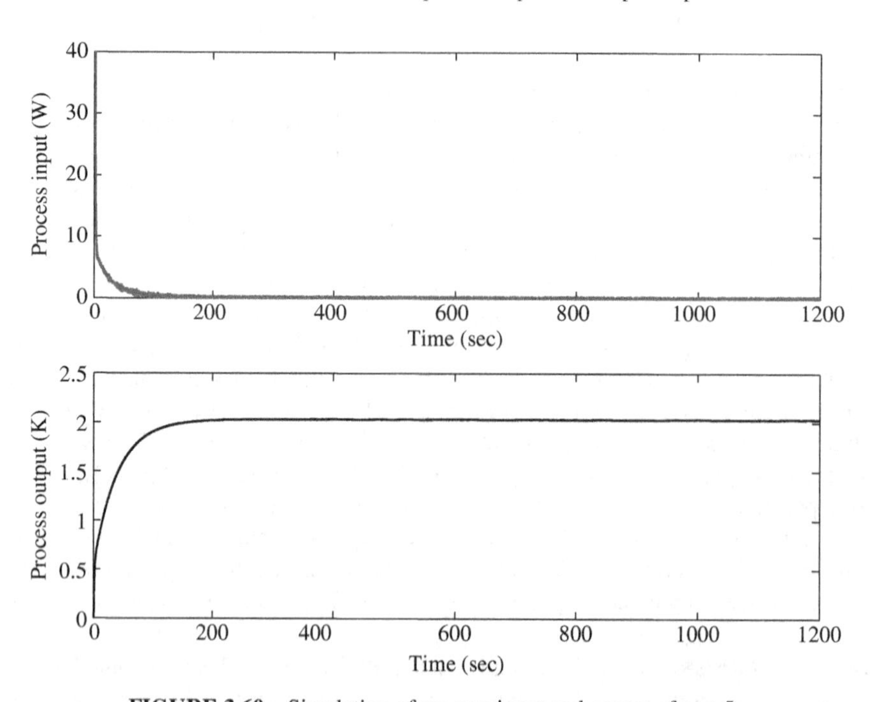

FIGURE 3.60 Simulation of process input and output of part 5.

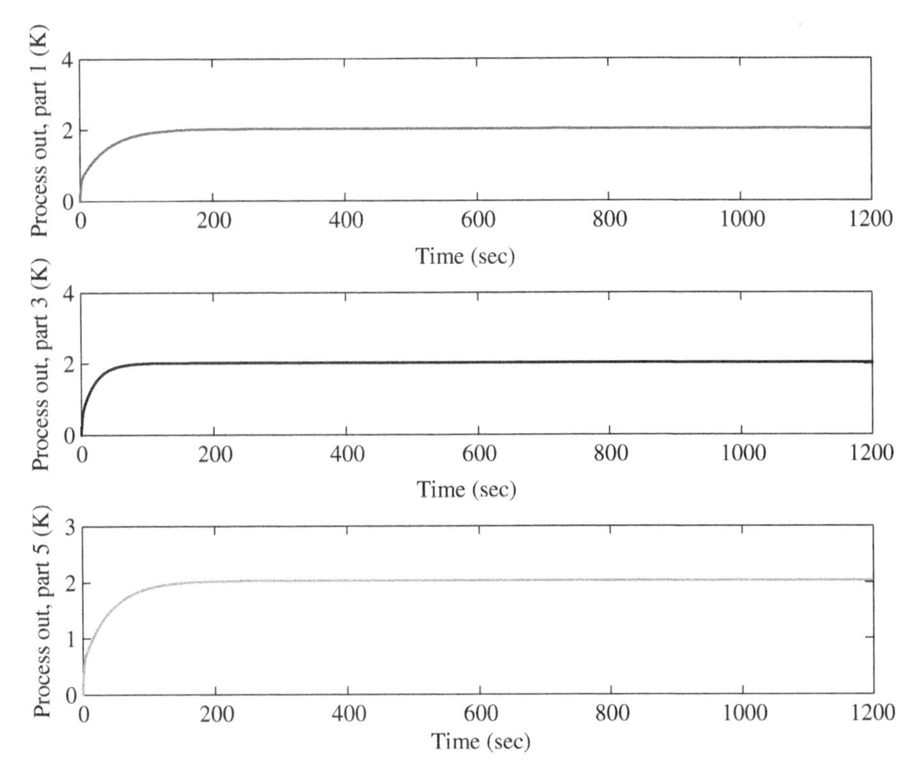

FIGURE 3.61 Simulation of process outputs of parts 1, 3, and 5.

In one two-input, two-output system, the well-posedness is defined as follows.

Consider the problem of stabilizing a nonlinear continuous-time process P by a controller K as shown in Figure 3.67, where the system is with real input spaces of continuous functions with continuous first derivative. For convenience, the feedback control system is denoted as $\{P, K\}$.

The system $\{P, K\}$ is well-posed if the closed-loop system input–output operator from u_1, u_2 to e_1, e_2 exists namely,

$$\begin{bmatrix} I & -K \\ -P & I \end{bmatrix}^{-1} \tag{3.205}$$

TABLE 3.4 Parameters of Experiment

Reference input	$r_1^* = r_3^* = r_5^* = 2.0$
Design parameters	$b_1 = 0.54, b_3 = 0.7$
	$b_5 = 0.4$
Experiment time	1200 sec
Sampling time	200 msec

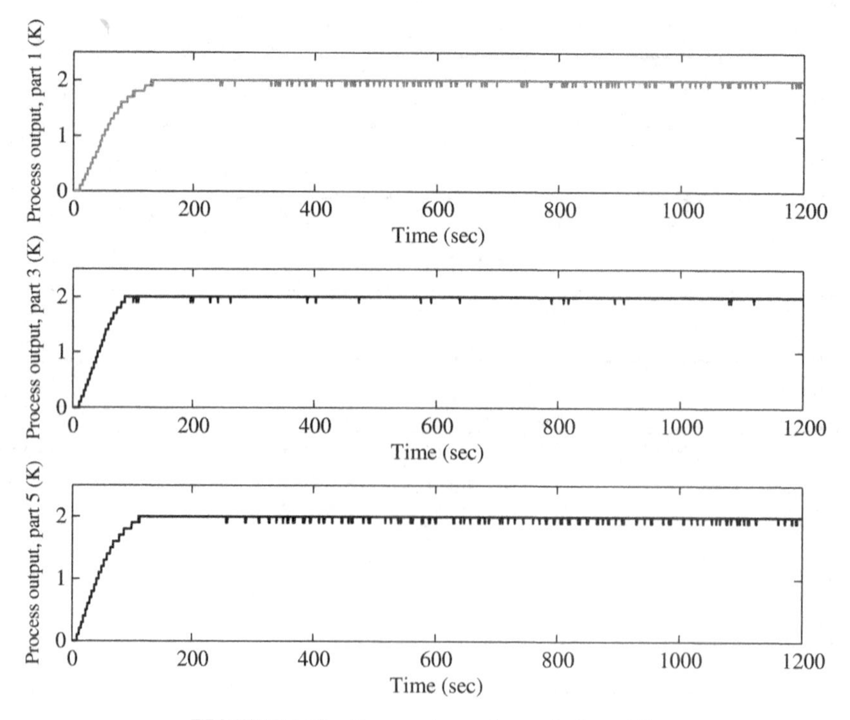

FIGURE 3.62 Temperature of parts 1, 3, and 5.

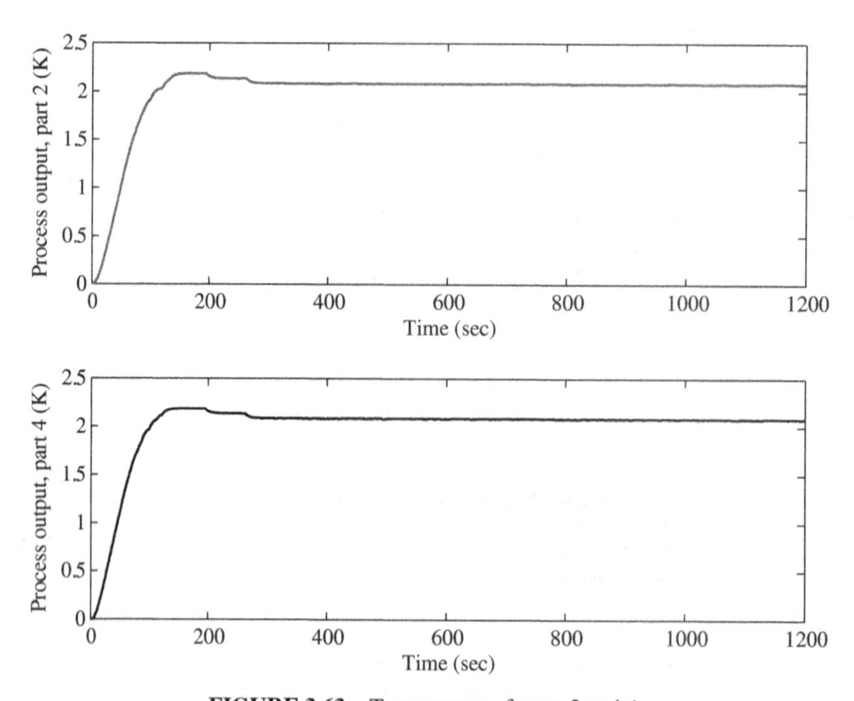

FIGURE 3.63 Temperature of parts 2 and 4.

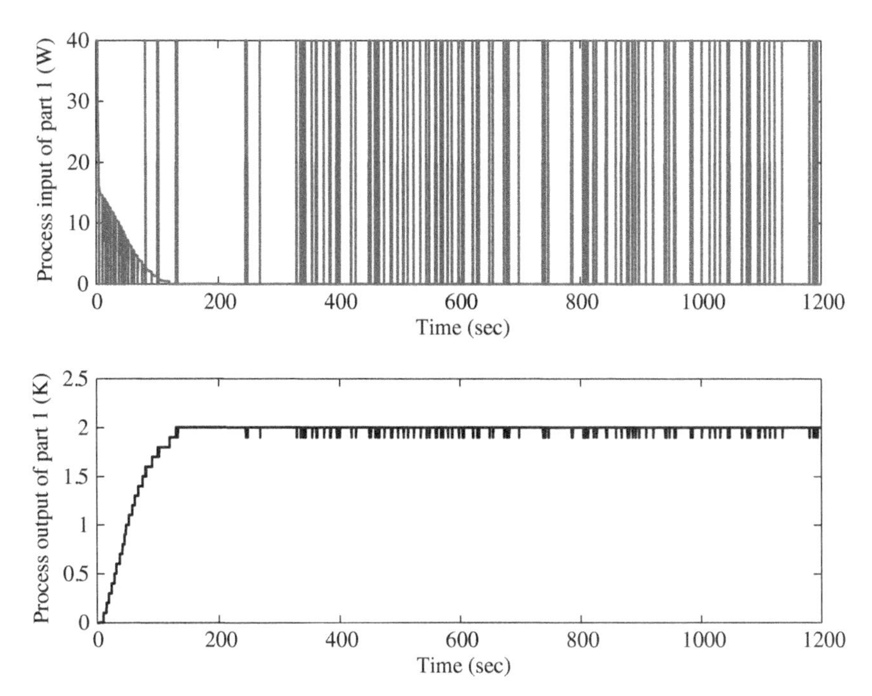

FIGURE 3.64 Experimental results of process input and output of part 1.

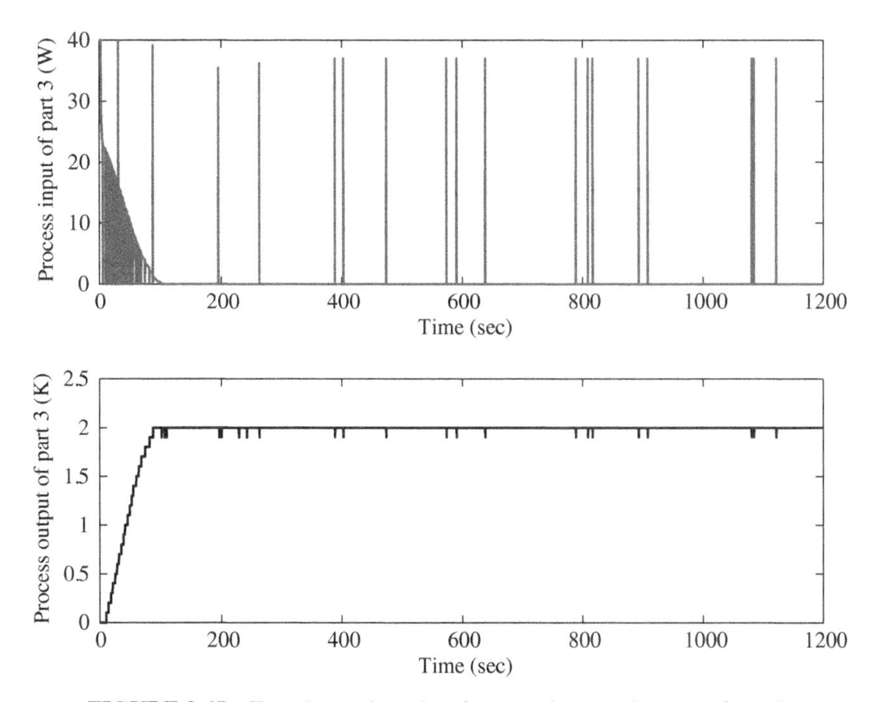

FIGURE 3.65 Experimental results of process input and output of part 3.

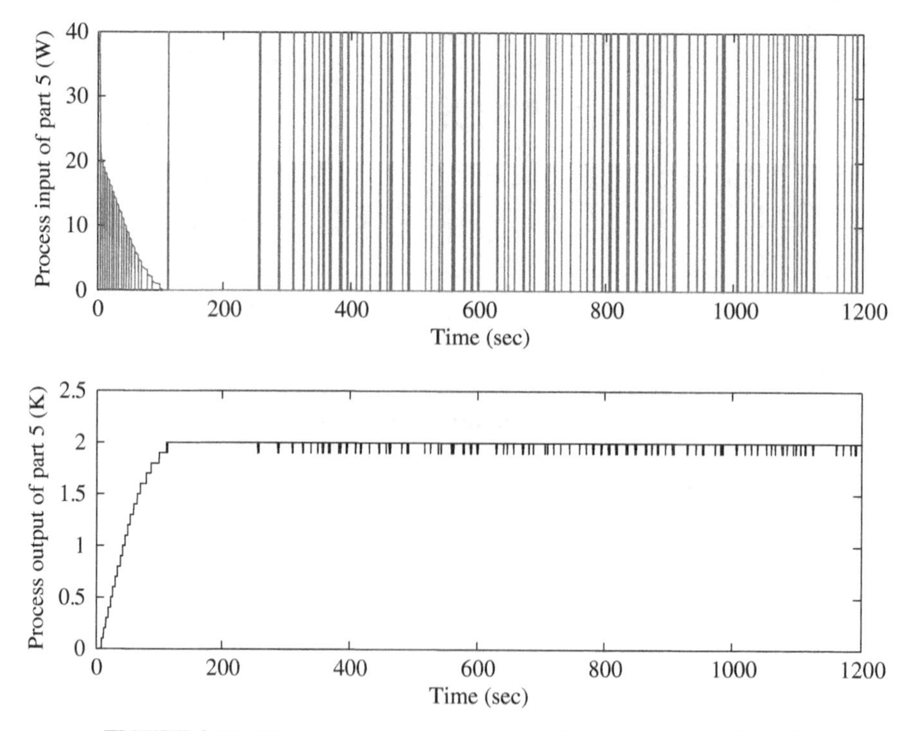

FIGURE 3.66 Experimental results of process input and output of part 5.

However, the condition (4.76) is difficult to be decided for nonlinear systems. The reason is that it is hard to determine the inverse of a matrix with nonlinear operators. To consider that issue, the contraction mapping theorem is effective. Consider the two-input, two-output system shown in Figure 3.68. The signals have the relations

$$\alpha_1 = \phi_1(\beta_1) - \psi_1(\beta_2) \qquad \alpha_2 = \phi_2(\beta_2) - \psi_2(\beta_1) \tag{3.206}$$

where $\alpha_i \in X$, $\beta_i \in Y$, $\phi_i \in \mathcal{U}(X, Y)$ is the unimodular operator, and $\psi_i : Y \rightarrow X$ $(i = 1, 2)$ is the stable operator. Then the following result is obtained.

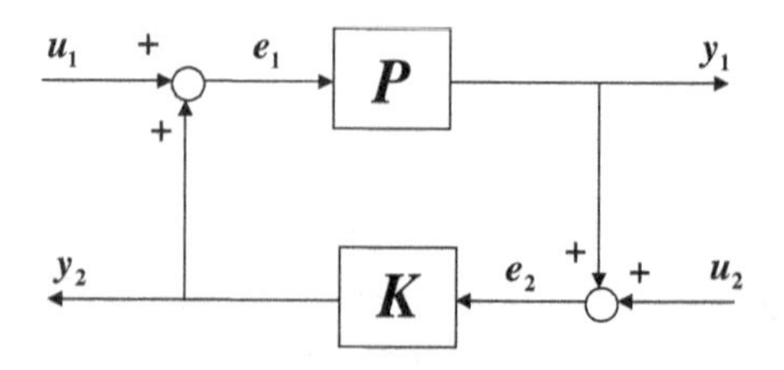

FIGURE 3.67 Feedback system $\{P, K\}$.

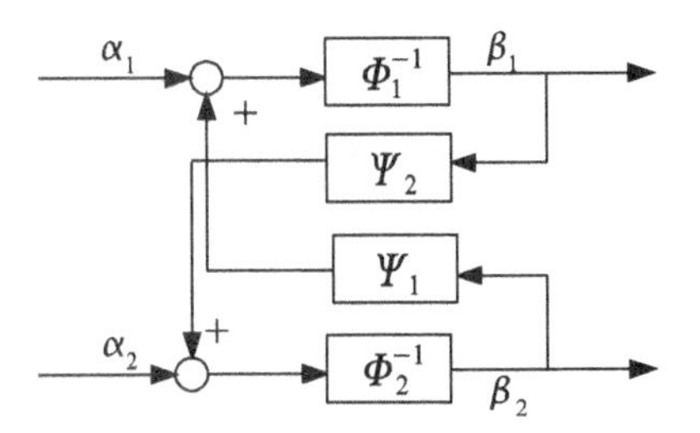

FIGURE 3.68 Two-input, two-output nonlinear system.

Lemma 3.8 Suppose that both ϕ_i^{-1} and ψ_i in the two-input, two-output system shown in Figure 3.68 are generalized Lipschitz operators with

$$\begin{aligned} \left\|[\phi_i^{-1}(\alpha)]_T - [\phi_i^{-1}(\tilde{\alpha})]_T\right\|_{Y_B} &\leq L_i \left\|\alpha_T - \tilde{\alpha}_T\right\|_{X_B} \\ \left\|[\psi_i(\beta)]_T - [\psi_i(\tilde{\beta})]_T\right\|_{X_B} &\leq K_i \left\|\beta_T - \tilde{\beta}_T\right\|_{Y_B} \end{aligned} \tag{3.207}$$

where $\alpha, \tilde{\alpha} \in X^e$, $\beta, \tilde{\beta} \in Y^e$, and L_i and K_i are Lipschitz constants such that $L_i K_i < 1$ $(i = 1, 2)$. Then, for any given $(\alpha_1, \alpha_2) \in X_s$, it follows that the signal pair $(\beta_1, \beta_2) \in Y_s$ and is uniquely determined by (α_1, α_2).

Proof Because ϕ_i $(i = 1, 2)$ is a unimodular operator, then

$$\beta_1 = \phi_1^{-1}[\alpha_1 + \psi_1(\beta_2)] \tag{3.208}$$

$$\beta_2 = \phi_2^{-1}[\alpha_2 + \psi_2(\beta_1)] \tag{3.209}$$

Consequently, substituting (3.208) and (3.209) into each other yields

$$\beta_1 = \phi_1^{-1}(\alpha_1 + \psi_1\phi_2^{-1}(\alpha_2 + \psi_2(\beta_1))) := f_1(\beta_1) \tag{3.210}$$

$$\beta_2 = \phi_2^{-1}(\alpha_2 + \psi_2\phi_1^{-1}(\alpha_1 + \psi_1(\beta_2))) := f_2(\beta_2) \tag{3.211}$$

For any given $\alpha_1, \alpha_2 \in X_s$, it follows that

$$\begin{aligned} &\|f_1(\beta_1) - f_1(\tilde{\beta}_1)\| \\ &\quad = \|\phi_1^{-1}\{\alpha_1 + \psi_1\phi_2^{-1}[\alpha_2 + \psi_2(\beta_1)]\} - \phi_1^{-1}\{\alpha_1 + \psi_1\phi_2^{-1}[\alpha_2 + \psi_2(\tilde{\beta}_1)]\}\| \\ &\quad \leq L_1\|\psi_1\phi_2^{-1}[\alpha_2 + \psi_2(\beta_1)] - \psi_1\phi_2^{-1}[\alpha_2 + \psi_2(\tilde{\beta}_1)]\| \\ &\quad \leq L_1 L_2 K_1 K_2 \|\beta_1 - \tilde{\beta}_1\| \end{aligned} \tag{3.212}$$

The condition $L_i K_i < 1\,(i = 1, 2)$ implies that f_1 is a contraction mapping on Banach space. According to the definition of the Lipschitz operator and the contraction mapping theorem, β_1 is uniquely determined by α_1 and α_2 and $\beta_1 \in Y_s$. The same result holds for β_2, completing the proof of the lemma. ■

Based on Lemma 3.8, the uniqueness of the signals is determined. That is, for the MIMO nonlinear feedback control system, if the controllers satisfy some restrictions, then the outputs are uniquely determined by a series of inputs. In other words, to a certain degree, the problem is transformed into a problem of points. However, Lemma 3.8 is effective when the amounts of the inputs and outputs are small, but it becomes complex for large amounts of inputs and outputs. To avoiding this, using the operator-based implicit function theorem is an effective method.

Suppose that one SISO feedback control system is written as

$$F(r, w) = 0 \tag{3.213}$$

where r and w denote the input and the quasi-state of operator-based control systems, respectively, and F represents the whole system.

Lemma 3.9 Assume that $F : U \times W \to U$ is Fréchet differentiable with $F(r_0, w_0) = 0$, and $(\partial/\partial w)F(r_0, w_0)$ is invertible. Then there exist neighborhood $U' \subseteq U$ of r_0 and a unique continuous operator $G : U \to W$ such that $F(r, G(r)) = 0$ and $w_0 = G(r_0)$. Moreover, if F is continuously differentiable, then the obtained G is also continuously differentiable.

This is the implicit function theorem in Banach space. In this book, the considered operators are generalized Lipschitz operators. As has been stated in Chapter 3, $\mathcal{A}(D, Y)$ is the family of operators defined by

$$\mathcal{A} = \big\{Q : D \to Y \big| Q'(x) \text{ exists for each } x \in X \tag{3.214}$$

$$\text{and} \sup_{x \in D} \| Q'(x)\|_Y < \infty\big\} \tag{3.215}$$

where $D \subseteq X$ is an open and convex subset. Moreover, let $\mathcal{A}_c = \mathcal{A}_c(D, Y)$ be the family of continuously differentiable operators defined by

$$\mathcal{A}_c = \big\{Q \in \mathcal{A} \big| Q'(x) \text{ is continuous on } D\big\} \tag{3.216}$$

Then, Lemma 3.5 still holds if $\mathcal{A}(D, Y)$ is replaced by $\mathcal{A}_c(D, Y)$. That is, $Q \in \mathcal{A}_c(D, Y)$ if and only if Q is differentiable on D and $Q \in \text{Lip}(D, Y)$. Moreover, $\mathcal{A}_c(D, Y)$ is a closed subspace of the Banach space $\text{Lip}(X, Y)$ under the Lipschitz norm. In summary, the relations of the tree spaces are

$$\mathcal{A}_c(D, Y) \subset \mathcal{A}(D, Y) \subset \text{Lip}(D, Y) \tag{3.217}$$

which implies that the condition of Lemma 3.9 is satisfied for the Lipschitz operator and continuous differentiability indicates BIBO stability.

The implicit function theorem and its variations are very important tools in the study of nonlinear control, mechanics and other engineering applications. Moreover, this theorem can be extended to the MIMO case. However, the existence and uniqueness of G are strongly dependent on the initial condition. Therefore, global existence for the implicit function problem is interesting and many versions of the global implicit function theorem have been given. In the following, one lemma on global existence for the implicit function of the MIMO system is introduced.

Lemma 3.10 [46] Considering system (3.213), assume that $F : R^n \times R^m \rightarrow R^1$ is continuous and continuously differentiable in the second variable $w \in R^m$. If

$$\left|\frac{\partial}{\partial w_i} F(r, w_1, \ldots, w_m)\right| - \sum_{j \neq i} \left|\frac{\partial}{\partial w_j} F(r, w_1, \ldots, w_m)\right| \geq d$$
$$\forall (r, w) \in R^n \times R^m$$

for some $i = 1, \ldots, m$, where $d > 0$ is a fixed constant. Then there exists a unique mapping $G : R^n \rightarrow R^m$ such that $F(r, G(r)) = 0$. Moreover, G is continuous. Additionally, if F is continuously differentiable, then the obtained G is also continuously differentiable.

For the temperature control of the three-input, three-output thermal process, for example, the operator-based MIMO nonlinear feedback control system is given as Figure 3.34, and the coupling effects can be summarized as the operator H_{ij} $(i, j = 1, 2, \ldots, n,\ j \neq i)$, which relates to the effect from the output of P_j to the input of P_i:

$$z_i(t) = u_i(t) + \sum_{\substack{j=1 \\ j \neq i}}^{n} H_{ij}(y_j)(t) \qquad i = 1, 2, \ldots, n \tag{3.218}$$

where u_i is the output of controller B_i^{-1}, z_i is the control input of the ith plant P_i, and y_j is the output of the jth plant P_j. Thus, the operator-based system can be described as

$$r_i(t) = M_i(w_i)(t) - \sum_{\substack{j=1 \\ j \neq i}}^{n} B_i H_{ij} N_j(w_j)(t) \tag{3.219}$$

where $M_i = A_i N_i + B_i D_i$ is the unimodular operator. As a consequence, to ensure the uniqueness of the operator-based MIMO system, the following result is obtained.

Theorem 3.11 Assume that the operators M_i, B_i, H_{ij}, and N_i in the MIMO system (3.219) are continuously differentiable. If

$$\left\| M_i'(w_i) \right\| - \sum_{j \neq i} \left\| B_i'(H_{ij} N_j(w_j)) \cdot H_{ij}'(N_j(w_j)) \cdot N_j'(w_j) \right\| \geq d \tag{3.220}$$

for some $i, j = 1, \ldots, n$, where $d > 0$ is a fixed constant, then there exists a unique operator $\mathbf{G} : \mathbf{W} \to \mathbf{U}$ such that $\mathbf{w} = \mathbf{G}(\mathbf{r})$. Moreover, G is continuously differentiable.

Proof Based on Lemma 3.10, the proof can be completed similarly. ■

From Theorem 3.11, it can be seen that under the condition of (3.220), for any given reference input r_i, the quasi-state w_i is determined uniquely, that it, well-posedness is guaranteed.

In Section 3.4.6, nonlinear control and tracking design for MIMO nonlinear systems with uncertainties and unknown coupling effects are considered using robust right coprime factorization. By the design scheme, a sufficient condition is obtained. Based on the derived condition, the interaction in the plant is decoupled by the feedback property, and the coupling effects caused by controllers and plant outputs are stabilized using the definition of the generalized Lipschitz operator and the contraction mapping theorem. Also, to ensure the desired output tracking performance, a control design structure is introduced. To demonstrate the control design, an example of the temperature control of the three-input, three-output aluminum plate thermal process is presented. Moreover, the sufficient condition for ensuring the uniqueness of the signals is also discussed.

3.4.7 Summary

Operator-based robust nonlinear control and tracking design for MIMO nonlinear systems with uncertainties and coupling effects were considered using robust right coprime factorization. By considering the coupling effects, sufficient conditions for MIMO nonlinear systems with uncertainties were given in detail. Based on the derived conditions, the MIMO system is robustly BIBO stable and the output tracking performance is realized. Moreover, a discussion of the uncertainty in realizing perfect tracking was also given.

3.5 OPERATOR-BASED TIME-VARYING DELAYED NONLINEAR FEEDBACK CONTROL SYSTEMS DESIGN

Recently, the modeling, analysis, and control of networked control systems have emerged as a topic of significant interest to science, technology, and our daily life. (See a brief survey in [47] and a special issue in [48].) One of the most important problems in networked control is the time-varying delay problem. Time-varying delays in a network have negative effects on the network's closed-loop stability and performance. For this challenging problem, a considerable number of studies have attempted to suppress the negative effects of the delays in a network. The used techniques include Ricatti equations and linear matrix inequalities. In this chapter, from a different point of view, operator-based robust right coprime factorization and the Bezout identity are considered for the design of networked control system. The merit of the presenting control system is that the output feedback error signal is not affected by the perturbed signal from time-varying delay and the uncertainties of the controlled process. That is, the perturbed signal from the time-varying delay and the uncertainties of the controlled process cannot be transmitted back to the error signal provided the presenting Bezout identity is satisfied, where robust stability of the networked control system is guaranteed. Then, an operator-based tracking controller is designed to ensure the controlled process output tracks the desired reference input.

Nonlinear control system design problems have been considered by many researchers in different fields. One of the approaches is based on coprime factorization [18,20,49]. The robust stability of operator-based nonlinear feedback control system design has been considered in Chapter 2. This approach is based on robust right coprime factorization. That is, for a right coprime factorization ND^{-1} of nonlinear plant P, suppose that P has a perturbation ΔP and its right coprime factorization can be described as $(N + \Delta N)D^{-1}$. If the Bezout identity of plant P satisfies $S(N + \Delta N) + RD = L$ and $S(N + \Delta N) - SN = 0$, the perturbed plant $P + \Delta P$ is stable, where L is a unimodular operator. That is, the output feedback error signal is not affected by the perturbed signal from uncertainties of the controlled process. A robust condition for the case of $S(N + \Delta N) - SN \neq 0$ is derived in Lemma 2.7. Further, the perturbed nonlinear plant output tracking problem is considered [21]. As a result, we can design a tracking controller based the needs of the engineer consideration of N or $N + \Delta N$. However, operator-based robust right coprime factorization has not been considered for the design of the networked control system. In this chapter, we model time-varying delays as ΔN. Then, by extending the result in Chapter 2, operator-based nonlinear feedback control is designed so that the output feedback error signal is not affected by negative effects from the delays. Further, a tracking controller is designed where the controller can cause the controlled process output to track the desired reference input.

3.5.1 Networked Experimental System

The experimental system (see Figure 3.69) has roughly two parts connected by a local area network (LAN): (1) computer PC1 as a controller and (2) computer PC2, I/O board, and aluminum plate setup as a controlled process. Computer PC1 (Pentium 4, 2.8 GHz, 512 MB, Windows XP) demands to process a networked nonlinear control by using the controller will be given in Section 3.5.2, where the software is Visual C++. The experimental part is composed of a computer PC2 for delivering the control command from PC1 to the I/O board and delivering the process output information (temperature data) from the I/O board to a digital input(I)/output(O)

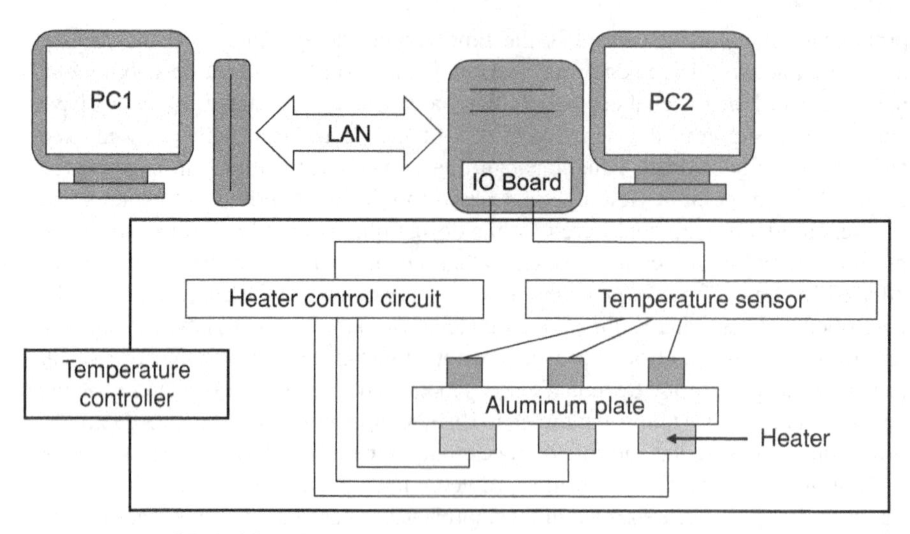

FIGURE 3.69 Experimental system connected by a LAN.

board, heater control circuit, thermometer circuit, and aluminum plate. The I/O board consists of A/D, D/A, and buffer boards. The heater control circuit is composed of a compensator, a counter oscillation circuit, and solid-state relay. The detailed parameters of the aluminum plate are shown in Table 3.5.

The input constraints are set as $u_{\max} = 40$ W and $u_{\min} = 0$ W. The schema of the experimental part is shown in Figure 3.70. As a result, the networked system is built using a web between PC1 and PC2. Furthermore, a STREAM socket is used when data are transmitted. This method establishes a connection between PC1 and PC2 using the Transmission Control Protocol/Internet Protocol (TCP/IP). With this method, although there is some time delay, loss of data does not take place. PC1 processes the following two tasks with a constant period:

1. Control of the temperature of the aluminum plate
2. Communication with PC2 to obtain the temperature data

TABLE 3.5 Parameters of Aluminum Plate

Density: 2700 kg/m^3
Specific heat: 0.917 kJ/kg K
Heat transfer coefficient: 20 W/m^2 K
Heat conductivity: 238 W/m K
Maximum calorific value: 40 W
Breadth of alminum plate: 250 mm
Thickness of alminum plate: 10 mm
Length of alminum plate: 100 mm

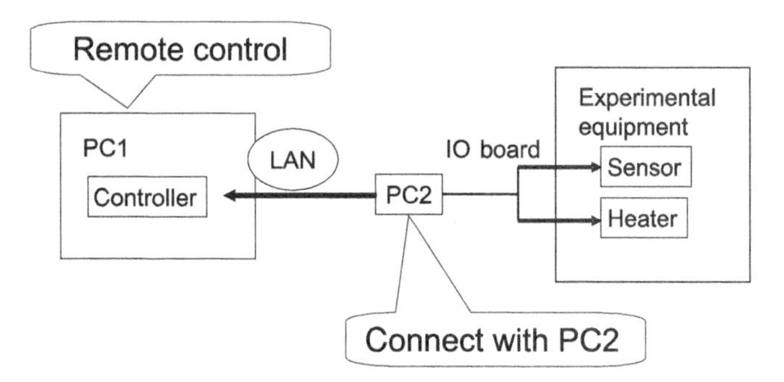

FIGURE 3.70 Schema of experimental part.

where PC1 commands the heater control circuit to turn on or off the switch between the heater and the AC power supply in PWM (pulse width modulation). The network-induced time delays are communication delays between PC1 and PC2 for delivering sensor (temperature) data as $A - B$ and between PC1 and PC2 for sending control data to the actuator (heater) as $C - E$, shown in Figure 3.71, where the delays $A - B$ and $C - E$ are measurable. The real control input includes the control signal being delivered over the network to the controlled process, and the measured signal being delivered over the network to the controller is given as

$$u' = u[t - \tau(t)] \tag{3.221}$$

where $\tau(t) = A - B + C - E$.

So far, to design a nonlinear feedback controller using operator-based right coprime factorization, some definitions were given in Chapter 2: operator, well-posedness, BIBO stability, and right coprime factorization.

Consider the aluminum plate thermal process described by the following right coprime factorization:

$$P = ND^{-1} \tag{3.222}$$

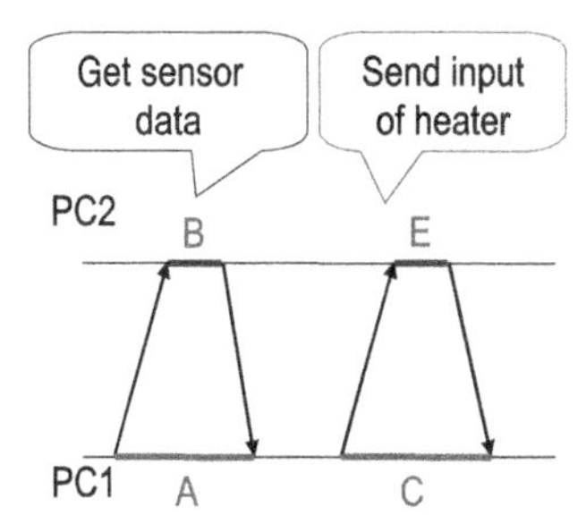

FIGURE 3.71 Fundamental relationship between delays.

Suppose that the plant P has a bounded perturbation ΔP caused by time-varying delays $\tau(t)$ and unmodeled uncertainties, that is, ΔD concerned with the effect from the delays and ΔN concerned with the uncertainties are bounded and the operators $D + \Delta D$ and $N + \Delta N$ are stable such that

$$\begin{aligned} P + \Delta P &= (N + \Delta N)D^{-1}[u(t - \tau)](t) \\ &= (N + \Delta N)(D + \Delta D)^{-1}(u)(t) \end{aligned} \tag{3.223}$$

where $D : W \to U$ and $N : W \to Y$ are stable operators, respectively, and the space W is called a *quasi-state* space [3] of P. Assume D and $D + \Delta D$ are invertible and $P : U \to Y$, where U and V are linear spaces over the field of real numbers, respectively. In this chapter, we consider the aluminum plate thermal process connected by a LAN as the application object of networked control. In the experimental system, since the temperature sensor and heater are not located in same position (see Figure 3.69), the heat transfer delay of the aluminum plate has to be considered, modeled in (3.230) and (3.231). The objective is to design a networked nonlinear feedback control for the aluminum plate thermal process.

3.5.2 Networked Nonlinear Feedback Control Design

We assume that the input signal u is in a subset U^* of U and output y is in a subset Y^* of Y. For nonlinear plant (3.221), under the condition of well-posedness, N and D are said to be right coprime factorizations if there exist two stable operators $S : Y \to U$ and $R : U \to U$ satisfying the Bezout identity

$$SN + RD = L \tag{3.224}$$

where R is invertible and L is a unimodular operator. In the same manner, under the condition of plant (3.222) being well-posed, we assume that there exist S and R satisfying the perturbed Bezout identity

$$S(N + \Delta N) + R(D + \Delta D) = \tilde{L} \tag{3.225}$$

where $\tilde{L}$ is a unimodular operator. Therefore, the perturbed plant retains a right comprime factorization. In this chapter, based on Lemma 2.7, we have the following theorem to guarantee the stability of the nonlinear feedback control system with perturbation, where the theorem is a general presentation of Lemma 2.9.

Theorem 3.12 Let D^e be a linear subspace of the extended linear space U^e associated with a given Banach space U_B, and let $L^{-1}(S(N + \Delta N) - SN + R(D + \Delta D) - RD) \in \text{Lip}(D^e)$. Let the Bezout identity of the nominal plant and the exact plant be $SN + RD = L \in \mathcal{U}(W, U)$, $S(N + \Delta N) + R(D + \Delta D) = \tilde{L}$, respectively.

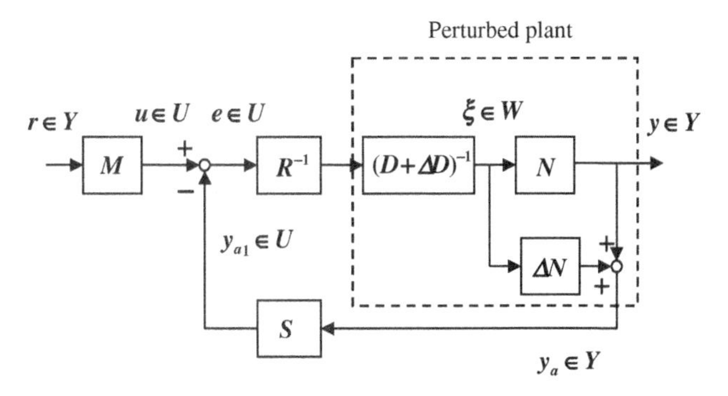

FIGURE 3.72 Tracking control system.

Under the condition of controller S to satisfying $S(N+\Delta N)+R(D+\Delta D)=L$, if

$$\|[S(N+\Delta N)-SN+R(D+\Delta D)-RD]L^{-1}\|<1 \tag{3.226}$$

the system shown in Figure 3.72 is stable, where $\|\cdot\|$ is a Lipschitz norm [1,2].

Proof If L is a unimodular operator, then L is invertible. From $SN+RD=M$, $S(N+\Delta N)+R(D+\Delta D)=\tilde{L}$, we have

$$\tilde{L}=L+[S(N+\Delta N)-SN+R(D+\Delta D)-RD]$$

Since $\tilde{L}=L+[S(N+\Delta N)-SN+R(D+\Delta D)-RD]=[I+(S(N+\Delta N)-SN+R(D+\Delta D)-RD)L^{-1}]L$ and $(S(N+\Delta N)-SN+R(D+\Delta D)-RD)L^{-1}\in \text{Lip}(D^e)$, $I+(S(N+\Delta N)-SN+R(D+\Delta D)-RD)L^{-1}$ is invertible, where I is an identity operator. Consequently, we have $\tilde{L}^{-1}=L^{-1}[I+(S(N+\Delta N)-SN+R(D+\Delta D)-RD)L^{-1}]^{-1}$. Meanwhile, since $\tilde{L}=L+[S(N+\Delta N)-SN+R(D+\Delta D)-RD]$, $L^{-1}(S(N+\Delta N)-SN+R(D+\Delta D)-RD)\in \text{Lip}(D^e)$ and $L\in\mathcal{U}(W,U)$, we have $\tilde{L}\in\mathcal{U}(W,U)$. That is, $\tilde{L}$ is a unimodular operator. Then, the system is overall stable. ■

Remark 3.3 The uncertain process retains a right coprime factorization if and only if (3.225) is satisfied, where (3.224) is the Bezout identity of process P and (3.225) is the perturbed Bezout identity of process $P+\Delta P$. Further, if (3.226) is satisfied, the uncertain process retains a robust right coprime factorization.

In the following, based on the result in [21], an output tracking system can be designed (see Figure 3.72), where $M: Y\to U$ is a designed stable operator and $r\in Y^*$ is a reference signal and the system is stable. Since Theorem 3.12 is satisfied, the effects from the delays cannot be transmitted back to error signal $e(t)$. The reason

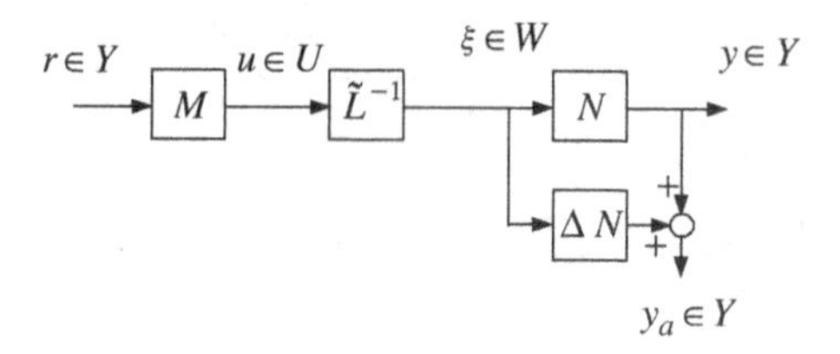

FIGURE 3.73 Equivalent diagram of Figure 3.72.

is that Figure 3.72 can be rewritten as Figure 3.73 [21], and the effect is removed by the robust control system in this chapter. From Figure 3.73, we know that we can design tracking operator M to consider the uncertain part ΔN. That is, based on the condition $N\tilde{L}^{-1}M(r)(t) = I(r)(t)$, process output $y(t)$ tracks to reference signal r, where y is observable. In practice, it is not unusual that operator L is designed as the identify operator. As a result, we have $\xi = \tilde{L}^{-1}(u)(t) = I(u)(t) = u(t)$.

3.5.3 Experimental Result

In this section, an experimental example to illustrate the efficacy of the design method is shown.

For the design of the controller, the nominal thermal process is described by the right coprime factorization:

$$y(t) = P(u_d)(t) = ND^{-1} \tag{3.227}$$

where

$$D(w)(t) = cmw(t) \qquad D^{-1}(u_d)(t) = \frac{1}{cm} \tag{3.228}$$

$$N(w)(t) = e^{-At}\int e^{A\tau}w(\tau)\,d\tau \tag{3.229}$$

Considering the real experimental system, the thermal process must deal with uncertainties and disturbances, and the above perturbation affects D and N in (3.227). In this chapter, suppose that the process P has a bounded perturbation ΔP, that is, ΔN is bounded and the operators D, N and $N + \Delta N$ is stable such that

$$\begin{aligned} P + \Delta P &= (N + \Delta N)D^{-1} \\ (N + \Delta N)(w)(t) &= (e^{-At} + \Delta_1)\int e^{A\tau}w(\tau)\,d\tau \end{aligned} \tag{3.230}$$

where $D{:}W \to U$ and $N{:}W \to Y$ are stable operators, respectively, and the space W is called a *quasi-state* space of P. Here D is invertible, and $P{:}U \to V$; U and V

are linear spaces over the field of real numbers, respectively. The uncertain factor is modeled as

$$\Delta_1 = e^{c-At}\left(\int e^{A\tau} w(\tau)\,d\tau\right)^{-1} \tag{3.231}$$

where $c = -6$. Further, time-varying delays are measurable only by experiment. The result is shown in Figure 3.74, where the y coordinate denotes the delay (in milliseconds). Based on the design scheme, to satisfying the Bezout identify (3.224), we design two operators R and S as

$$S(y_a)(t) = K_p(1 - B)\left(\frac{dy(t)}{dt} + Ay(t)\right) \tag{3.232}$$

$$R(u)(t) = \frac{K_pB - K_p + 1}{cm}u(t) \tag{3.233}$$

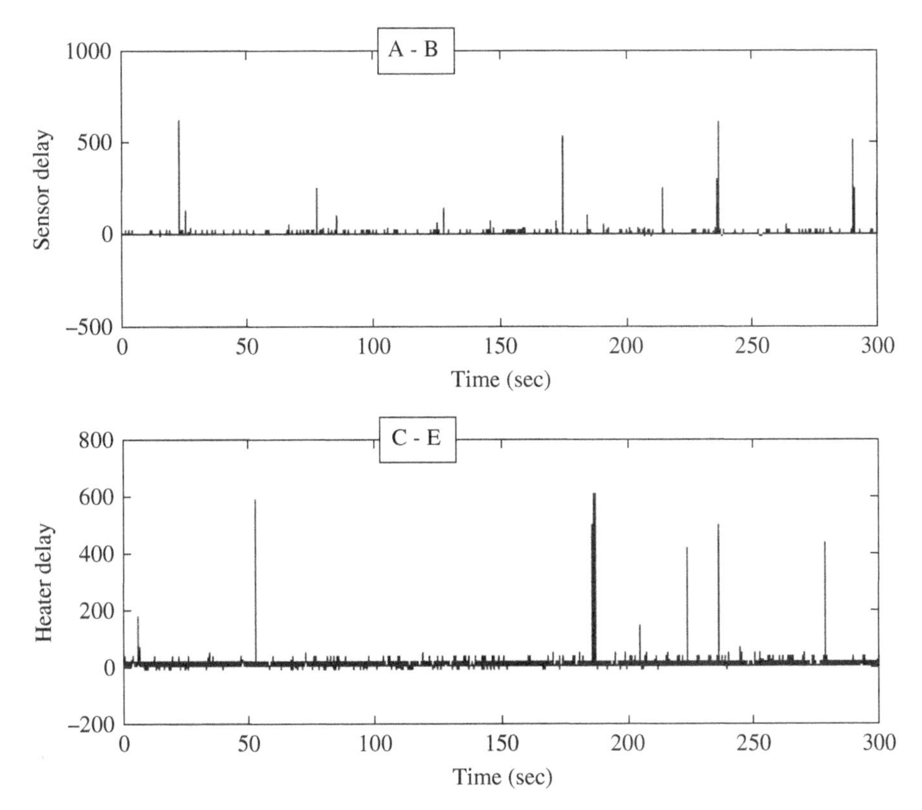

FIGURE 3.74 Delay variation between PC1 and PC2 for delivering sensor data and sending control data.

where $A = 0.013$, $B = 0.7$, $L = I$, and $u(t)$ is the process input without the delay, namely, $D(w)(t)$, and K_p is proposal gain and B is constant. Then, considering the process with the uncertain factor shown in (3.231), from (3.229) and (3.230) we have

$$S(N + \Delta N)(w)(t) = K_p(1 - B)w(t) \tag{3.234}$$

Then, for the case of the process without delays, the following relationship is satisfied based on the design condition of (3.224) and (3.225):

$$S(N + \Delta\tilde{N}) - S(N + \Delta N) = 0 \tag{3.235}$$

That is, $S(N + \Delta\tilde{N}) - RD = I(\xi)(t)$. For the process with delays shown in Figure 3.74, we evaluate the perturbed Bezout identity (3.225) and (3.226) in Theorem 3.12. Since (3.225) and (3.226) are satisfied, the control system without considering the tracking performance is stable. The tracking operator is designed as

$$M(r)(t) = Ar + (1 - A)re^{-t} \tag{3.236}$$

Figure 3.75 shows the experimental result of networked temperature control with delays. The desired temperature control result has been obtained where the upper figure is of the process output and the lower figure is of process input.

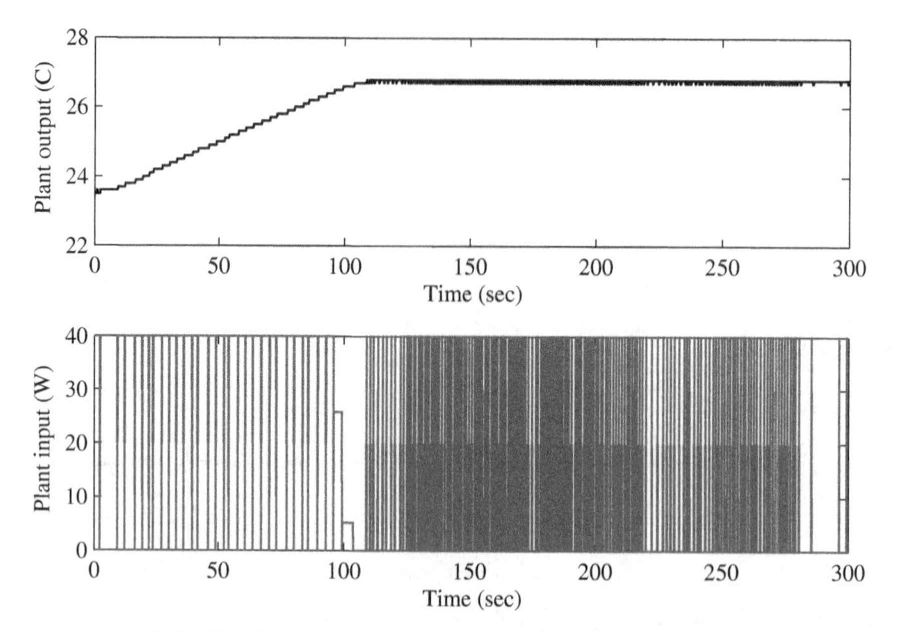

FIGURE 3.75 Experimental result based on networked control.

3.5.4 Summary

A networked robust nonlinear feedback control system was considered by using operator-based robust right coprime factorization. The problems of time-varying delays and unmodeled uncertainties were discussed. The effectiveness of the method was also confirmed by experiment.

CHAPTER 4

Tracking and Fault Detection Issues in Nonlinear Control Systems

Operator-based tracking controller design schemes and fault detection methods are developed in this chapter as an extension of robust right coprime factorization. Sections 4.1 and 4.2 discuss tracking controller design schemes and Sections 4.3 and 4.4 summarize fault detetion methods.

4.1 OPERATOR-BASED TRACKING COMPENSATOR IN NONLINEAR FEEDBACK CONTROL SYSTEMS DESIGN

4.1.1 Introduction

Control design problems of linear systems have made considerable and comprehensive progress in the last century [50]. At the same time, nonlinear control design systems have shown a slower progress. Many researchers entered the field of nonlinear control systems and have achieved remarkable results. However, the effect of nonlinear control systems is far from profound. One reason is that practical and effective methods in linear control systems could not be directly applied to nonlinear control systems owing to their complex structures and nonlinear characteristics. For nonlinear control systems, there are many problems that need further studies. Among them, robust issues play an important role and have attracted much attention from researchers in almost every field of control design systems [1–3, 51] (see Chapters 2 and 3). Besides the robustness of systems, output tracking is a significant problem in nonlinear systems.

Since the 1980s, coprime factorization has been proposed to deal with the robust issues in nonlinear control systems. Due to the contributions of many researchers, coprime factorization has been a promising method to deal with the control problems of nonlinear feedback control systems.

Coprime factorization supplies a convenient framework for researching the input–output stability properties of a nonlinear feedback control system, where the given plant is usually unstable. Many researchers have developed methods from various

Operator-Based Nonlinear Control Systems: Design and Applications, First Edition. Mingcong Deng.

aspects and various fields. Youla–Kucera parameterization and left coprime factorization for nonlinear systems were considered in [51, 52]. Robust right coprime factorization and robust stabilization for nonlinear feedback control systems were studied in [3] and the authors gave a condition to guarantee overall stability for right coprime factorization under the assumption that the control system is well-posed. The initial conditions of nonlinear systems and a new version of the method named dynamic right coprime factorization were proposed in [53]. However, the two papers [3, 53] did not consider the tracking problem in nonlinear control systems. A tracking design of perturbed nonlinear plants using robust right coprime factorization is considered by Deng et al. [54]. They gave a nonlinear operator-based design method for nonlinear plant output tracking to a reference input. Moreover, the tracking problem and designed tracking controller based on the exponential iteration theorem are shown in Chapter 3. Yet the condition existed only for $t \leq T$ large enough. Usually, we request that the output should track to the reference input within a finite time.

This chapter considers the tracking problem of nonlinear control systems with perturbations based on robust right coprime factorization, that is, a universal design scheme. Different from the former design schemes, the universal scheme not only can guarantee the robust stability of the nonlinear control systems with perturbations but also can realize the output tracking to the reference input. That is, with only a design scheme, we solve two problems and the tracking problem can be better realized. Simulations are given to demonstrate the effectiveness of the method.

4.1.2 Tracking Controller Design Scheme Using Unimodular Operator

Some important definitions are recalled before we adressing the tracking controller design issue.

The concept of right coprime factorization is as follows, where $D(T)$ represents the domain of the operator T, and similarly $R(T)$ stands for the range of T: Let $T : D(T) \to R(T)$ be a causal and stabilizable nonlinear operator, where T is said to have a *right coprime factorization* on $D(T)$ over the space of finite-gain stable and causal operators if it has a right factorization $T = ND^{-1}$ on $D(T)$ over the space of finite-gain stable and causal operators. Moreover there exist two operators $A : R(N) \to D(T)$ and $B : R(D) \to D(T)$, and for the unimodular operator $M : D(T) \to D(T)$, we have the Bezout identity

$$AN + BD = M \tag{4.1}$$

Additionally, $D(N)$ refers to *quasi-state space*.

Definition 4.1 If the following condition is satisfied, $y(t)$ is said to be high order and infinitely smaller than $x(t)$, written in the form $y = \circ(x)$:

$$\lim_{t \to \infty} \frac{y(t)}{x(t)} = 0 \tag{4.2}$$

Definition 4.2 If a function $\alpha(t)$ satisfies

$$\lim_{t\to\infty} \alpha(t) = 0 \tag{4.3}$$

then the function $\alpha(t)$ is said to be strongly stable.

Otherwise:

Definition 4.3 If a function $\beta(t)$ satisfies

$$\lim_{t\to\infty} \beta(t) = m \leq \infty \tag{4.4}$$

where $m \neq 0$, then the function $\beta(t)$ is said to be weakly stable. Moreover, if $m = \infty$, $\beta(t)$ is unstable.

The section considers two problems in nonlinear control systems with perturbation: the robust stability of the system and the output tracking to the reference input. The objective of this chapter is to give a universal design scheme to solve the two problems based on the approach of robust right coprime factorization. We first consider the robust right coprime factorization problem for the nonlinear system is shown in Figure 4.1, where the nominal plant P is unstable. The nominal plant is in the form $P = ND^{-1}$, where $N : W \to Y$, $D : W \to U$ are both stable operators and the operator D is invertible. Here U, W, and Y are input space, output space, and quasi-state space, respectively.

Before giving the main results, we introduce a definition and a theorem from [3]. The main work in this part of stabilizing the unstable plant is to design operators $A : Y \to U$, $N : W \to Y$, $D : W \to U$, and $B : U \to U$, where B and D are invertible for the given unimodular M to get the Bezout identity $AN + BD = M$.

Next the output tracking performance of the perturbed plant based on the right coprime factorization for the nonlinear control system is considered. The framework is shown in Figure 4.2, which is simplified to be equivalent to the one of Figure 4.3 by the Bezout identity.

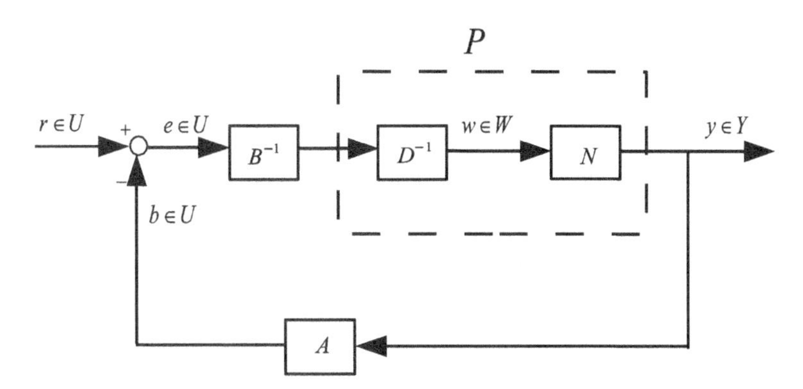

FIGURE 4.1 Nonlinear control system.

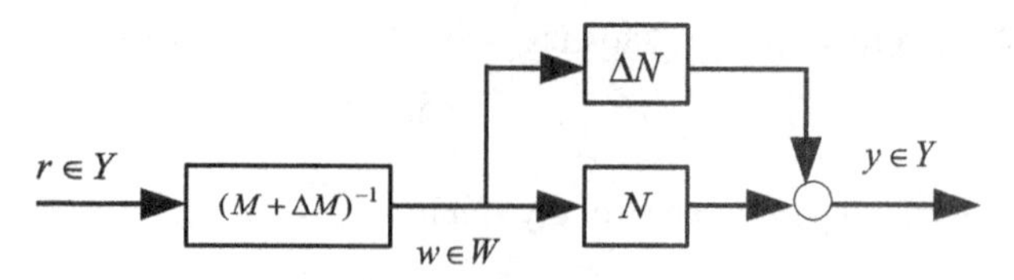

FIGURE 4.2 Output tracking of perturbed plant.

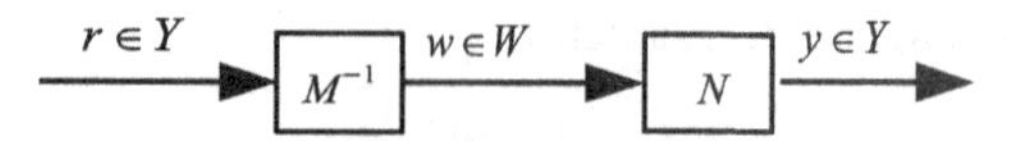

FIGURE 4.3 Output tracking of perturbed plant.

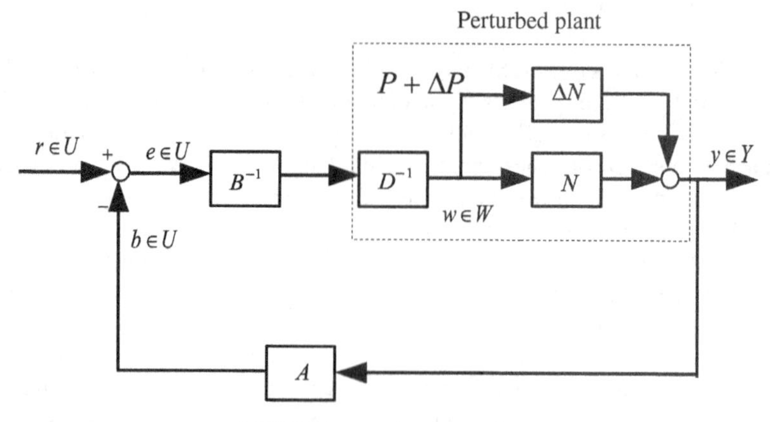

FIGURE 4.4 Perturbed plant.

Theorem 4.1 Assume that the system shown in Figure 4.4 is well-posed, consider the right factorization of the unstable plant $P + \Delta P = (N + \Delta N)D^{-1}$, and assume the designed two operators A, B satisfy the Bezout identity

$$A(N + \Delta N) + BD = M + \Delta M \in \mu(W, U) \tag{4.5}$$

If the equation

$$(N + \Delta N)(M + \Delta M)^{-1} = I \tag{4.6}$$

is satisfied, then the output tracks to the reference input.

Proof The system shown in Figure 4.4 is stable by the result in Chapter 2. From Figure 4.2,

$$y = (N + \Delta N)(M + \Delta M)^{-1}(r) = I(r) = r \tag{4.7}$$

so the problem of output tracking to the reference input is resolved. ■

Similarly, the method can be applied to the nominal plant.

Corollary 4.1 Suppose that the system shown in Figure 4.1 is well-posed, consider the right factorization of the unstable plant $P = ND^{-1}$, and design two operators A, B so they satisfy the Bezout identity $AN + BD = M \in \mu(W, U)$. If

$$NM^{-1} = I \tag{4.8}$$

then output tracking to the reference input is guaranteed.

Remark 4.1 Perfect tracking is always a hot topic in problems in control systems. However, there are seldom research results realizing it, especially for nonlinear control systems. To some extent, this chapter has done such a job.

4.1.3 Simulation

In this section, two simulations for different reference inputs $r_1(t)$, $r_2(t)$ are given to show the effectiveness of the method.

Let $U = C_{[0,T]}$ be a space of continuous functions and its subspace $C^1_{[0,T]}$ is constructed with all functions having a continuous first derivative, where $T < \infty$. Let U and Y be two linear spaces:

$$U = C_{[0,T]}$$
$$Y = \left\{ e^{-t} \int_0^t u(\tau)\,d\tau | u \in C^1_{[0,T]} \right\} \subset U$$

Consider the system in Figure 4.4, and let the input space, output space, quasi-state space be $U, Y, W = U$, where the plant $P + \Delta P = (N + \Delta N)D^{-1}$ with $D : W \to U$, $N + \Delta N : W \to Y$, respectively. The operators $A : Y \to U$, $B : U \to U$ are stable with $B \in \mu(U, U)$, and A, B^{-1} represent the feedback and feedforward controllers, respectively.

Consider the given plant $P + \Delta P : U \to Y$ defined as

$$P[\tilde{u}(t)] = e^{-t} \int_0^t [e^{2\tau}\tilde{u}(\tau) + g(\tau)]\,d\tau$$

where $\tilde{u} \in W$, and $g(t)$ is an unknown but bounded disturbance with $|g(t)| \le \epsilon$.

Then we design the right factorization of the perturbed plant $P + \Delta P = (N + \Delta N)D^{-1}$ with $N + \Delta N : W \to Y$ and $D : W \to U$ defined by

$$(N + \Delta N)[w(t)] = e^{-t} \int_0^t [w(\tau) + g(\tau)]\,d\tau$$
$$D[w(t)] = e^{-2t}w(t)$$

where $w(t) = \circ(e^t)$ and

$$D^{-1}[\tilde{u}(t)] = e^{2t}\tilde{u}(t) \tag{4.9}$$

Now let us verify the stability of the two operators $N + \Delta N$, D:

$$\begin{aligned}\lim_{t\to\infty}(N + \Delta N)[w(t)] &= \lim_{t\to\infty}\frac{\int_0^t [w(\tau) + g(\tau)]\,d\tau}{e^t} \\ &= \lim_{t\to\infty}\frac{w(t) + g(t)}{e^t} \\ &= 0\end{aligned} \tag{4.10}$$

With equation (4.2), we can find that the above limit is established. So is the operator D. And $(N + \delta N)$ and D are the right factorization of P:

$$\begin{aligned}(N + \Delta N)D^{-1}[\tilde{u}(t)] &= N[e^{2t}\tilde{u}(t)] \\ &= e^{-t}\int_0^t [e^{2\tau}\tilde{u}(\tau) + g(\tau)]\,d\tau \\ &= (P + \Delta P)[\tilde{u}(t)]\end{aligned} \tag{4.11}$$

We can find that

$$\begin{aligned}\lim_{t\to\infty}(P + \Delta P)[\tilde{u}(t)] &= \lim_{t\to\infty}\frac{\int_0^t [e^{2\tau}\tilde{u}(\tau) + g(\tau)]\,d\tau}{e^t} \\ &= \lim_{t\to\infty}\frac{e^{2t}\tilde{u}(t) + g(t)}{e^t} \\ &= \lim_{t\to\infty} e^t\tilde{u}(t) \\ &= \infty\end{aligned} \tag{4.12}$$

So the perturbed plant $P + \Delta P$ is unstable.

Next we design the two operators A, B to obtain the Bezout identity with the given unimodular:

$$(M + \Delta M)[w(t)] = \int_0^t e^{-\tau}[w(\tau) + g(\tau) - e^{\tau}r(\tau)]\,d\tau \tag{4.13}$$

Here $r(0) = 0$, and the inverse of $M + \Delta M$ is

$$(M + \Delta M)^{-1}[r(t)] = e^t\dot{r}(t) - g(t) + e^t r(t) \tag{4.14}$$

It is clear that $(M + \Delta M)[w(t)]$ and $(M + \Delta M)^{-1}[r(t)]$ are stable with $t \in [0, T]$.

Here, we choose the operators $B = I$ and A is defined as

$$A(N + \Delta N)[w(t)] = A\left(e^{-t}\int_0^t [w(\tau) + g(\tau)]\,d\tau\right)$$
$$= \int_0^t e^{-\tau}[w(\tau) + g(\tau)]\,d\tau - \int_0^t r(\tau)\,d\tau - e^{-2t}w(t) \quad (4.15)$$

Therefore, we have

$$\begin{aligned}[A(N + \Delta N) + BD][w(t)] &= \int_0^t e^{-\tau}[w(\tau) + g(\tau)]\,d\tau \\ &\quad - \int_0^t r(\tau)\,d\tau - e^{-2t}w(t) + e^{-2t}w(t) \\ &= \int_0^t e^{-\tau}[w(\tau) + g(\tau) - e^{\tau}r(\tau)]\,d\tau \\ &= (M + \Delta M)[w(t)] \end{aligned} \quad (4.16)$$

Therefore, the Bezout identity is satisfied so that the system is overall stable.

Then the output tracking problem is verified as follows:

$$\begin{aligned} y(t) &= (N + \Delta N)(M + \Delta M)^{-1}[r(t)] \\ &= (N + \Delta N)[e^{t}\dot{r}(t) - g(t) + e^{t}r(t)] \\ &= e^{-t}\int_0^t [e^{\tau}\dot{r}(\tau) - g(\tau) + e^{\tau}r(\tau) + g(\tau)]\,d\tau \\ &= e^{-t}\int_0^t [e^{\tau}\dot{r}(\tau) + e^{\tau}r(\tau)]\,d\tau \\ &= e^{-t}\left(e^{t}r(t) - \int_0^t e^{\tau}r(\tau)\,d\tau + \int_0^t e^{\tau}r(\tau)\,d\tau\right) \\ &= r(t) \end{aligned} \quad (4.17)$$

that is, $y(t) = r(t)$. So the problem of output tracking to the reference input can be solved with this method. Figures 4.5–4.8 demonstrate the effectiveness of the method. The reference inputs in Figures 4.5 and 4.7 for the two simulations are, respectively,

$$r_1(t) = 0.6e^{-t} - 0.5e^{-3t} - 0.1e^{-11t}$$
$$r_2(t) = 1 - 0.1e^{-11t} - 0.9e^{-t}$$

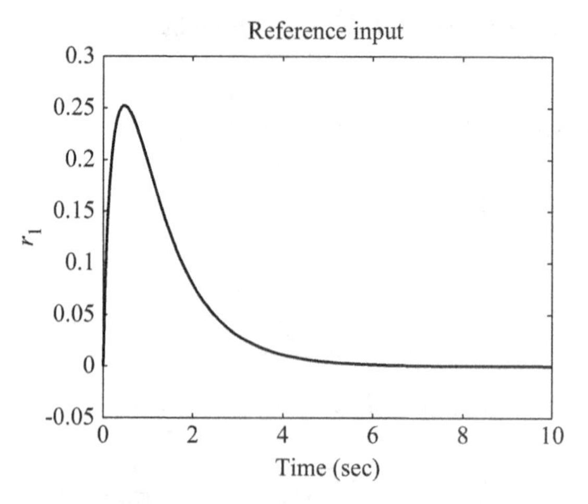

FIGURE 4.5 Reference input r_1 (simulation 1).

4.1.4 Summary

This section considered the tracking problem of nonlinear control systems with perturbations based on robust right coprime factorization. The robust stability of the system and the output tracking problems were considered. Based on the approach of right coprime factorization, a universal design scheme was introduced and, with it, we can guarantee the robust stability of nonlinear control systems with perturbations and realize the output tracking to the reference input. The simulations confirm the effectiveness of the method.

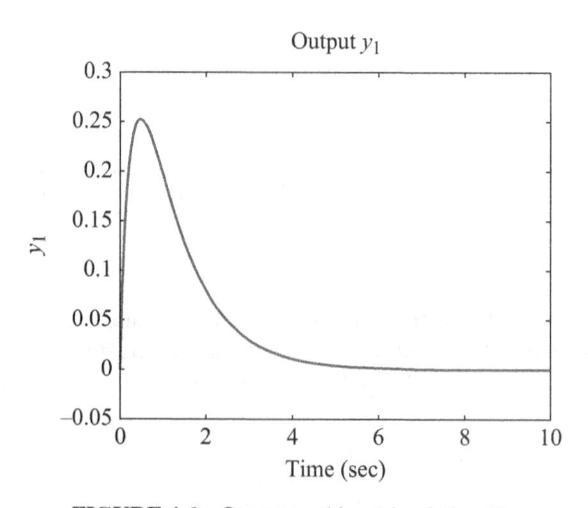

FIGURE 4.6 Output tracking (simulation 1).

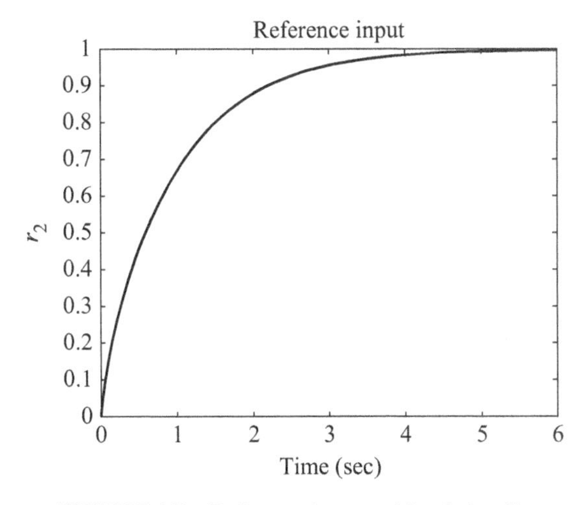

FIGURE 4.7 Reference input r_2 (simulation 2).

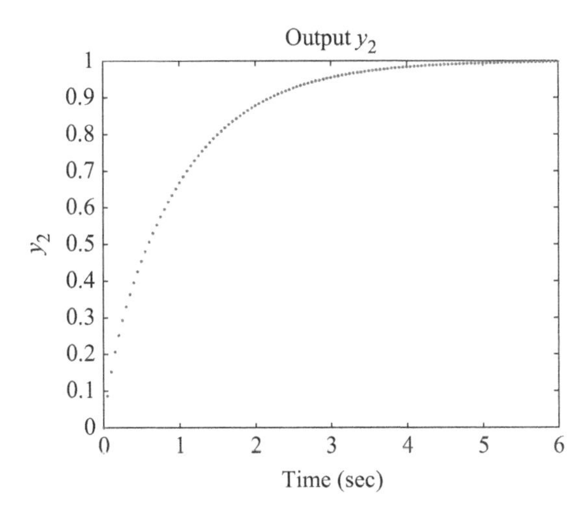

FIGURE 4.8 Output of plant (simulation 2).

4.2 ROBUST CONTROL FOR NONLINEAR SYSTEMS WITH UNKNOWN PERTURBATIONS USING SIMPLIFIED ROBUST RIGHT COPRIME FACTORIZATION

4.2.1 Introduction

Uncertainties usually exist in real systems. The method of decreasing or offsetting the effects produced by uncertainties is critical, namely, the robust control, which entails finding a controller to stabilize the nonlinear system with the nominal plant as well as the perturbed plant if the perturbations are bounded [16, 55–57]. Robust control has shown increasing interest in many fields and many methods have been used, such as the linear matrix inequality method (LMI) ([58]), sliding-mode control (SMC) ([59, 60]), and robust right coprime factorization (RRCF) ([5, 7, 18, 20, 51, 61, 66] (See also Chapters 2 and 3.). Among these methods, robust right coprime factorization is effective in dealing with the robust issues, and it has been proved to be a promising method in the control and design of nonlinear systems with perturbations.

Among the existing results using right coprime factorization, operator-based robust right coprime factorization has been proved to be effective to deal with the robust control and tracking problem in nonlinear feedback control systems. Recently, the robust stabilization problem, output tracking problem, and factorization method were considered (see [2], [3], and [5], respectively). Robust stability and robust stabilization are considered in [3], where a sufficient condition guaranteeing the robust stability of the nonlinear feedback system is based on the definition of a null set. Another condition guaranteeing the robust stability of perturbed nonlinear systems is shown in [2] based on the Lipschitz norm. Moreover, the output tracking problem is also considered in [2] and a sufficient condition is used to improve the tracking performance based on the exponential iteration theorem. Another fundamental problem of right coprime factorization is proposed in [5], that is, where the factorization of the plant is realized using the isomorphism approach, and a quantitative design scheme of the controllers is proposed. Detailed explanations on the isomorphism approach have been given in Chapter 2. In this chapter, for nonlinear systems with unknown perturbations, the problem of robust control is considered. That is, a simplified design of the robust controllers as well as the tracking problem will be considered for a class of nonlinear systems with unknown perturbations. For the designed controllers, existence of the inverses of the operators M and $\tilde{M}$ is obtained using a roots formula and the inverse is also guaranteed to be bounded. Then, the operators are guaranteed to be unimodular operators. Therefore, the Bezout identities of the nominal nonlinear system $(AN + BD = M)$ and the perturbed nonlinear system $[A(N + \Delta N) + BD = \tilde{M}]$ are respectively satisfied, which means the robust stability of the nonlinear system with unknown perturbation is guaranteed. That is, a more quantitative design of the robust controllers is given to guarantee robust stability and the asymptotic tracking on nonlinear systems with unknown perturbations.

4.2.2 Robust Design of Tracking Controller

First, some definitions and notation are recalled.

A vector space $\boldsymbol{X}$ is closed under addition and scalar multiplication and its subset $\boldsymbol{X}_s$ is said to be normed if each element $x \in \boldsymbol{X}_s$ has a norm $\|x\|$ which is defined such that the following three properties are fulfilled:

1. $\|x\|$ is a real, positive number and is different from zero unless x is identically zero,
2. $\|ax\| = |a|\|x\|$, and
3. $\|x_1 + x_2\| \leq \|x_1\| + \|x_2\| (x_1, x_2 \in \boldsymbol{X}_s)$.

Consider $\boldsymbol{M}^*$ to be the family of real-valued measurable functions defined on $[0, \infty)$, which is a linear space. For each constant $T \in [0, \infty)$, let P_T be the *projection operator* mapping from $\boldsymbol{M}^*$ to another linear space, $\boldsymbol{M}_T^*$, of measurable functions such that

$$f_T(t) := P_T(f)(t) = \begin{cases} f(t) & t \le T \\ 0 & t > T \end{cases} \tag{4.18}$$

where $f_T(t) \in \boldsymbol{M}_T^*$ is called the *truncation* of $f(t)$ with respect to T. Then, for any given Banach space $\boldsymbol{X}_B$ of measurable functions, set

$$\boldsymbol{X}^e = \{f \in \boldsymbol{M} : \|f_T\|_{\boldsymbol{X}_B} < \infty \text{ for all } T < \infty\} \tag{4.19}$$

Obviously, $\boldsymbol{X}^e$ is a linear subspace of $\boldsymbol{M}^*$. The space $\boldsymbol{X}^e$ so defined is called the *extended linear space associated with the Banach space* $\boldsymbol{X}_B$. Two linear spaces $\boldsymbol{U}$ and $\boldsymbol{Y}$ are denoted to be subspaces of $\boldsymbol{X}^e$, and their normed subspaces are respectively denoted as $\boldsymbol{U}_S$ and $\boldsymbol{Y}_S$ and are called *stable subspaces* of $\boldsymbol{U}$ and $\boldsymbol{Y}$, respectively.

Let $P : \boldsymbol{U} \to \boldsymbol{Y}$ be an operator, and the domain and range of P are denoted as $\mathfrak{D}(P)$ and $\mathfrak{R}(P)$, respectively. If for the operator $P : \mathfrak{D}(P) \to \boldsymbol{Y}$ the rules $P : \alpha u_1 + \beta u_2 \to \alpha P(u_1) + \beta P(u_2)$ for $\forall u_1, u_2 \in \mathfrak{D}(P)$ and $\forall \alpha, \beta \in \boldsymbol{C}$ ($\boldsymbol{C}$ denotes the set of the complex numbers) are satisfied, then the operator is a linear operator. Otherwise, it is nonlinear. In this chapter, we consider the nonlinear cases and assume that $\mathfrak{D}(P) = \boldsymbol{U}$ and $\mathfrak{R}(P) \subseteq \boldsymbol{Y}$.

Definition 4.4 Let $P : \boldsymbol{U} \to \boldsymbol{Y}$ be a nonlinear operator that describes an input–output stable control system with the input and output spaces $\boldsymbol{U}$ and $\boldsymbol{Y}$, respectively. If, furthermore, an operator norm $\|P\|$ for $P : \boldsymbol{U} \to \boldsymbol{Y}$ is well defined and finite, $\|P\| < \infty$, then the system is said to be finite-gain input-output stable.

Definition 4.5 Let $P : \mathfrak{D}(P) \to \mathfrak{R}(P)$ be a causal and stabilizable nonlinear operator. Assume P has a right coprime factorization on $\mathfrak{D}(P)$ over the space of finite-gain stable and causal operators if it has a right factorization $P = ND^{-1}$ on $\mathfrak{D}(P)$ over the space of finite-gain stable and causal operators and moreover there exist two operators $A : \mathfrak{R}(N) \to \mathfrak{D}(P)$ and $B : \mathfrak{R}(D) \to \mathfrak{D}(P)$ and for the unimodular operator $M : \mathfrak{D}(N) \to \mathfrak{D}(P)$ we have the Bezout identity

$$AN + BD = M \tag{4.20}$$

Additionally, $W = \mathfrak{D}(N)$ is named for quasi-state space.

It is worth mentioning that the initial state should also be considered, that is, $AN(w_0, t_0) + BD(w_0, t_0) = M(w_0, t_0)$ should be satisfied. Based on the satisfaction of the Bezout identity, the nonlinear feedback system shown in Figure 4.9 can be equivalently transformed to the system shown in Figure 4.10.

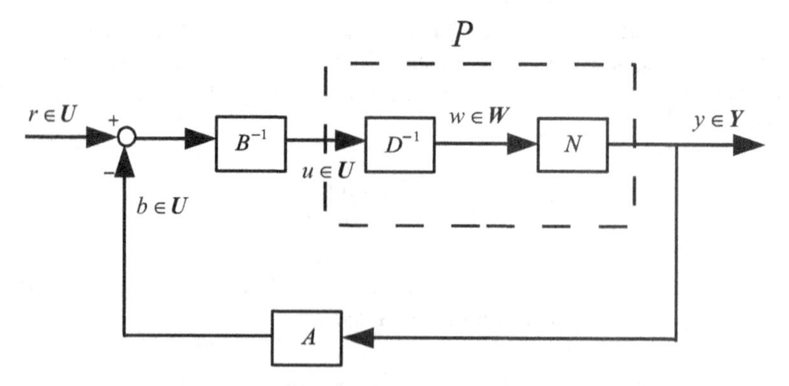

FIGURE 4.9 Nonlinear feedback control system.

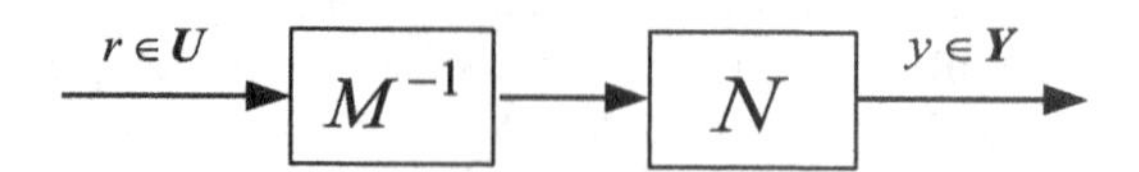

FIGURE 4.10 Equivalent system.

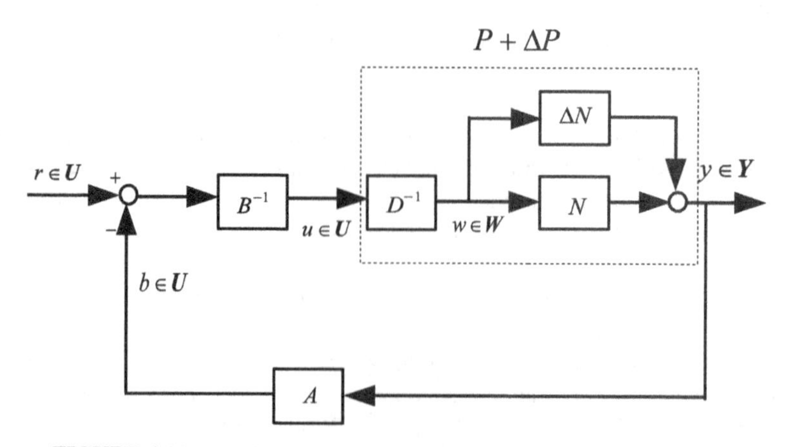

FIGURE 4.11 Nonlinear feedback system with unknown perturbations.

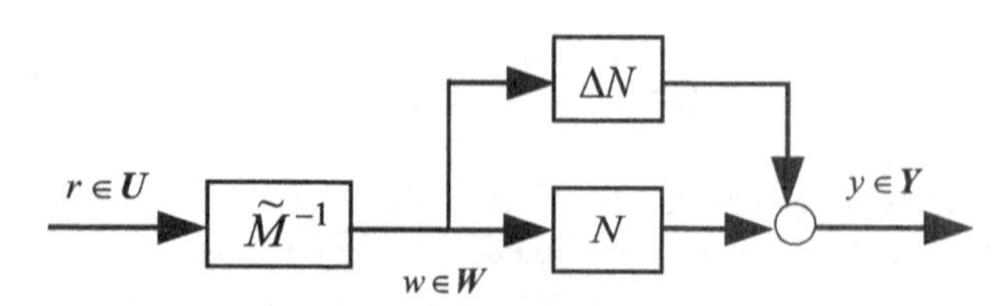

FIGURE 4.12 Perturbed equivalent system.

Since the uncertainties usually exist in real systems, the robustness of right coprime factorization is considered wherein the sufficient condition of offsetting the effects of the uncertainties is shown in Chapter 2 and the perturbed nonlinear control system is shown in Figure 4.11. Similarly, based on the satisfaction of the perturbed Bezout identity, the perturbed nonlinear control system can be equivalently transformed into the system shown in Figure 4.12. Some necessary definitions of the sufficient condition are given again as follows.

Definition 4.6 Let $\boldsymbol{U}^e$ and $\boldsymbol{Y}^e$ be two extended linear spaces which are associated respectively with two given Banach spaces $\boldsymbol{U}_B$ and $\boldsymbol{Y}_B$ of measurable functions defined on the time domain $[0, \infty)$. Let $\boldsymbol{D}^e$ be a subset of $\boldsymbol{U}^e$. A nonlinear operator $P : \boldsymbol{D}^e \to \boldsymbol{Y}^e$ is called a generalized Lipschitz operator on $\boldsymbol{D}^e$ if there exists a constant L such that

$$\|[P(x)]_T - [P(\tilde{x})]_T\|_{Y^e} \leq L\|x_T - \tilde{x}_T\|_{U^e} \tag{4.21}$$

for all $x, \tilde{x} \in \boldsymbol{D}^e$ and for all $T \in [0, \infty)$. $Lip(\boldsymbol{D}^e)$ denotes the family of nonlinear generalized Lipschitz operators that map $\boldsymbol{D}^e$ to itself.

Note that the least such constant L is given by

$$\|P\| := \sup_{T\in[0,\infty)} \sup_{\substack{x,\tilde{x}\in D^e \\ x_T\neq\tilde{x}_T}} \frac{\|[Px]_T - [P\tilde{x}]_T\|_{Y^e}}{\|x_T - \tilde{x}_T\|_{U^e}} \tag{4.22}$$

Lemma 4.1 Let $\boldsymbol{D}^e$ be a linear subspace of the extended linear space $\boldsymbol{U}^e$ associated with a given Banach space $\boldsymbol{U}_B$, and let $[A(N + \Delta N) - AN]M^{-1} \in \text{Lip}(\boldsymbol{D}^e)$. If the robust conditions

$$AN + BD = M \in \mu(\boldsymbol{W}, \boldsymbol{U}) \tag{4.23}$$

$$A(N + \Delta N) + BD = \tilde{M} \tag{4.24}$$

$$\left\|[A(N + \Delta N) - AN]M^{-1}\right\| < 1 \tag{4.25}$$

are satisfied, then the system shown in Figure 4.11 is stable, where $\|\cdot\|$ is defined in (4.22).

According to the results of [3], if (4.23) is satisfied, then nonlinear systems with the nominal plant shown in Figure 4.9 are stable. Meanwhile, the nonlinear systems with unknown perturbations shown in Figure 4.11 are also stable by the satisfaction of (4.23) and (4.25). Moreover, the perturbed plant $P = (N + \Delta N)D^{-1}$ can be said to have a robust right coprime factorization since the robustness of the right coprime factorization of the perturbed plant is linked to the robust stability of the perturbed nonlinear systems.

Similarly, based on the satisfaction of the perturbed Bezout identity $A(N + \Delta N) + BD = \tilde{M}$, the system shown in Figure 4.11 is stable and can be equivalently transformed into the one shown in Figure 4.12. From the equivalent system, the relationship between the plant output and the reference input can be simply obtained: $y = (N + \Delta N)\tilde{M}^{-1}(r)$.

Considering the perfect tracking property, the robust stabilizing controllers were designed (see [5]), but in some cases, the plant output space is not the as same as the reference input space. Even if they are the same, the perfect tracking problem is difficult to be realized. Therefore, for the nonlinear systems with unknown perturbations, the stabilizing robust controllers (A, B) are designed by which both the nominal nonlinear system and the perturbed nonlinear system can be stabilized. Meanwhile, the plant output can asymptotically track the reference input.

Consider the nonlinear feedback system with unknown perturbations shown in Figure 4.11, where the system is assumed to be well-posed and the factors N, $N + \Delta N$, and D are assumed to be obtained from the isomorphism factorization method shown in Chapter 2. That is, the nonlinear parts are included in the factor $N + \Delta N$ while the linear parts are contained in D^{-1}. Also, the existence domain of the controllers (A, B) are given, but the design scheme is only effective for the nonlinear system where perfect tracking is guaranteed. In most cases, perfect tracking cannot be realized, especially for unknown perturbations. Therefore, in the following, we will consider a design scheme of robust controllers (A, B) for the nonlinear feedback system with unknown perturbations.

Theorem 4.2 If the following assumptions are satisfied, and the two controllers (A, B) are designed to be as $A(y)(t) = \eta(t)y(t)$ and $B(u)(t) = (1/K)u(t)$, then the designed controllers (A, B) are robust stabilizing controllers for the nonlinear system shown in Figure 4.11.

- Assumption I: The nonlinear feedback system with unknown perturbations shown in Figure 4.11 is well-posed.
- Assumption II: The factors are obtained from the isomorphism factorization method.
- Assumption III: The unknown perturbations are assumed to be positive and bounded.
- Assumption IV: $\eta(t) \to 1$ as $t \to \infty$.

Proof According to assumptions I and II, the operators should be designed as

$$N(w)(t) = \phi(t)w^{\gamma}(t)$$

$$(N + \Delta N)(w)(t) = [\phi(t) + \lambda(t)]w^{\gamma}(t)$$

$$D(w)(t) = \psi(t)w(t)$$

where $\gamma = 1, \ldots, 4$, $\phi(t) \nrightarrow 0$, and $\psi(t)$ are BIBO stable, the unknown parameter $\lambda(t) > 0$, $|\lambda(t)| < \beta$. Then the following conditions are satisfied since assumption IV is satisfied:

$$
\begin{aligned}
M(w)(t) = (AN + BD)(w)(t) &= \eta(t)\phi(t)w^{\gamma}(t) + \frac{1}{K}\psi(t)w(t) \\
&\to \phi(t)w^{\gamma}(t) + \frac{1}{K}\psi(t)w(t)
\end{aligned} \tag{4.26}
$$

According to assumption II and the design of B, we can find that the operator M is bounded and it does not tend to zero. Then M is a unimodular operator (for the case of $\gamma = 1, \ldots, 4$, the inverse of M can be obtained by the roots formula). Therefore, the Bezout identity for the nominal nonlinear system is satisfied. Therefore, the nonlinear system with the nominal plant is stabilized by controllers A and B. Next we need to verify whether the controllers can stabilize the perturbed nonlinear system:

$$
\begin{aligned}
\tilde{M}(w)(t) &= [A(N + \Delta N) + BD](w)(t) \\
&= \eta(t)[\phi(t) + \lambda(t)]w^{\gamma}(t) + \frac{1}{K}\psi(t)w(t) \\
&\to [\phi(t) + \lambda(t)]w^{\gamma}(t) + \frac{1}{K}\psi(t)w(t)
\end{aligned} \tag{4.27}
$$

According to assumptions III and IV, $\lambda(t)$ is positive and bounded while $\eta(t) \to 1$ as $t \to \infty$. The operator $\tilde{M}$ is similarly proved to be a unimodular operator. Then (4.27) is proved to be a Bezout identity, and based on the results (see [3]), the stability of the nonlinear system with unknown perturbations is guaranteed. The designed controllers (A, B) guarantee that both the nonlinear system with nominal plant and the nonlinear system with unknown perturbations are stable. Therefore, the designed controllers (A, B) are robust stabilizing controllers. This completes the proof. ■

In this proof, we can find that whether D^{-1} is stable or not, the results are still established $[\phi(t) \nrightarrow 0]$, which means that for the stable or unstable plant, the stability of the nonlinear feedback control systems can be guaranteed by the designed robust controllers. Next, the plant output tracking the reference input problem will be discussed. For simplification, only the case of $\gamma = 2$ is considered in the following. Then Theorem 4.3 can be obtained for the nonlinear system shown in Figure 4.11, which is equivalent to the nonlinear feedback system shown in Figure 4.10.

Theorem 4.3 Considering the nonlinear equivalent system shown in Figure 4.11, if there exists $T > 0$, when $t > T$, $K \gg \psi(t)$, then the plant output can be found to asymptotically track the reference input.

Proof Since the operator $\tilde{M}$ is a unimodular operator, then its inverse is obtained by completing the square of some corresponding parameters as follows:

$$
\tilde{M}^{-1}(r)(t) = \left[\frac{1}{\eta(t)[\phi(t) + \lambda(t)]}\right]^{1/2} F(r)(t) \tag{4.28}
$$

where

$$F(r)(t) = \frac{\sqrt{\psi^2(t) + 4r(t)K^2\eta(t)[\phi(t) + \lambda(t)]} - \psi(t)}{2K\sqrt{\eta(t)[\phi(t) + \lambda(t)]}} \tag{4.29}$$

Then based on the satisfaction of (4.27), the following condition is satisfied:

$$\begin{aligned}(N + \Delta N)\tilde{M}^{-1}(r)(t) &= \left[\phi(t) + \lambda(t)\right]\left(\frac{1}{\eta(t)[\phi(t) + \lambda(t)]}\right)F^2(r)(t) \\ &= \frac{1}{\eta(t)}\left[r(t) - \frac{\psi(t)[\sqrt{\psi^2(t) + 4r(t)K^2\eta(t)[\phi(t) + \lambda(t)]} - \psi(t)]}{2K^2\eta(t)[\phi(t) + \lambda(t)]}\right] \\ &= \frac{1}{\eta(t)}\left[r(t) - \frac{2r(t)}{\sqrt{1 + 4r(t)\dfrac{K^2}{\psi(t)}\eta(t)[\phi(t) + \lambda(t)]} + 1}\right]\end{aligned} \tag{4.30}$$

If $K \gg \psi(t)$ when $t > T$, then (4.30) tends to $[1/\eta(t)]r(t)$. Moreover, by the assumption $1/\eta(t) \to 1$, the following condition is satisfied:

$$y(t) = (N + \Delta N)\tilde{M}^{-1}(r)(t) \to r(t) \tag{4.31}$$

Therefore, the plant output asymptotically tracks the reference input. This completes the proof. ∎

In this discussion, the design of the robust stabilizing controllers A and B are given in Theorem 4.2. That is, for the general case when the plant output cannot track the reference input, robust stabilizing controllers (A, B) are designed to guarantee the stability of the nonlinear system with nominal plant and perturbed plant. Moreover, by the robust stabilizing controllers, the plant output asymptotically tracks the reference input, which is shown in Theorem 4.3.

The main results, which are different from the results given in Chapter 2, are as follows: The robust stability of the perturbed nonlinear systems are directly guaranteed by the satisfaction of the perturbed Bezout identity according to the designed simplified robust controllers instead of satisfying the *robust conditions* (which means calculating the complex Lipschitz norm). Meanwhile, the plant output can be guaranteed to asymptotically track the reference input. Especially, based on the existing results [5, 7], the design scheme can be extended to the case of perturbations existing in both the numerator and the denominator.

4.2.3 Illustrative Examples

In this section, a numerical example and a simulation of the temperature control in the spirit of the aluminum plate thermal process will be given to show the effectiveness of the design scheme.

Simulation 1 First, two spaces $\boldsymbol{U}$ and $\boldsymbol{Y}$ are given:

$$\boldsymbol{U} = \boldsymbol{C}^*_{[0,\infty)}$$

$$\boldsymbol{Y} = \{y(t)|y(t) = \beta(t)u^2(t), u(t) \in \boldsymbol{U}\}$$

where $\boldsymbol{C}^*_{[0,\infty)}$ is a set of positive and continuous functions defined in $t \in [0, \infty)$ and $\beta(t)$ is also a continuous function.

Let $\|\cdot\|$ be the sup-norm defined by $\|u\|_\infty = \sup_{t\in[0,\infty)} |u(t)|$. Define

$$\boldsymbol{U}_s = \{u(t)|u(t) \in \boldsymbol{U}, \|u\|_\infty < \infty\}$$

$$\boldsymbol{Y}_s = \{y(t)|y(t) \in \boldsymbol{Y}, \|y\|_\infty < \infty\}$$

It can be verified that both of the above spaces are linear normed. Then they can be used to be the stable subspaces of $\boldsymbol{U}$ and $\boldsymbol{Y}$, respectively.

Consider the nonlinear feedback system with unknown perturbations shown in Figure 4.11. By using isomorphism factorization (see [5]), the nominal plant and real plant are respectively factorized as follows:

$$\begin{aligned} N(w)(t) &= (1 + e^{-t})w^2(t) \\ (N + \Delta N)(w)(t) &= \left\{1 + [1 + \lambda(t)]e^{-t}\right\} w^2(t) \\ D^{-1}(u)(t) &= e^t u(t) \end{aligned} \tag{4.32}$$

where the uncertain factor $\lambda(t)$ is bounded with $|\lambda(t)| < \beta_1$ and the operator D can be obtained as $D(w)(t) = e^{-t}w(t)$. Then we find that the factors N and $N + \Delta N$ are unimodular operators, D is stable, and D^{-1} is unstable. Next, we will design the nominal plant as well as the one with unknown perturbations.

For simplification, the controllers A and B are respectively designed according to Theorem 4.2 as follows:

$$\begin{aligned} A(y)(t) &= \frac{1}{1 + e^{-t}} y(t) \\ B(u)(t) &= \frac{1}{K} u(t) \end{aligned} \tag{4.33}$$

Then we will verify the satisfaction of the Bezout identity for the nonlinear feedback system with a nominal plant and a perturbed plant, respectively.

Given $M = AN + BD$, the following is obtained:

$$M(w)(t) = (AN + BD)(w)(t) = w^2(t) + \frac{e^{-t}}{K} w(t) \tag{4.34}$$

Then we can find that M is a unimodular operator with

$$M^{-1}(r)(t) = \frac{\sqrt{e^{-2t} + 4r(t)K^2} - 1}{2K}$$

Similarly,

$$\tilde{M}(w)(t) = [A(N + \Delta N) + BD](w)(t) = \frac{1 + [1 + \lambda(t)]e^{-t}}{e^{-t} + 1} w^2(t) + \frac{e^{-t}}{K} w(t)$$

According to the definition of the unimodular operator and the method of completing the square, we can find that $\tilde{M}$ is unimodular with

$$\tilde{M}^{-1}(r)(t) = \frac{\sqrt{\pi^2(t)e^{-2t} + 4r(t)\pi(t)K^2} - \pi(t)e^{-t}}{2K} \tag{4.35}$$

where

$$\pi(t) = \frac{1 + e^{-t}}{1 + [1 + \lambda(t)]e^{-t}}$$

Then, based on the equivalent system, we can find that

$$\begin{aligned} y(t) = (N + \Delta N)\tilde{M}^{-1}(r) &= \frac{e^t + 1 + \lambda(t)}{e^t} \left(\frac{\sqrt{\pi^2(t)e^{-2t} + 4r(t)\pi(t)K^2} - \pi(t)e^{-t}}{2K} \right)^2 \\ &= \frac{e^t + 1 + \lambda(t)}{e^t} \left(\frac{\pi^2(t)e^{-2t} + 2r(t)\pi(t)K^2 - \pi(t)e^{-t}\sqrt{\pi^2(t)e^{-2t} + 4r(t)\pi(t)K^2}}{2K^2} \right) \end{aligned} \tag{4.36}$$

Designing K according to Theorem 4.3 that is,

$$\frac{\pi^2(t)e^{-t} - \pi(t)e^{-t}\sqrt{\pi^2(t)e^{-2t} + 4r(t)\pi(t)K^2}}{2K^2} \to 0 \quad \text{as} \quad t \to \infty \tag{4.37}$$

$y(t) \to r(t)$. Therefore, the plant output asymptotically tracks the reference input.

The simulation results are given in Figures 4.13–4.15, where $r(t) = 10 + e^{-t}$, $K = \frac{1}{4}$, and $\lambda(t) = 0.02$ rand(1). From them, we can find that the plant output asymptotically tracks the reference input while the quasi-state plant output and reference input are all BIBO stable from about 7 sec.

By designing the controllers A and B, it has been proved that the plant output can asymptotically track the reference input without knowing the value of the perturbations $\lambda(t)$. In many cases, the asymptotic tracking can be guaranteed in a finite time interval instead of the case $t \to \infty$ which is verified in the simulation results.

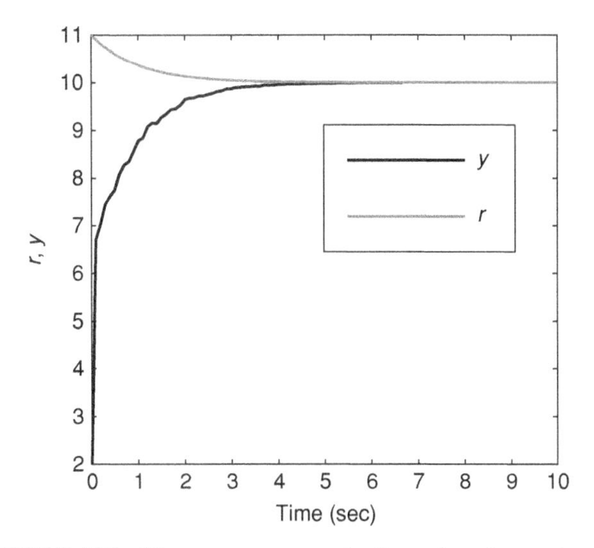

FIGURE 4.13 Plant output asymptotically tracks reference input.

Simulation 2 Similar to the aluminum plate thermal process this simulation shows the effectiveness of the method where the temperature $y(t)$ of the aluminum plate is controlled to asymptotically track the reference input $r(t)$ while the stability of the nonlinear system is guaranteed. The experimental apparatus shown in Figure 4.16 as well as its photograph in Figure 4.17 is used to control the temperature of the aluminum plate using a computer. This apparatus system is composed of a Peltier

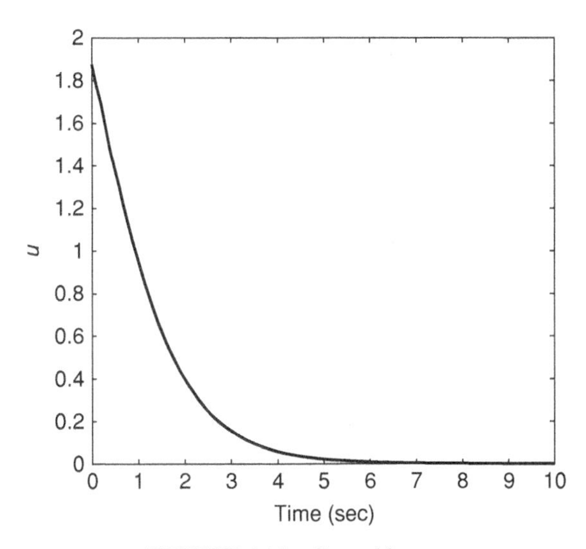

FIGURE 4.14 Control input.

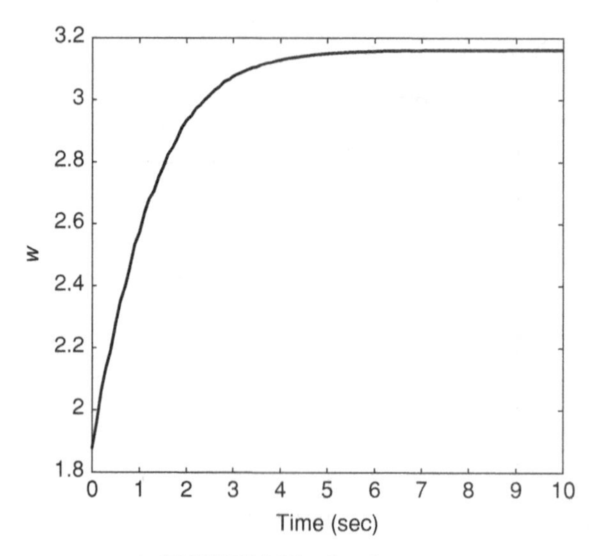

FIGURE 4.15 Quasi-state.

actuated process to cool the aluminum plate, and the serial communication (RS232C) provides an interface between the personal computer and the microcomputer. The aluminum plate thermal process is shown in Figure 4.18. Then according to Fourier's law of thermal conduction, Newton's cooling law, the specific heat capacity equation, the electrothermal amount by the Peltier effect, thermal conduction by temperature

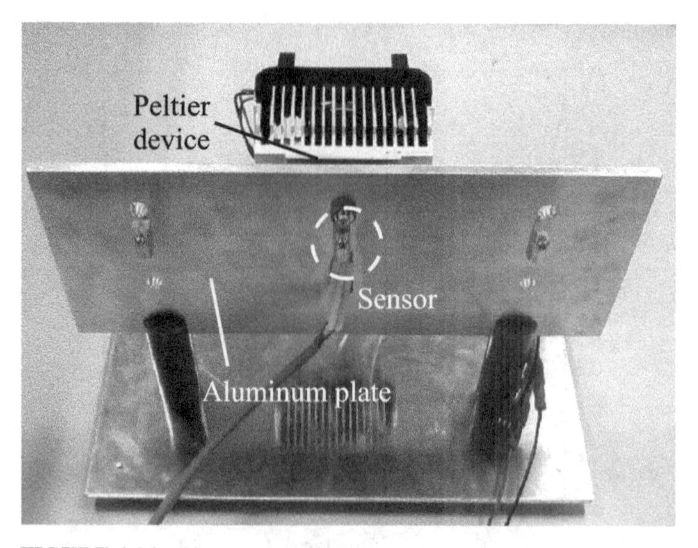

FIGURE 4.16 Photograph of experiment system of aluminum plate.

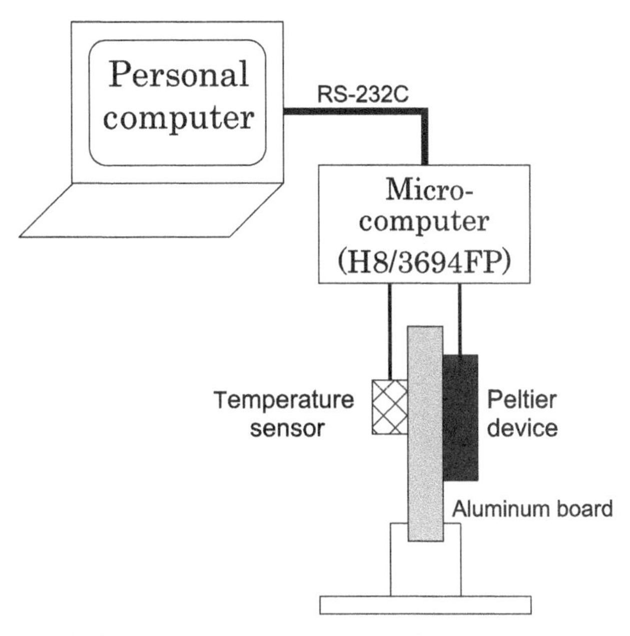

FIGURE 4.17 Experiment system of aluminum plate.

difference, and Joule exothermic heat by a current, a differential equation for heat conduction is obtained as

$$\begin{aligned}\frac{d(T_0 - T_x)mc}{dt} &= S_p T_1 i - K(T_h - T_1) - \frac{1}{2}R_p i^2 \\ &\quad - \alpha(T_0 - T_x)(2S_1 + 2S_2 - S_3) - \frac{2\lambda S_4(T_0 - T_x)}{d_1}\end{aligned} \tag{4.38}$$

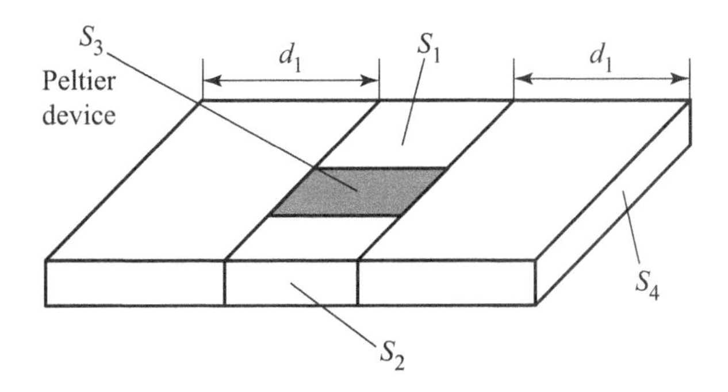

FIGURE 4.18 Aluminum plate thermal process.

TABLE 4.1 Parameters of Aluminum Plate Thermal Process

c	900 J/kg/K	S_1	6.8×10^{-3} m^2
α	15 W/m^2 K	S_2	3.4×10^{-4} m^2
λ	238 W/m K	S_3	9.0×10^{-4} m^2
d_1	0.095 m	S_4	5.0^{-4} m^2
K	0.63 W/K	R_p	4.2 Ω
m	0.81 kg	S_p	0.053 V/K

In [21], $T_0 - T_x$ is defined as process output $y(t)$ and the total endothermic amount $S_p T_1 i - K(T_h - T_1) - \frac{1}{2} R_p i^2$ is defined as control input $u(t)$. Then, the model of the thermal process is reexpressed in the form

$$y(t) = \frac{1}{cm} e^{-Ht} \int e^{H\tau} u(\tau)\, d\tau \tag{4.39}$$

where

$$H = \frac{\alpha(2S_1 + 2S_2 - S_3) + 2\lambda S_4 / d_1}{cm}$$

According to the parameters in Table 4.1, the nonlinear mathematical model with unknown perturbations can be estimated as follows:

$$(P + \Delta P)(u)(t) = \frac{1}{291.6} e^{-0.01t} \int_0^t e^{0.01\tau} u(\tau)\, d\tau + \delta^* \tag{4.40}$$

where δ^* denotes the unknown bounded perturbation. By the isomorphism approach, the nonlinear plant can be factorized into two parts:

$$\begin{aligned} (N + \Delta N)(w)(t) &= \frac{1}{291.6} w(t) + \delta^* \\ D^{-1}(u)(t) &= e^{-0.01t} \int_0^t e^{0.01t} u(\tau)\, d\tau \end{aligned} \tag{4.41}$$

Then we get the form

$$D(w)(t) = \frac{1}{100} w(t) + \dot{w}(t) \tag{4.42}$$

According to Theorem 4.2, the controllers A and B can be designed as

$$\begin{aligned} A(y)(t) &= y(t) \\ B(u)(t) &= \frac{1}{291.6 \times 10^2} u(t) \end{aligned} \tag{4.43}$$

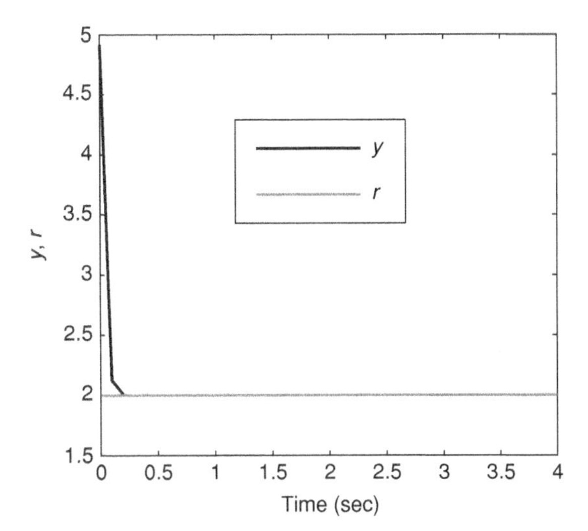

FIGURE 4.19 Plant output y asymptotically tracks r.

Thus the system is guaranteed to be BIBO stable and the plant output asymptotically tracks the reference input from about 0.2 sec. The simulation results are shown in Figures 4.19–4.21, where the environment temperature $T_0 = 21°\text{C}$, reference input $r = 2$ K, and $\delta^* = 0.2$ rand(1).

In this simulation, the controller A is specially designed to be I to show the main difference from the design scheme in [5] wherein the controller A cannot be designed to tend to I due to the unimodular property of the controller[B. Further, $B(u)(t)$ can be designed as a fixed-gain controller provided that assumption III is satisfied.

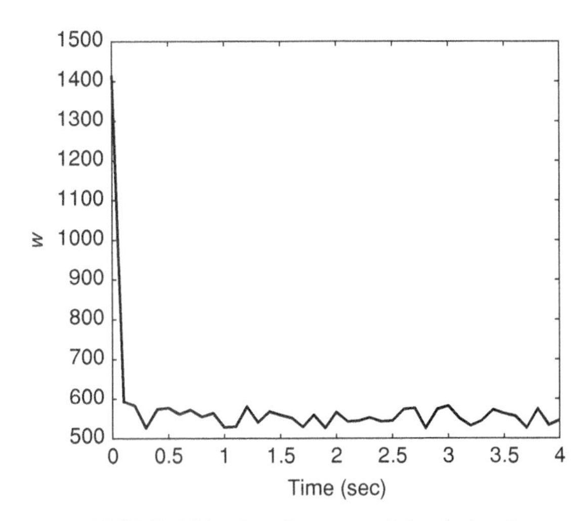

FIGURE 4.20 Quasi-state w of simulation 2.

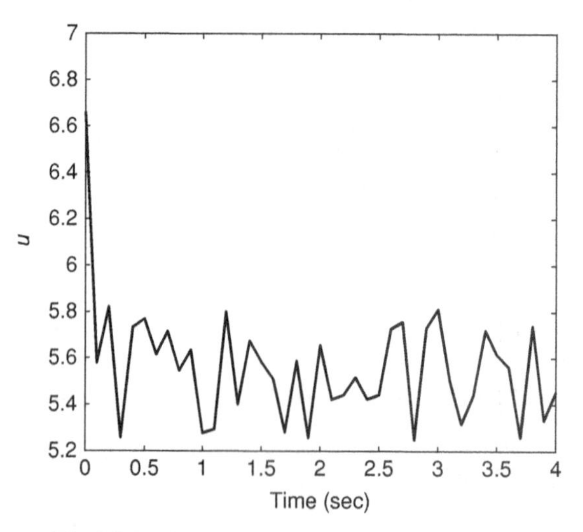

FIGURE 4.21 Control input u of simulation 2.

4.2.4 Summary

The robust control of the nonlinear feedback system with unknown perturbations was considered. The robust stabilizing controllers (A, B) guarantee not only the robust stability of the nonlinear perturbed system but also that the plant output asymptotically tracks the reference input. The effectiveness of the design scheme was confirmed by simulation results.

4.3 OPERATOR-BASED ACTUATOR FAULT DETECTION METHODS

4.3.1 Introduction

Process control is very popular in real control systems. The main purposes of the process control system are to stabilize the process and minimize the negative effect of disturbances, noises, and process input constraints. Examples of the process are tanks, compressors, and heat exchangers. These processes control water level, pressure, and temperature, respectively. To stabilize such real control systems, which contain some negative factors mentioned above, the process models seem to be developed based on a few physical laws or system identification. Following this, the controllers for the processes can be designed. However, a real process is likely to contain a fault signal owing to various environmental factors (see [62, 63]). Besides, the process control system needs to deal with some input constraints, such as limited power of heater or compressor. Furthermore, a real process has uncertainties whose linearity is unknown. Therefore, the uncertain process with an input constraint is a considerable issue in the real process. As for fault detection, a large number of interesting design methods have been researched (see [45, 64, 65]). In general, fault signals are classified as three kinds, namely sensor fault, actuator fault, and process fault. From the viewpoint of process safety, the control systems should have the capability of detecting the fault signal, because the process will be corrupted unless the system has the capability. To detect the fault signal, one approach is to locate many sensors. On the other hand, in this approach the costs of detecting and monitoring the sensor data are high. For this reason, analytical methods that utilize measurable information in the process have been considered.

In this chapter, different from the tracking filter fault detection in [45], operator-based robust right coprime factorization (see [2, 3, 66], also see Chapter 2) is applied to an uncertain thermal process of an aluminum plate with process input constraints. First, the thermal process is modeled as a right coprime factorization description. Next, a robust tracking control system is shown. Then, the Bezout identity can be given for stabilizing the system by robust right coprime factorization based on operator theory. Finally, the fault signal on an actuator is analyzed using two sorts of operators. An experimental result is presented to support the designed control system and fault detection system. In what follows, first modeling and problem setup are presented in Section 4.3.2. The following section discusses how to design operators. An algorithm of the fault detection system is given in Section 4.3.3. An experimental result is shown in Section 4.3.4, and its discussion is also given. Finally, Section 4.3.5 draws conclusions.

4.3.2 Actuator Fault Detetion Method in Nonlinear Systems

Most definitions and lemmas concerned with operators and bounded input and bounded output stability were given in Chapter 2. In this setion, first we recall a few significant definitions. Then the process model is explained briefly. See Chapter 2 for detailed information in regards to the modeling.

Let U and Y be linear spaces over the field of real numbers and let U^* and Y^* be normed subspaces, called the stable subspaces, of U and Y, respectively. An operator $Q: U \to Y$ is said to be BIBO stable or, simply, stable if $Q(U^*) \subseteq Y^*$.

Let $\mathcal{S}(U, Y)$ be the set of stable operators from U and Y. Then $\mathcal{S}(U, Y)$ contains a subset defined by

$$\mathcal{U}(U, Y) = \{M : M \in \mathcal{S}(U, Y)\} \tag{4.44}$$

where M is invertible with $M^{-1} \in \mathcal{S}(U, Y)$. Elements of $\mathcal{U}(U, Y)$ are called unimodular operators.

For convenience, the feedback control system is denoted as $\{P, K\}$.

The following lemmas of a right coprime factorization (rcf) are employed.

Lemma 4.2 Given $\{P, K\}$ and $P = ND^{-1}$ and $K = SR^{-1}$, the rcf's of the process and controller, respectively. Then $\{P, K\}$ is well-posed if and only if

$$\begin{bmatrix} D & -S \\ -N & R \end{bmatrix}^{-1} \tag{4.45}$$

exists and is internally stable if and only if

$$\begin{bmatrix} D & -S \\ -N & R \end{bmatrix}^{-1} \tag{4.46}$$

is BIBO stable.

Hence the stability and well-posedness of the system depend on the existence and stability of the operator $\begin{bmatrix} D & -S \\ -N & R \end{bmatrix}^{-1}$.

Lemma 4.3 Suppose $P = ND^{-1}$ and $K = SR^{-1}$ such that the operators D, N, S, R are BIBO stable. Then these are rcf's for P and K if they satisfy (4.46).

Lemma 4.4 Suppose that Lemma's 4.2 and 4.3 are satisfied. Then the system is overall stable if and only if the operator M is a unimodular operator, namely, $M \in \mathcal{U}(W, U)$.

There are three main parts for the experimental setup:

- Aluminum part
- Interface part
- Computer part

The first part consists of an aluminum board divided into three blocks hypothetically, that is, three thermal sensors and three heaters whose maximum outputs are 40 W. The sensors and heaters are both fixed on three spots of those three blocks.

To measure the temperature of an aluminum plate, thermal sensors are used and the measured temperatures are output as voltages. Then, the binary data are converted to base 10 at the computer. The model is shown in Figure 3.55, and a picture of the process is shown in Figure 3.56.

The configuration of the aluminum plate thermal process is shown in Figure 4.22. Fourier's law of heat conduction, Newton's law of cooling, and the equation between heat capacity and objects and their specific heat are used in the development of the mathematical model. From these three laws, the thermal process is described as

$$y(t) = \frac{1}{cm} e^{-At} \int e^{A\tau} u_d(\tau)\, d\tau \tag{4.47}$$

where c is specific heat, m is mass, α is the heat transfer coefficient, and A is defined as

$$A = \frac{\alpha(4s_1 + 2s_2 + 4s_3 + s_4 + 2s_5 - s_6)}{cm} \tag{4.48}$$

In addition, u_d is defined as process input.

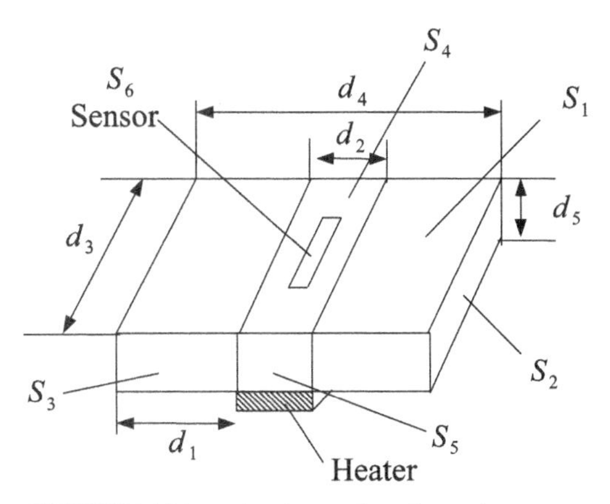

FIGURE 4.22 Aluminum plate thermal process.

Consider the nominal thermal process described by the right coprime factorization

$$y(t) = P(u_d)(t) = ND^{-1}(u_d)(t) \tag{4.49}$$

where D^{-1} is invertible. The equations are given as

$$D(w)(t) = cmw(t),\ D^{-1}(u_d)(t) = \frac{1}{cm}u_d(t) \tag{4.50}$$

$$N(w)(t) = e^{-At}\int e^{A\tau}w(\tau)\,d\tau \tag{4.51}$$

In a real system, a thermal process is necessary to deal with uncertainties or disturbances, and the above perturbations should affect D or N. Therefore, suppose that the process P has a bounded perturbation ΔP. That is, assume that only N has a bounded perturbation ΔN and the operators $N + \Delta N$ is stable such that

$$P + \Delta P = (N + \Delta N)D^{-1} \tag{4.52}$$

$$(N + \Delta N)(w)(t) = (e^{-At} + \Delta_1)\int e^{A\tau}w(\tau)\,d\tau \tag{4.53}$$

where $D : W \to U$ and $N : W \to Y$ are stable operators, respectively, and the space W is a quasi-state space (see [3, 21]) of P. Assume D is invertible and Δ_1 in (4.54) is regarded as an uncertainty created by the approximation of the aluminum plate in modeling. The modeling of the uncertain factor is written in [45].

Here, the process input $u_d(t)$ is subject to the following constraint on its magnitude:

$$u_d(t) = \sigma(u_1(t))$$
$$\sigma(v) = \begin{cases} u_{\max} & \text{if} \quad v > u_{\max} \\ v & \text{if} \quad u_{\min} \le v \le u_{\max} \\ u_{\min} & \text{if} \quad v < u_{\min} \end{cases} \tag{4.54}$$

where $u_1(t)$ is the control input before the constraint.

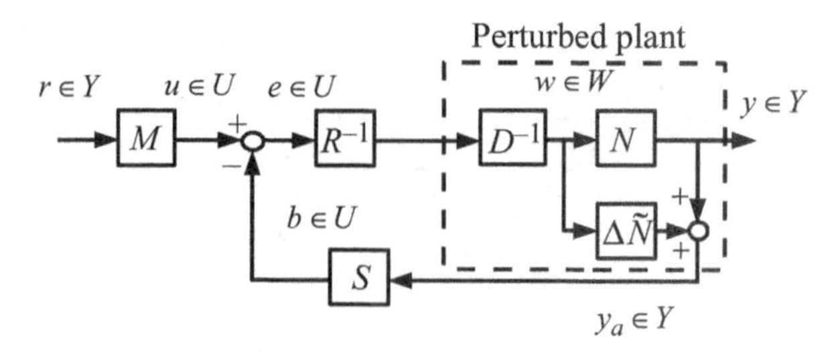

FIGURE 4.23 Tracking system [45].

To control the thermal process, two stable operators $S : Y \rightarrow U$ and $R : U \rightarrow U$ are required to be designed under the condition of well-posedness and that of N and D being said to have a right coprime factorization. If these two operators exist and satisfy the Bezout identity

$$(SN + RD)(w)(t) = I(w)(t) \tag{4.55}$$

where R is invertible and I is the identity operator, the process can be controlled. As for the case containing additive perturbation ΔP, the process is likely to be represented as

$$P + \Delta P = (N + \Delta N)D^{-1} \tag{4.56}$$

Assume that two stable operators S and R exist satisfying the perturbed Bezout identity

$$(S(N + \Delta N) + RD)(w)(t) = I(w)(t) \tag{4.57}$$

When the process input is limited, the process uncertainty is described as the operator

$$\Delta\tilde{N} : W \rightarrow Y$$

The tracking system shown in Figure 4.23 can be designed to satisfy the two conditions

$$(N + \Delta\tilde{N})M(r)(t) = r(t) \tag{4.58}$$

$$(S(N + \Delta\tilde{N}) + RD)(w)(t) = I(w)(t) \tag{4.59}$$

where $M : Y \rightarrow U$ is a stable tracking operator and $r \in Y$ is a reference signal. Furthermore, if there is no process input constraint, (4.55) and (4.57) and the following conditions should be satisfied:

$$(N + \Delta N)M(r)(t) = r(t) \tag{4.60}$$

$$NM(r)(t) = r(t) \tag{4.61}$$

The system is internally stable due to Lemmas 4.2 and 4.3 and the fact that the three Bezout identities mentioned above are satisfied, where all operators are BIBO stable.

Eventually, operators S, R, and M are designed as

$$S(y_a)(t) = K_p(1 - B)[\frac{dy(t)}{dt} + Ay(t)] \tag{4.62}$$

$$R(u_d)(t) = \frac{K_pB - K_p + 1}{cm}u_d(t) \tag{4.63}$$

$$M(r)(t) = Ar + (1 - A)re^{-t} \tag{4.64}$$

where K_p is the gain and B is a design parameter.

4.3.3 Algorithm of Fault Detection System

In [45], a fault detection system for the tracking operator M was proposed. One advantage of the method is that the fault signal in the tracking operator can be obtained without using a large number of sensors. The fault detection system can be meaningful provided the tracking operator works on hardware. Besides, the fault detection system in this chapter may be more useful than the one in [45], since it is applied to an actuator fault and does not depend on the kind of operator, that is, software or hardware.

To detect the fault signal, first the three operators R_0, S_0, and D represented in Figures 4.24 and 4.25 are designed.

In Figure 4.23, the Bezout identities

$$(SN + RD)(w)(t) = I(w)(t) \tag{4.65}$$

$$(S(N + \Delta N) + RD)(w)(t) = I(w)(t) \tag{4.66}$$

$$(S(N + \Delta \tilde{N}) + RD)(w)(t) = I(w)(t) \tag{4.67}$$

are satisfied.

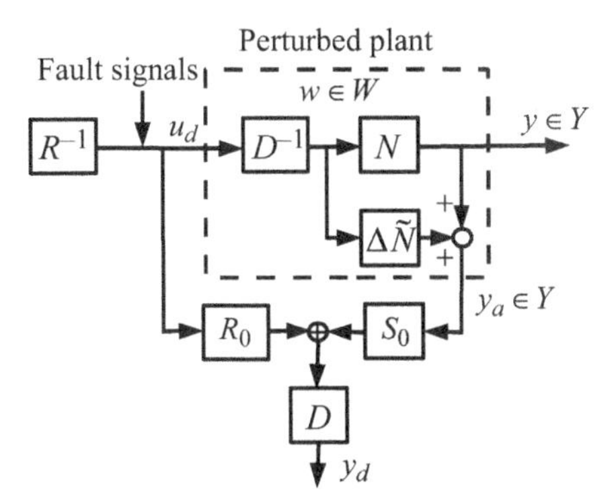

FIGURE 4.24 Fault detection system.

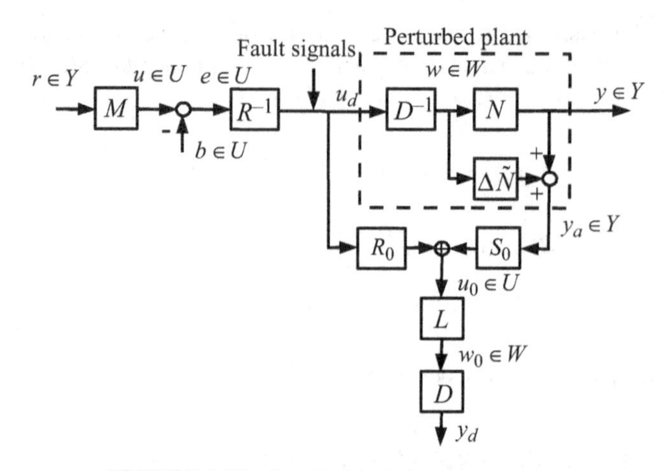

FIGURE 4.25 Detailed fault detection system.

When R_0 and S_0 are designed to satisfy the Bezout identity

$$(S_0 N + R_0 D)(w)(t) = I(w)(t) \tag{4.68}$$

S_0 and R_0 are obtained as

$$S_0(y_a)(t) = (1 - K_0)(\frac{dy(t)}{dt} + Ay(t)) \tag{4.69}$$

$$R_0(u_d)(t) = \frac{K_0}{cm} u_d(t) \tag{4.70}$$

where K_0 is a constant. Assume that the sum of the outputs of S_0 and R_0 is a mapping from space W to U as well as (4.68) according to Figure 4.24. That is, the sum $u_0 \in U$ is represented as

$$\begin{aligned} u_0 &= R_0(u_d)(t) + S_0(y_a)(t) \\ &= R_0 D(\omega)(t) + S_0(y_a)(t) \end{aligned} \tag{4.71}$$

where $\omega(t)$ is obtained by the other method. Moreover, process input u_d becomes

$$u_d = R^{-1}(e)(t) + \text{fault} \tag{4.72}$$

It may be understandable from (4.65) that signal w is equivalent to u because the Bezout identity is the identity mapping from $W \to U$. Similarly, (4.68) implies that signal w equals u_0. However, we have not ensured that $u = u_0$ for general case. If so, (4.68) can be written as

$$(S_0 N + R_0 D)(w)(t) = M(w)(t) \tag{4.73}$$

TABLE 4.2 Parameters for Experiment

Maximum power of heater	40 W
Reference input	$r = 1.5$
Constant	$B = 0.7$
Designed gain	$K_p = 3.2$
Gain for fault detection	$K_0 = 0.95$
Simulation time	1800 sec

where $M(w)(t)$ is unimodular, although in this chapter $M(w)(t)$ is decided as $I(w)(t)$. As for operator D, a mapping exists from signal w to u_d from Figure 4.23. Therefore, y_d in Figure 4.24 is equal to the signal in an actuator affected by the fault signal as long as w is equivalent to w_0. However, connecting u_0 to D directly like Figure 4.24 is impossible due to the difference of spaces to which u_0 and w_0 belong. One way to solve this problem is to design the space-change operator L shown in Figure 4.25, which is required to be mapped from U to W. The operator is described as

$$L(R_0 D + S_0 N) = I \tag{4.74}$$

where N, S_0, and R_0 are invertible. Eventually, the difference between the actuator's signal before it is affected by the fault signal, that is, the output of operator R^{-1} and y_d, results in the actuator faults. In other words, the following equation enables us to detect the fault signal:

$$\text{Detected fault signal} = \text{abs}[R^{-1}(e)(t) - y_d] \tag{4.75}$$

If there are no actuator faults,

$$\text{Detected fault signal} = 0$$

4.3.4 Experiments and Discussion

An experiment to show the effectiveness of the fault detection system is conducted. Table 4.2 presents the parameters utilized in the simulation, and the initial temperature and desired temperature of the aluminum plate are in Table 4.3. Other detailed information is in [45].

Figure 4.26 shows the process input and output, and Figure 4.27 expresses the output of R^{-1}, y_d, and detected fault signal in $0 < t \leq 1800$ sec (see Table 4.4). Moreover, Figures 4.28 and 4.29 show the comparison of y_d and u_d and detected

TABLE 4.3 Temperatures of Aluminum Plate

Desired temperature	28.3°C
Initial temperature	26.8°C

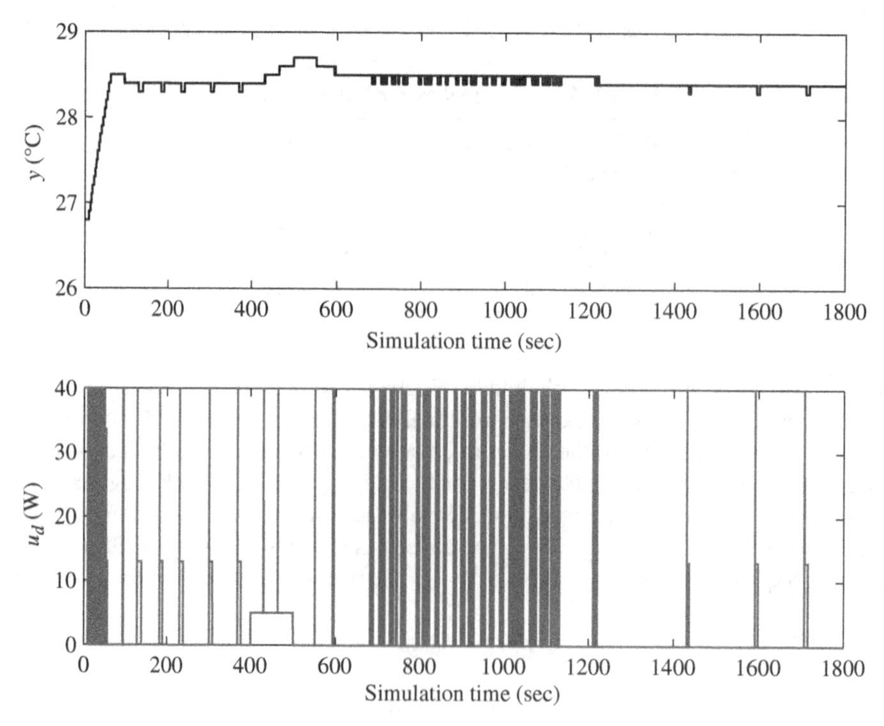

FIGURE 4.26 Plant output y and u_d.

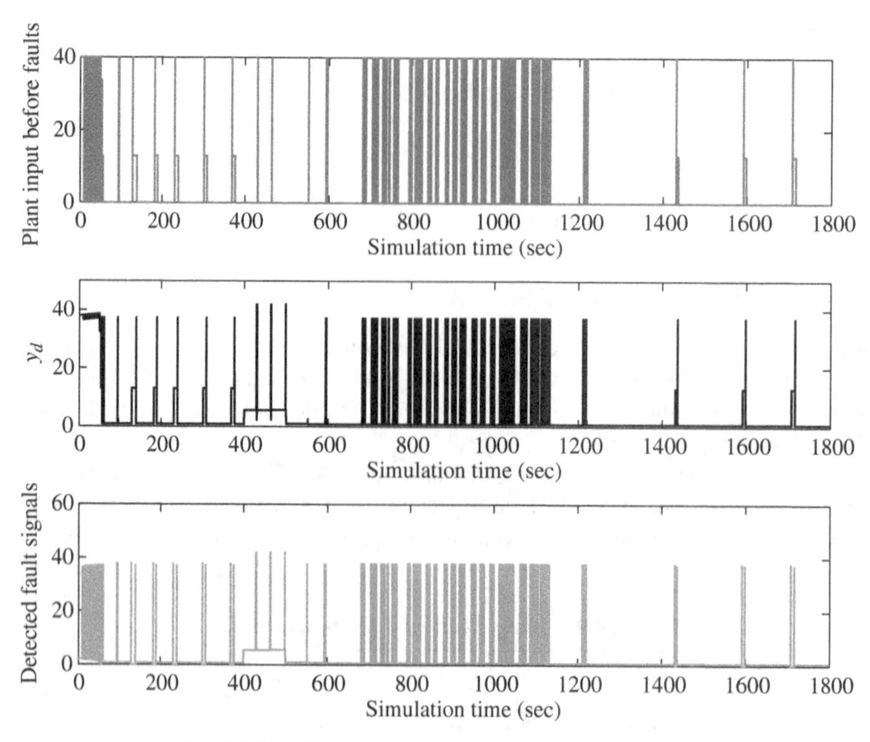

FIGURE 4.27 Before fault, y_d, and detected faults.

TABLE 4.4 Fault Signal in Actuator

$0 < t \leq 400$	0 W
$400 < t < 500$	5.0 W
$500 \leq t \leq 1800$	0 W

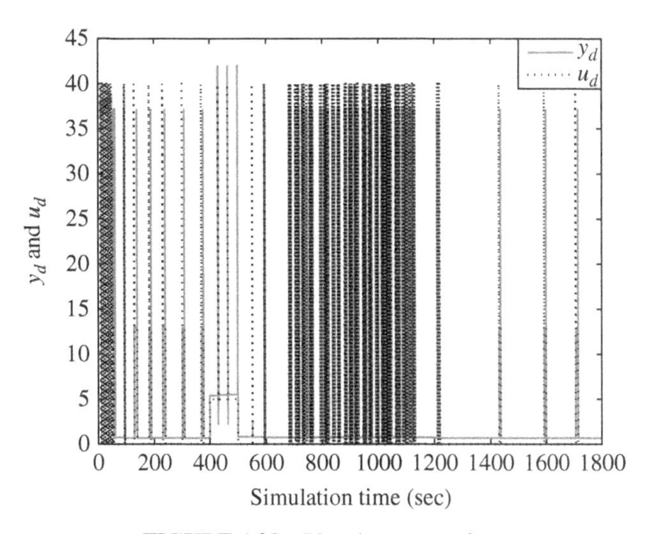

FIGURE 4.28 Plant input y_d and u_d.

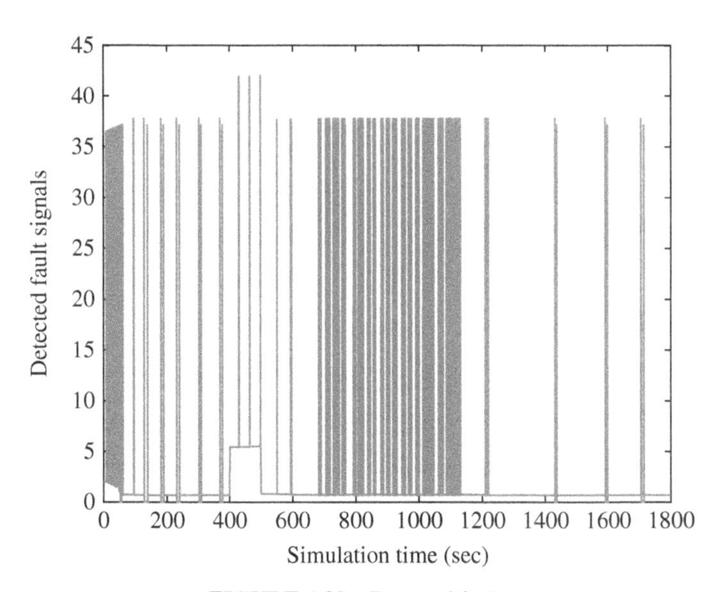

FIGURE 4.29 Detected faults.

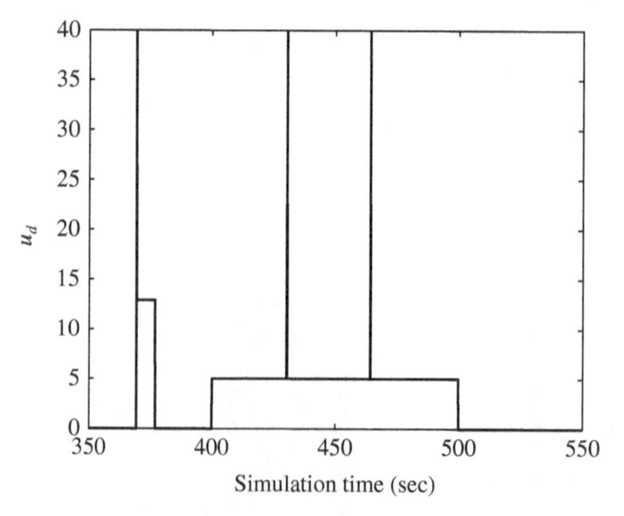

FIGURE 4.30 u_d when $350 \leq t \leq 550$.

fault signal, respectively. Figures 4.30 and 4.31 are magnified figures of the process input and the detected fault signal, respectively. Finally, the output of the tracking operator u (thick one) is compared with the sum of R_0 and S_0, that is, u_0 in Figure 4.32.

From Figure 4.26, process output tracks to the almost desired temperature before the fault signal happened, and after the actuator fault is fixed, the temperature is back to the desired temperature. Further, the designed fault detection system can detect a similar fault signal added according to Figures 4.29 and 4.31. Detailed discussion in regard to the simulation will be given in next section.

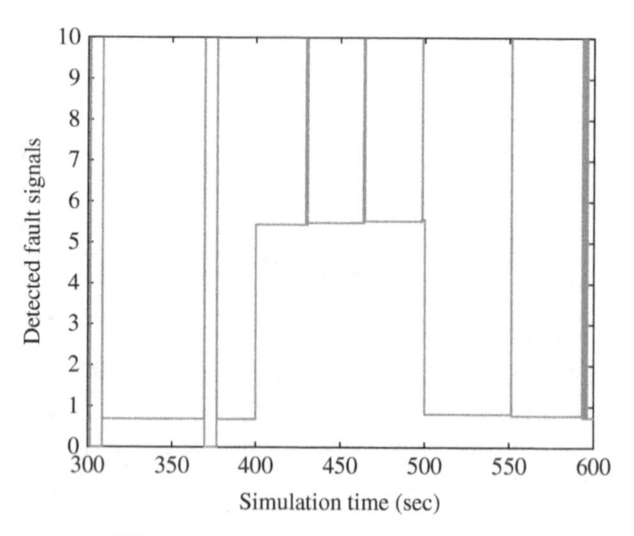

FIGURE 4.31 Detected faults when $300 \leq t \leq 600$.

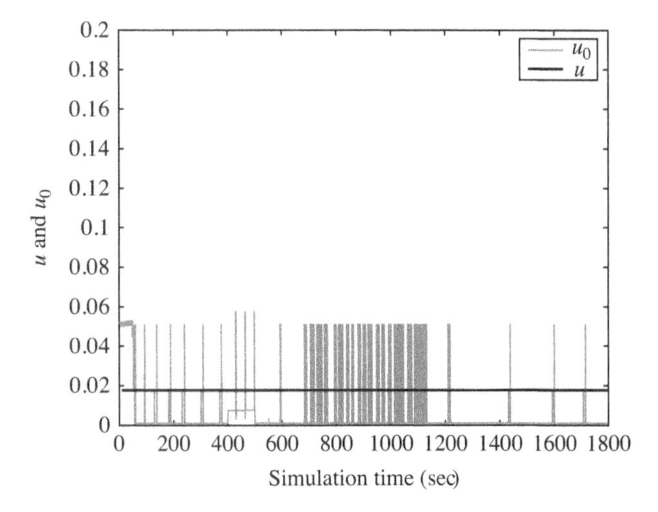

FIGURE 4.32 Tracking operator u and u_0.

From Figures 4.26 and 4.30, process output y tracks to the desired temperature approximately 100 sec, and after adding the fault signal, its output is back to the desired temperature. As for process input u_d, when $400 < t < 500$, the value of its input increases owing to the influence from the fault signal, and its value is 5 W. This result is similar to the simulation result shown in Figure 4.33, where the other simulation result including control is omitted for brevity. According to the simulation result, due to feedback from the process output, including an actuator fault, the impact from an actuator fault seems to be mitigated. However, the actuator fault affects the actuator directly in this experiment. This may be because of process input constraints. The steady-state value of process input is 0 W, which is equivalent to minimum input

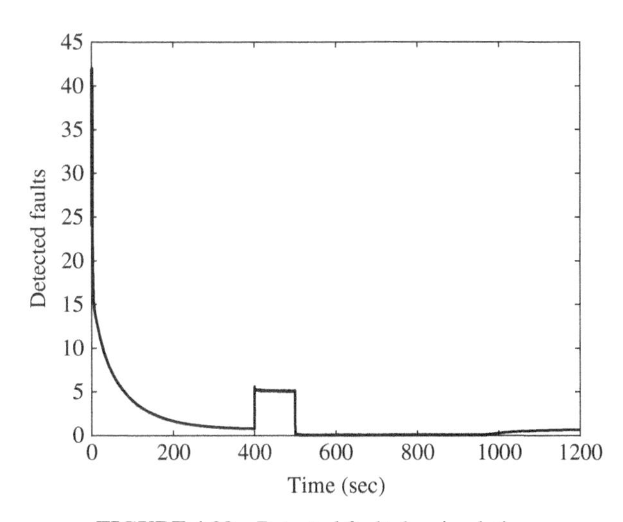

FIGURE 4.33 Detected faults by simulation.

constraints. Consequently, the actuator faults cannot be reduced. Furthermore, a differentiator in the operator S affects the signal, whose size is 40 W, shown in the center graph in Figure 4.27.

The bottom graph for detected tault signals in Figure 4.27 and Figure 4.29 shows the effectiveness of the designed fault detection system, since the fault signal detected and the one added are similar, though the former is slightly larger (see Figure 4.31). This might be because gain K_0 is not a proper number. Another possible reason will surface when we discuss the difference between y_d and u_d. From Figure 4.27, the designed system can detect the fault signal, as signal y_d can be the same as the signal u_d if $w = w_0$ is satisfied. However, according to Figure 4.28, these two signals are not exactly the same. This is likely to come from the effect from process input constraints and (4.74). If the input constraints are considered somehow and the equation is satisfied as $L(R_0 D + S_0 N) = I$, where L is unimodular, this problem would be solved.

Finally consider the error between u and u_0. It is clear from Figure 4.25 that u_0 is the sum of $S_0(y)(t)$ and $R_0(u_d)(t)$. From Figure 4.32, the signal u does not equal the signal u_0. The impulse signals that occur when the temperature of the aluminum plate changes are created by a differentiator in the operator S_0, which originally comes from low-temperature resolution of complementary metal–oxide–semiconductor (CMOS) sensors. Besides, the output of operator R_0 is restricted by process input constraints. For these two reasons, the value of u_0 is different from that of u, and this also could lead to data mismatching between y_d and u_d.

During the experiment, the environment temperature is fixed. As a result, the heat loss to the environment can be calculated. Based on the result in [45], including the heat loss, the uncertain factor in (4.54) is modeled as

$$\Delta_1 = e^{c-At} \left(\int e^{A\tau} w(\tau)\, d\tau \right)^{-1}$$

The uncertain effect can be controlled and removed by stability condition (4.58) and fault detecting conditions (4.68) and (4.74). However, the issue of an unknown uncertain factor will be considered in the future.

4.3.5 Summary

In this section, using the operator-based robust right coprime factorization, a fault detection system to an actuator was considered. The effectiveness of the fault detection system was confirmed by the experimental result. By extending the MIMO design techniques in Chapter 3, a fault detection system can also be designed for the MIMO system [85].

4.4 OPERATOR-BASED INPUT COMMAND FAULT DETECTION METHOD IN NONLINEAR FEEDBACK CONTROL SYSTEMS

4.4.1 Introduction

As shown in Section 4.3, the well-known process includes reactors, heat exchangers, pumps, compressors, and tanks. The control variables have temperature, pressure, water level, position, and reactor speed. The process model can be derived based on first principles and system estimation techniques. As a result, the process controller can be designed using linear or nonlinear design schemes. However, in applying the designed control schemes to the real process systems, operator failures happen due to various environmental factors, and the process control system must also deal with some constraints on input variables such as pressure and temperature limits. Further, a real process has uncertainties and is nonlinear, and the problem of the uncertain nonlinear process with input constraint is of considerable practical importance. A nonlinear process with a fault will be corrupted by the unknown fault signal. For the purpose of engineering safety, the control system must have the capability of distinguishing the fault signal. Concerning fault detection, many interesting design schemes have been considered [62–65, 86–90]. In general, fault diagnosis technology covers two main research areas. One is a mechanical method that uses an increased number of sensors. In this approach, each sensor parameter represents a respective working part and has a static value in the normal working state. Therefore it is simple to depict the abnormal signal by detecting whether the present sensory data deviate from the standard value or not. Another one is an analytical method that uses measurable process information. That is, measurable signal analysis schemes and process analysis using process models together with parameter and state estimation methods have been considered. The analytical methods are studied because the mechanical method costs more than the analytical method. The most commonly used analytical method is the method that uses model-based estimation methods. However, these known results using the analytical method are related only to the case of state feedback and transfer function presentations. In view of the input and output nature of the nonlinear process concept, it seems useful to establish computer-oriented approaches to nonlinear fault diagnosis system analysis and design caused by the enormous computational power of the computer. Motivated by the above consideration, the fault diagnosis analysis of nonlinear processes based on an operator-theoretic approach is considered Section 4.4.3. That is, based on operator theory, a nonlinear fault diagnosis system is designed. The theoretic basis is the concept of robust right coprime factorization described in Chapter 2. This method uses the relationships between faults and symptoms which are at least partially known so that this a priori knowledge can be represented in causal relations. However, considering nonlinear elements in real processes and in comparison with the inherently difficult problems of nonlinearity, the increasing demand for a mathematical theory is desired to fault diagnosis. Addressing this problem, in this chapter, an operator-based robust fault detection nonlinear tracking control system using robust right coprime factorization is considered. In practice, it is not unusual that the tracking controller operator can be realized by hardware as a command trajectory unit. However, the consistency of the output data of the designed unit with the desired tracking filter realized by a software as a model block needs to be checked. Instead of a mechanical method that uses an increased number of sensors, Section 4.4.3 gives a robust fault detection scheme for output tracking in processes with uncertainties and input constraints. Applying operator-based robust right coprime factorization from Chapter 2 to uncertain nonlinear processes with input constraints, an uncertain thermal process with input constraints is analyzed in this section. First, the thermal process is modeled as a right coprime factorization. Next, robust stability using right coprime factorization based on operator theory for the thermal process with control input constraints is obtained. Further, a robust tracking control system is also shown. Then, a Bezout identity can be given for the above class of nonlinear processes. Using two operators in the Bezout identity, the fault signal for the tracking operator system is analyzed. Simulation and experimental examples are presented to support the theoretical analysis.

Section 4.4.2 reviews the definitions for operator, coprime factorization, and stability. Modeling of a thermal process and problem setup are also discussed. Section 4.4.3 gives details of the design of robust fault detection. Simulation for drawing the theorectical description and the experimental result for the thermal process are illustrated in Section 4.4.4 to show the effectiveness of the method.

4.4.2 Modeling and Problem Setup

For the operator-based right coprime factorization of the thermal process, several definitions of operator [2, 3, 17, 20, 21] are first reviewed. Then, the process model and problem setup are given.

In the following, well-posedness and the stability of nonlinear feedback systems are described, and only those systems which are well-posed will be considered. Consider the problem of stabilizing a nonlinear continuous-time process P by a controller K, where the system is with real input spaces of continuous functions with continuous first derivative. For convenience, the feedback control system is denoted as $\{P, K\}$.

Definition 4.7 The system $\{P, K\}$ is well-posed if the closed-loop system input–output operator from $[u_1 u_2]^T$ to $[e_1 e_2]^T$ exists, namely,

$$\begin{bmatrix} I & -K \\ -P & I \end{bmatrix}^{-1} \tag{4.76}$$

Definition 4.8 The given process operator $P : U \to Y$ is said to have a right factorization if there exist a linear space W and two stable operators $D : W \to Y$ and $N : W \to Y$ such that D is invertible from U to W and $P = ND^{-1}$ on U. Such a factorization of P is denoted (N, D), and the space W is called a quasi-state space of P.

Definition 4.9 Let N and D be a right factorization for $P : U \to Y$:

$$P = ND^{-1} \qquad \begin{matrix} N : W \to Y \\ D : W \to U \end{matrix} \tag{4.77}$$

where N and D are stable operators from the quasi-state space W to the input and output spaces. Then (N, D) is a right coprime factorization of P if and only if, for any unbounded input $w \in W$, Nw or Dw is unbounded.

Definition 4.10 Let (N, D) be a right factorization of P. The factorization is said to have a right coprime factorization if there exist two stable operators $S : Y \to U$ and $R : U \to U$, where R is invertible, satisfying the Bezout identity

$$SN + RD = M \quad \text{for some } M \in \mathcal{U}(W, U) \tag{4.78}$$

The Bezout identity is often used in the following equation for simplicity:

$$SN + RD = I \qquad I : \text{identity operator} \tag{4.79}$$

The following lemmas of a right coprime factorization are employed.

Lemma 4.5 Given $\{P, K\}$ and $P = ND^{-1}$ and $K = SR^{-1}$, the rcf's of the process and controller, respectively, then $\{P, K\}$ is well-posed if and only if

$$\begin{bmatrix} D & -S \\ -N & R \end{bmatrix}^{-1} \quad \text{exists} \tag{4.80}$$

and is internally stable if and only if

$$\begin{bmatrix} D & -S \\ -N & R \end{bmatrix}^{-1} \quad \text{is BIBO stable} \tag{4.81}$$

Proof For the proof see [17]. ■

Hence the stability and well-posedness of the system depend on the existence and stability of the operator $\begin{bmatrix} D & -S \\ -N & R \end{bmatrix}^{-1}$. In fact, the relationship is somewhat stronger, and coprimeness also results from the stability of this operator, as is explored in the following lemma.

Lemma 4.6 Suppose $P = ND^{-1}$ and $K = SR^{-1}$ such that the operators D, N, S, R are BIBO stable. Then these are rcf's for P and K if they satisfy (4.81).

Proof For the proof see [17]. ■

Lemma 4.7 Suppose that Lemmas 4.5 and 4.6 are satisfied. Then the system is overall stable if and only if the operator M is a unimodular operator, namely, $M \in \mathcal{U}(W, U)$.

Proof The proofs of sufficiency and necessity can be made based on Lemmas 4.5 and 4.6 [3]. ■

In the following, the above definitions and lemmas are summarized based on the viewpoint of engineering. We consider the design of operator-based nonlinear tracking control systems and solve the fault diagnosis problem for the tracking control system from a normed linear space to another normed linear space of complex-valued functions defined on the time domain. These functions are the fundamental entities on which the results of the nonlinear tracking control system and fault diagnosis in the following sections are based. The definitions of operator and unimodular operator in Chapter 2 are mathematical preliminaries. The BIBO stable operator and the description of the normed subspace, called the stable subspace, are also given in Chapter 2. Operators can be extended to a more general setting to be suitable for nonlinear systems theory and applications. That is, the definition of operator theory given previously from a standard normed linear space setting is very useful for nonlinear systems theory and engineering in the consideration of stability and Lemma 4.5 and causality, robustness, and coprime factorizations in Definitions 4.8, 4.9, and 4.10. An advantage of the operator approach is that the nonlinear feedback system is well-posed (see Definition 4.7 and Lemma 4.5). This section considers the case for processes with right coprime factorization. The coprime factorization approach gives a convenient framework for researching input–output stability properties of nonlinear feedback control systems, where overall stability, namely the internal stability shown in Lemma 4.5 for the right coprime factorization of the controlled process, is considered under the assumption that the control system is well-posed. In the following, the process model and problem setup are given.

The configuration of the aluminum plate thermal process is shown in Figure 4.34.

Fourier's law of heat conduction, Newton's law of cooling, and the equation between heat capacity and objects and their specific heat are given as follows:

$$q_1 = -\lambda \left(\frac{d\theta}{dn} \right) \tag{4.82}$$

$$q_2 = \alpha(\theta_s - \theta_f) \tag{4.83}$$

$$cmd\theta = dQ \tag{4.84}$$

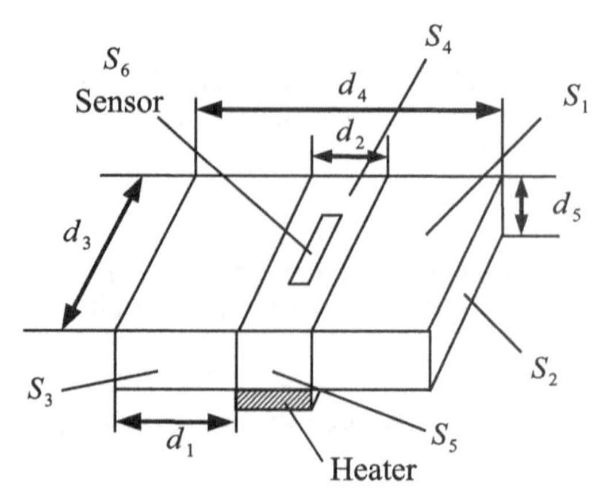

FIGURE 4.34 Aluminum plate thermal process.

From equations (4.82), (4.83), and (4.84),

$$\frac{cm(\theta_s - \theta_f)}{dt} = u_d - \alpha(\theta_s - \theta_f)(4s_1 + 2s_2 + 4s_3 + s_4 + 2s_5 - s_6) - \frac{2\lambda s_2}{d_1}(\theta_s - \theta_f)$$

Define $y(t) = \theta_s - \theta_f$, and the above equation is transformed as follows:

$$y(t) = \frac{1}{cm} e^{-At} \int e^{A\tau} u_d(\tau)\, d\tau \tag{4.85}$$

where

$$A = \frac{\alpha(4s_1 + 2s_2 + 4s_3 + s_4 + 2s_5 - s_6) + 2\lambda s_2/d_2}{cm} \tag{4.86}$$

Consider the nominal thermal process described by the following right coprime factorization:

$$y(t) = P(u_d)(t) = ND^{-1} \tag{4.87}$$

where

$$D(w)(t) = cmw(t) \qquad D^{-1}(u_d)(t) = \frac{1}{cm} \tag{4.88}$$

$$N(w)(t) = e^{-At} \int e^{A\tau} w(\tau)\, d\tau \tag{4.89}$$

Considering the real experimental system, the thermal process must deal with uncertainties and disturbances, and the above perturbation affects D and N in equation (4.14). In this case, suppose that the process P has a bounded perturbation ΔP, that is, ΔN is bounded and the operators and $N + \Delta N$ are stable such that

$$P + \Delta P = (N + \Delta N)D^{-1}$$

$$(N + \Delta N)(w)(t) = (e^{-At} + \Delta_1) \int e^{A\tau} w(\tau)\, d\tau \tag{4.90}$$

where $D : W \to U$ and $N : W \to Y$ are stable operators, respectively, and the space W is called a quasi-state space [3, 21] of P. Assume D is invertible, $P : U \to V$, and U and V are linear spaces over the field of real numbers, respectively. The uncertain factor is modeled as

$$\Delta_1 = e^{c-At} \left(\int e^{A\tau} w(\tau)\, d\tau \right)^{-1} \tag{4.91}$$

Here, assume that input signal u_1 is in a normed subset U^* of U and output $y(t)$ is in a normed subset Y^* of Y. Considering the input limitation of the heater (Figure 4.34), control input u_d is subject to a constraint on its magnitude, $u_d(t) = \sigma(u_1(t))$, where

$$\sigma(v) = \begin{cases} u_{\max} & \text{if} \quad v > u_{\max} \\ v & \text{if} \quad u_{\min} \leq v \leq u_{\max} \\ u_{\min} & \text{if} \quad v < u_{\min} \end{cases}$$

As a result, the modeled process has the modeling error given in (4.91) and an input constraint from the heater. For the above process, the objective of the section is to demonstrate how robust right coprime factorization theory can be utilized to achieve this control objective y tracking r and to detect the fault signal of $M(r)$ in tracking the operator system given in Section 4.4.3.

4.4.3 Robust Input Command Fault Detection Method

For nonlinear process (4.87), under the condition of well-posedness, N and D are said to be right coprime factorization if there exist two stable operators $S : Y \to U$ and R: $U \to U$ satisfying the Bezout identity

$$SN + RD = I(w)(t) \tag{4.92}$$

where R is invertible and I is the identify operator. In the same manner, under the condition that process (4.90) is well-posed, assume that there exist two operators S and R satisfying the perturbed Bezout identity

$$S(N + \Delta N) + RD = I(w)(t) \tag{4.93}$$

Therefore, the perturbed process retains a right coprime factorization. Consider the case when there is the input limit described in equation (4.91). When the control input is limited, the process uncertainty is represented as the operator

$$\Delta \tilde{N} : W \to Y \tag{4.94}$$

Based on the result in [21], an output tracking system can be designed by satisfying the conditions (see Figure 4.35)

$$(N + \Delta \tilde{N})M(r)(t) = r(t) \tag{4.95}$$

$$S(N + \Delta \tilde{N}) + RD = I(w)(t) \tag{4.96}$$

where $M : Y \to U$ is a designed stable operator and $r \in Y^*$ is a reference signal. Based on Lemmas 4.5 and 4.6, since equations (4.92), (4.93), and (4.96) are satisfied, the system is internally stable, where the operators are BIBO stable. Further, in the process without control constraints, equations (4.92) and (4.93) and the following

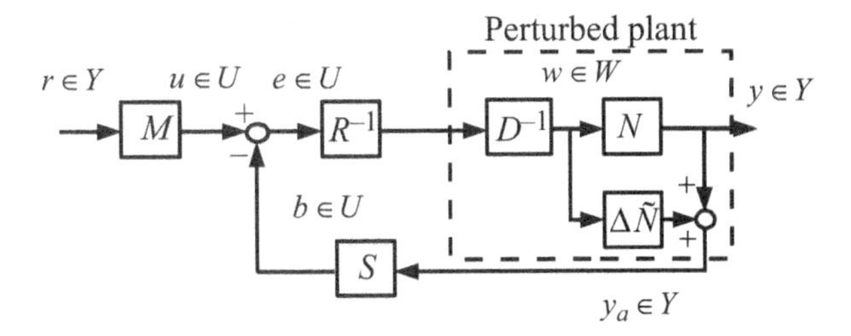

FIGURE 4.35 Robust tracking operator system.

conditions should be satisfied:

$$(N + \Delta N)M(r)(t) = I(r)(t) = r(t) \tag{4.97}$$

$$NM(r)(t) = I(r)(t) = r(t) \tag{4.98}$$

The merit of the tracking control system shown in Figure 4.35 is that the error signal $e(t)$ is not affected by the perturbed signal. That is, the perturbed signal $\Delta\tilde{N}$ cannot be transmitted back to the error signal $e(t)$ provided that equations (4.92), (4.93), and (4.96) are satisfied [3]. In practice, in this sense the undesired influence from uncertainties, output disturbances, and feedback sensor error is avoided. Meanwhile, for the perturbed process, the tracking problem of the real process output $y_a(t)$ is undertaken by tracking operator M, namely, $y_a(t)$ tracks to the reference signal $r(t)$ under the perturbations of ΔN or $\Delta\tilde{N}$. Here, the normed linear subspaces of U^* and Y^* are designed based on the control objective [21].

In practice, it is not unusual that the tracking operator M can be realized by hardware as a command trajectory unit. However, a check of the consistency of output data of the designed unit with the desired tracking filter realized by software as a transfer function block is necessary. Instead of a mechanical method that uses an increased number of sensors, the task of this chapter is to give a robust fault detection scheme for output tracking in processes with bounded perturbation $\Delta\tilde{N}$.

In the following, the task is to give a robust fault detection scheme for an output tracking system in a process with bounded perturbations $\Delta\tilde{N}$. If the perturbed Bezout identity can be satisfied, a system of robust fault detection can be given to check the output of the tracking operator M using S, R, and the desired tracking operator M', which is designed as well as M. The diagram of the detecting system is shown in Figure 4.36 for the case of y being observable. For the case of y_a being observable, if $\mathbf{S}_0 = S$ and $\mathbf{R}_0 = R$, then $\mathbf{u}_0 = u$ is obtained from (4.96). The robust fault detecting system is designed using the robust right coprime factorization condition. That is, if under the uncertainties and input constraints, the relationship $u = I(w)(t)$ remains unchanged for the identify operator I, then the uncertain process maintains a right coprime factorization which satisfies equations (4.93) and (4.96) [3]. Thus, the output signal of tracking operator M can be observed by using signal $w(t)$. The meaning of the above explanations is clear from the relationship between Figures 4.36 and 4.37. Further, signal $w(t)$ is observed based on the Bezout identify. In practice, the designed

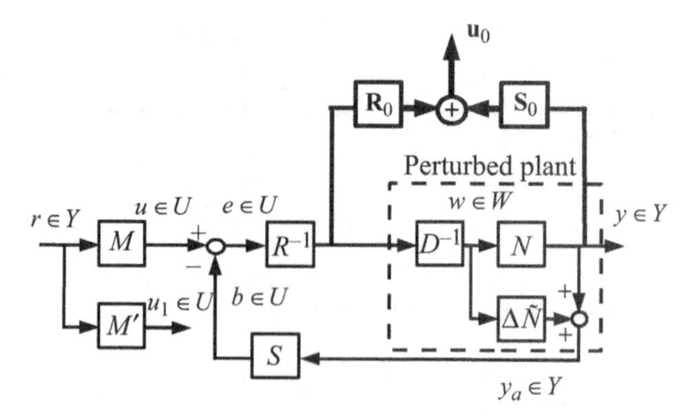

FIGURE 4.36 Robust fault detecting system.

controller is realized by the actuator. So the observed $\mathbf{R}_0$ and $\mathbf{S}_0$ of R and S can describe the real actuator output if the observing unit is asymptotic. In this chapter, the following theorem is given to guarantee the observed signal $\mathbf{u}_0$ generated by the outputs of $\mathbf{R}_0$ and $\mathbf{S}_0$ equaling the output u of the tracking operator M.

Theorem 4.4 Suppose that the perturbed Bezout identity equations (4.93) and (4.96) are satisfied and the designed tracking operator M is stable. Then, the output $\mathbf{u}_0$ of the detecting system tracks to the output u of M provided that $\mathbf{S}_0 = S$ and $\mathbf{R}_0 = R$.

Proof For the case of the process with input constraints, based on the result in [21], the nonlinear system depicted in Figure 4.36 can be regarded as a system with perturbation depicted in Figure 4.37.

From Figure 4.37, $\mathbf{u}(t) = w(t) = I(u)(t) = u(t)$. The input signal of $\mathbf{S}_0$ is $D(w)(t)$. Then, from equation (4.96) and Figure 4.36

$$\begin{aligned}\mathbf{u}_0(t) &= \mathbf{R}_0 D(w) + \mathbf{S}_0(y) \\ &= \mathbf{R}_0 D(w) + \mathbf{S}_0(N + \Delta\tilde{N})(w) \\ &= I(w)(t) \\ &= w(t)\end{aligned} \tag{4.99}$$

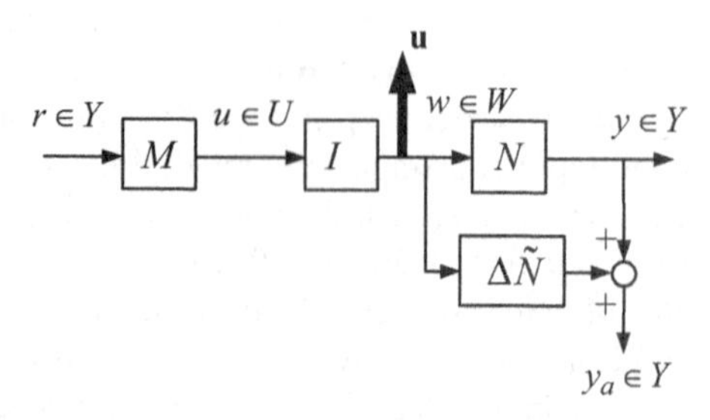

FIGURE 4.37 Equivalent diagram of Figure 4.36 without desired tracking operator M'.

provided that $S(N + \Delta\tilde{N}) + RD = I$, $\mathbf{S}_0 = S$, and $\mathbf{R}_0 = R$. That is, from Figure 4.37, $\mathbf{u}(t) = w(t) = u(t)$ is obtained. Then, from equation (4.99), $\mathbf{u}_0 = w(t) = u(t)$. For the case of process without input constraint, the desired result can be obtained by the same argument. ■

Theorem 4.4 is for the case of $y(t)$ being observable. According to the same manner, if $y_a(t)$ is observable and equations (4.93) and (4.96) are satisfied, then $\mathbf{u}_0 = u$ is also satisfied.

From Figure 4.37, the signal $u(t)$ is not affected by the perturbed signal. That is, the perturbed signal from $\Delta\tilde{N}$ cannot be transmitted back to the error signal $e(t)$ in Figure 4.36. In practice, in this sense the undesired influence from uncertainties, output disturbances, and feedback sensor error is avoided to the robust fault detection system. In the following, some examples by simulation and experiment illustrate the efficacy of the fault detecting method.

4.4.4 Simulation and Experimental Results

To explain the design procedure and to demonstrate the effectiveness of the above design scheme, simulation studies are conducted by using the thermal process and robust fault detection system shown in Figures 4.34 and 4.36. Parameters of the simulations are given in Table 4.5. The desired temperature is 25.9°C and the initial temperature is 22.9°C.

For the process without control constraints, from equation (4.92)

$$S(y_a)(t) = K_p(1 - B)\left(\frac{dy(t)}{dt} + Ay(t)\right) \tag{4.100}$$

$$R(u_d)(t) = \frac{K_pB - K_p + 1}{cm}u_d(t) \tag{4.101}$$

where u_d is process input, namely, $D(w)(t)$: K_p is proposal gain, and B is constant. Then, considering the process with uncertain factor shown in Section 4.4.2, from equations (4.89) and (4.90)

$$S(N + \Delta N)(w)(t) = K_p(1 - B)w(t) \tag{4.102}$$

TABLE 4.5 Parameters for Simulations

Integral constant	$c = -6.0$
Maximum power of heater	40 W
Reference input	$r = 3.0$
Constant	$B = 0.7$
Proposal gain	$K_p = 1.5$
Simulation time of tracking	600 sec
Simulation time of fault detection	1800 sec

The uncertain factor in equation (4.102) is considered using equation (4.91), namely,

$$\Delta_1 = e^{c-At}\left(\int e^{A\tau} w(\tau)\, d\tau\right)^{-1}$$

where c is constant. Then, equation (4.93) is satisfied. Further, for the process with control constraints, the following relationship should be satisfied based on the design condition of equations (4.93) and (4.96):

$$S(N + \Delta\tilde{N}) - S(N + \Delta N) = 0 \tag{4.103}$$

Since equation (4.103) is satisfied, the control system without the tracking performance is internally stable. The tracking operator is designed as

$$M(r)(t) = Ar + (1 - A)re^{-t} \tag{4.104}$$

where the tracking operator design condition of equation (4.95) is satisfied, and the above-mentioned operators are BIBO stable. In with real temperature control, the designed tracking operator is realized by a hardware unit including some electric parts, for example, a diode, a condenser, and IC chip. However, the parts may be destroyed by undesired high-voltage signals during the control period. Addressing this problem, the fault detecting system for checking the tracking operator is considered using the design scheme. To satisfy the design condition of the fault detecting system, the problem of determining a precise mathematical description of $\mathbf{S}_0$ and $\mathbf{R}_0$ is considered based on some a priori knowledge of the control system from data representative of input and output behavior. Here, the main difference between the operator-based scheme and the known results [62–65, 86–90] is that the existing methods are related only to the case of state feedback and transfer function presentations and the scheme is based on the input and output nature of the nonlinear process.

Figures 4.38 and 4.39 show the simulation results of tracking control. The upper part in Figure 4.38 shows the process output y_a, and the lower part shows the process input u_d. Figure 4.39 is the result of comparing u_1 with $\mathbf{u}_0$ in the case without a fault signal. From Figure 4.38, the output of the tracking operator u tracks to the desired temperature. In addition, the process input is constrained in 40 W, namely, the maximum of the process input is 40 W.

The fault signal ($250 < t < 350$) mixed with the output of the tracking operator normal signal, namely, tracking operator output u, is given in Table 4.6.

Figures 4.40 and 4.41 show the simulation result of fault diagnosis. Figure 4.40 shows the output of the tracking operator M with a fault signal when $250 < t < 350$. Figure 4.40 is the result of $u_1 - \mathbf{u}_0$. Figure 4.40 shows that the tracking controller had a fault, where $\mathbf{u}_0$ describes tracking operator output u, which includes the tracking operator normal signal and the fault signal.

Two kinds of experiments, namely, without a fault signal and with a fault signal, are conducted. The results and experimental analyses are presented in this section. The notation and parameters used in the two experiments are shown in Table 4.7.

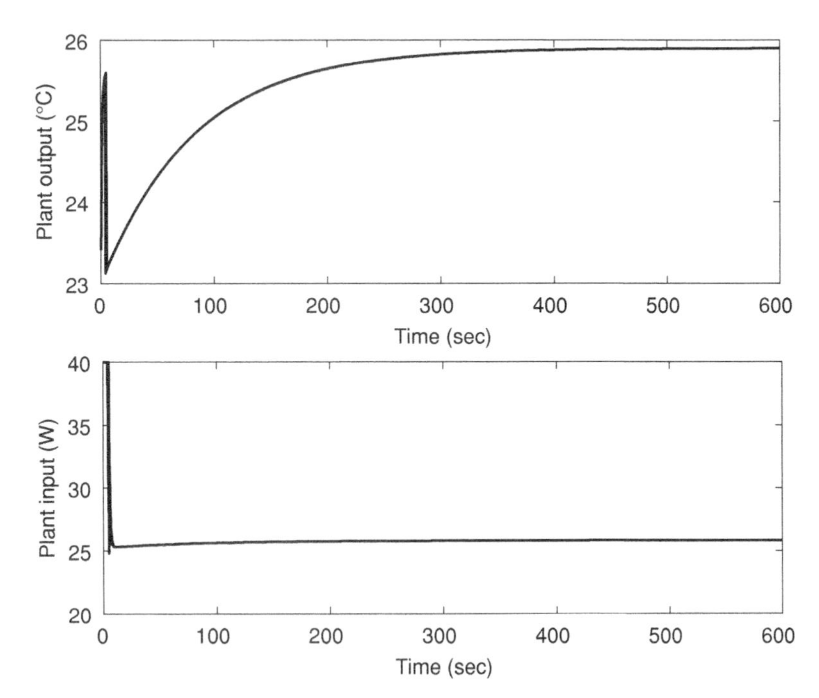

FIGURE 4.38 Process input $u_d(t)$ and output $y_a(t)$.

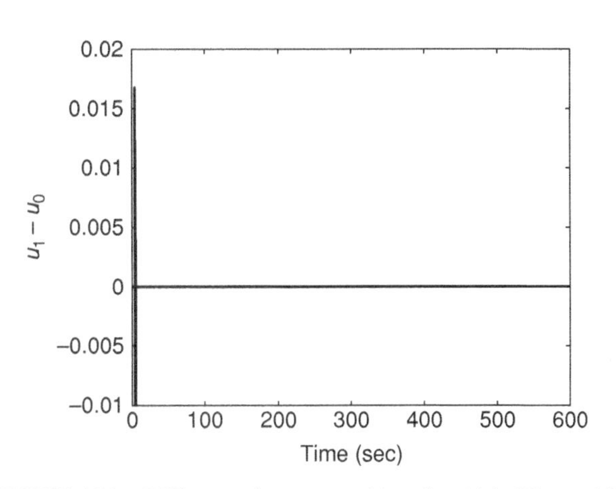

FIGURE 4.39 Difference between $u_1(t)$ and $\mathbf{u}_0(t)$ in Figure 4.36.

TABLE 4.6 Tracking Operator Output Signal u

$0 < t \leq 250$	$u = 3A + 3(1 - A)e^{-t}$
$250 < t < 350$	$u = 1.0$
$350 \leq t \leq 1800$	$u = 3A + 3(1 - A)e^{-t}$

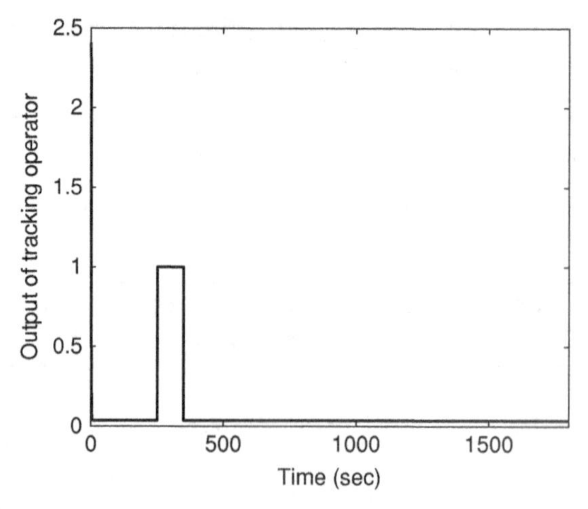

FIGURE 4.40 Output of tracking operator u with fault signal.

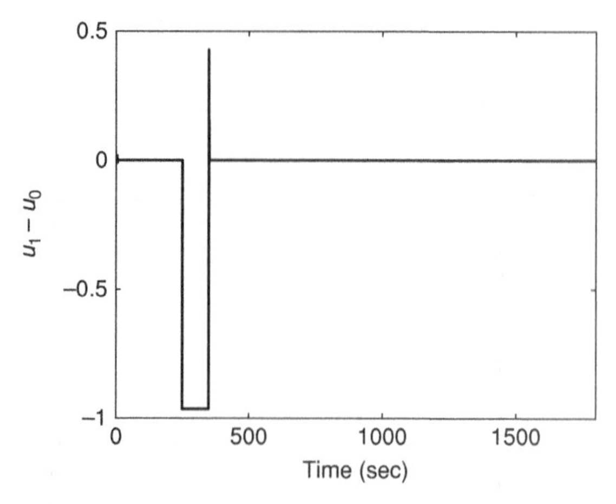

FIGURE 4.41 Difference between u_1 and $\mathbf{u}_0$ with fault signal.

TABLE 4.7 Parameters for Experiments

Maximum power of heater	40 W
Reference input	$r = 3.0$
Constant	$B = 0.7$
Proportional gain	$K_p = 3.2$
Experiment time of tracking	600 sec
Experiment time of fault detection	1800 sec

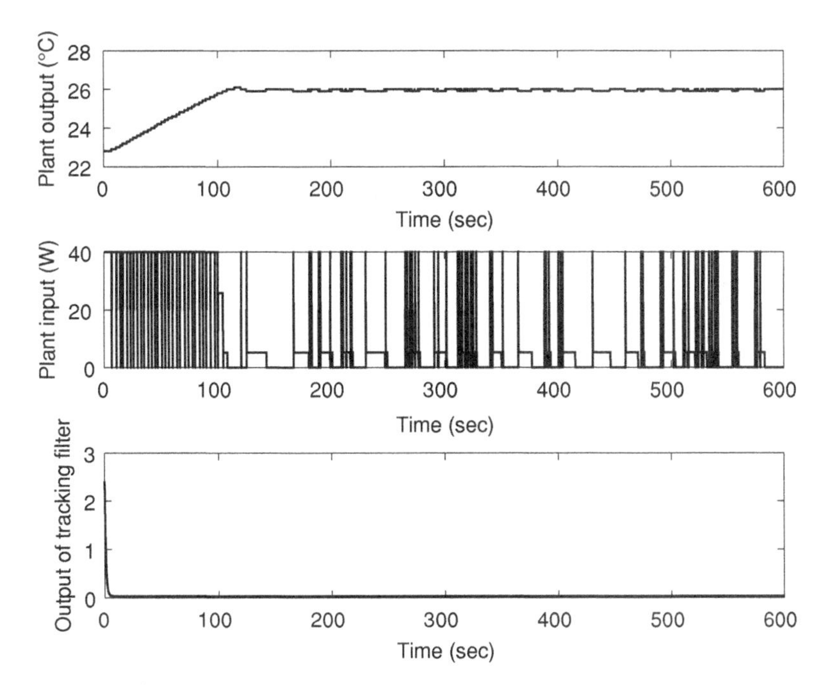

FIGURE 4.42 Process output $y_a(t)$, process input u_d, and tracking operator output u.

The initial temperature and the desired temperature are the same with the above simulation. Figures 4.42 and 4.43 show the experimental result for the case without fault signal. According to top graph in Figure 4.42 (without the fault signal), the desired tracking control has been obtained. Furthermore, Figure 4.43 recognizes the result $u_1 - \mathbf{u}_0 = 0$. Therefore, the process does not have a fault signal.

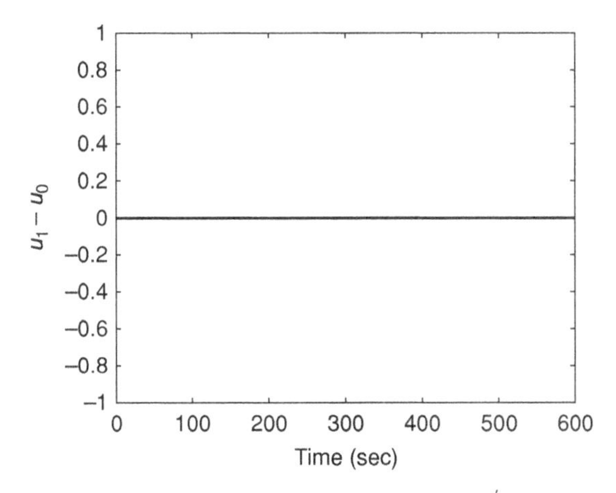

FIGURE 4.43 Difference between output u_1 of M' and output $\mathbf{u}_0$.

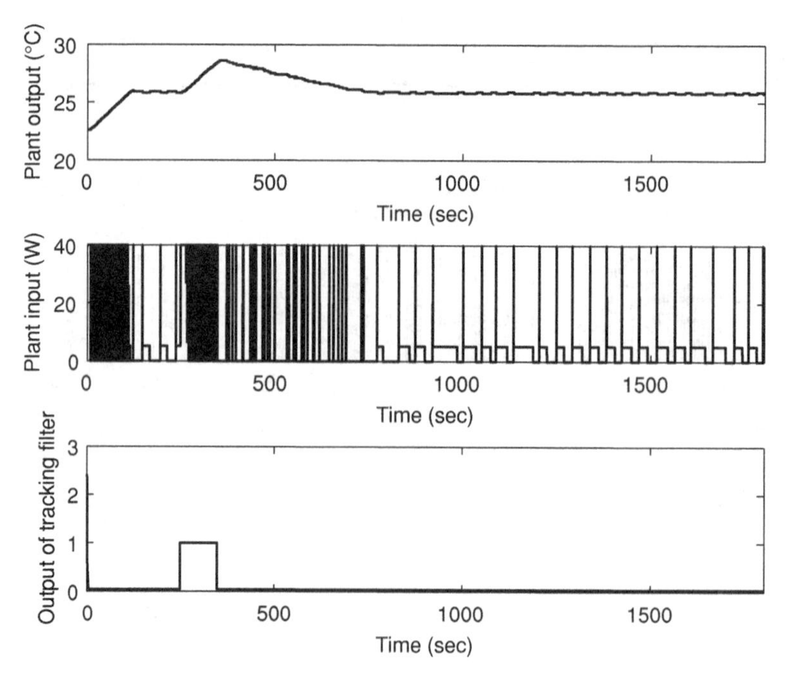

FIGURE 4.44 $y_a(t)$, u_d, and u with fault signal.

In the following, similar to the above simulation, the fault signal ($250 < t < 350$) concerned with the output of the tracking operator normal signal, generated as u, is given in Table 4.6. The experimental results for the process with fault detection are shown in Figures 4.44–4.47. From Figure 4.44, the temperature starts increasing under the influence of the fault signal in the tracking operator ($250 < t < 350$) after

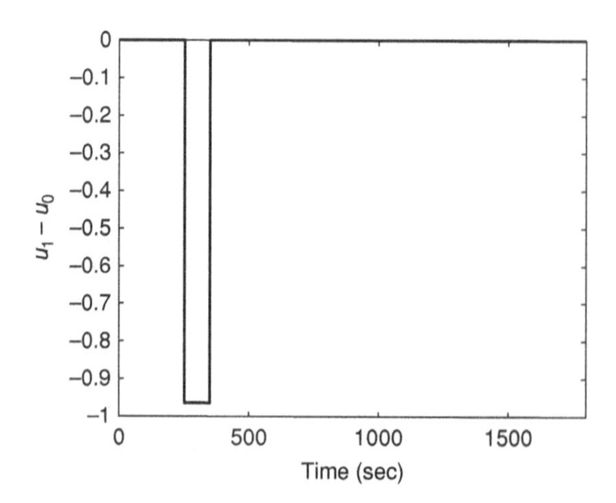

FIGURE 4.45 Difference between u_1 and $\mathbf{u}_0$.

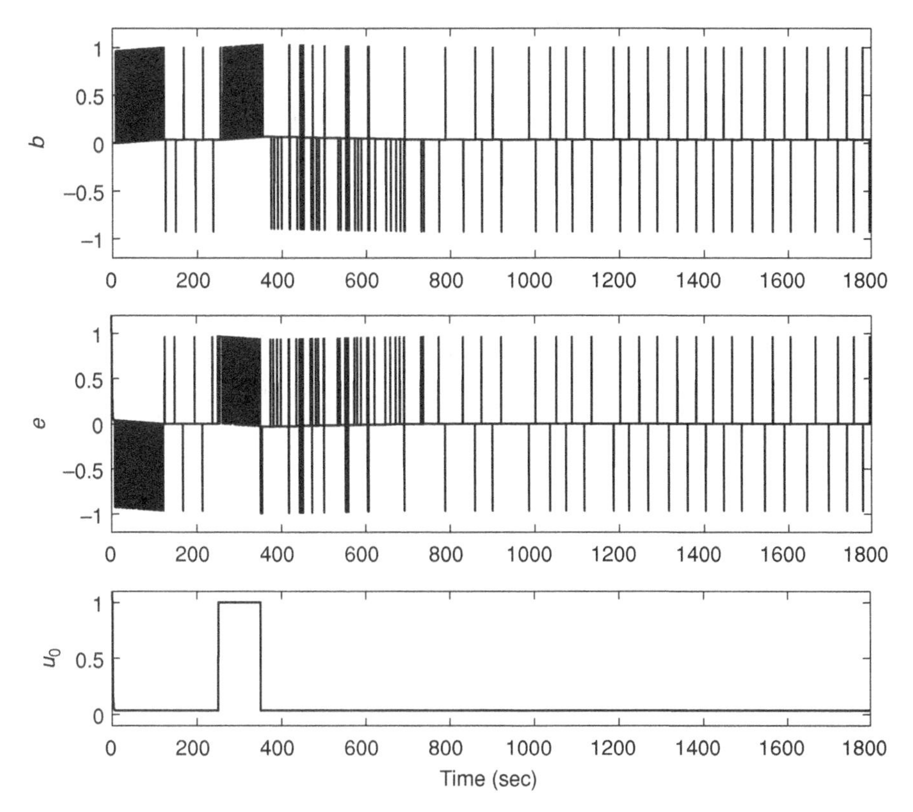

FIGURE 4.46 Output b and e of S_0 and R_0 respectively and $\mathbf{u}_0$.

the process is in a steady state. Then, process output returns to a steady state after the fault signal of the tracking operator is repaired. Figure 4.45 shows the value of $u_1 - \mathbf{u}_0$, where u_1 is the normal tracking signal. From the bottom part of Figure 4.46, the system could detect the fault signal when $250 < t < 350$. Figure 4.47 is a part from 200 sec to 400 sec of Figure 4.46, form Figure 4.47, the relationship $\mathbf{u}_0 = b + e$ is obtained, where the output b of operator S_0 and the output e of operator R_0 again describe the tracking output signal u provided that Theorem 4.4 is satisfied.

4.4.5 Summary

Using operator-based robust right coprime factorization, a fault detecting tracking operator was considered for an uncertain aluminum plate thermal process control system with input constraints. The Bezout identity was used to ensure that the observed fault signal equals the real fault signal. When the Bezout identity is satisfied, the perturbed signal of the system cannot be transmitted back to the controlled system.

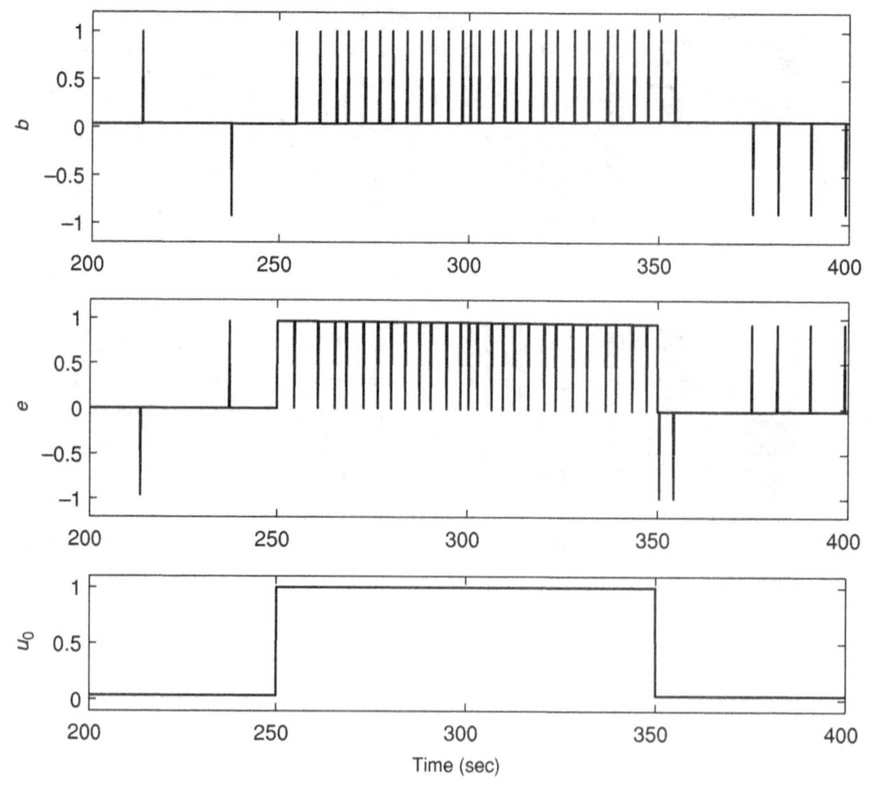

FIGURE 4.47 Outputs b, e and output of tracking operator $\mathbf{u}_0$ when $200 \leq t \leq 400$.

The effectiveness of the fault detecting method was confirmed by simulation and experiment. Two kinds of experiments, without a fault signal and with a fault signal, were conducted. Comparing the experimental results, the method achieved detection of the fault signal.

CHAPTER 5

Operator-Based Nonlinear Control Systems with Smart Actuators

Smart materials such as piezoelectric elements, shape memory alloys (SMAs), and ionic polymer–metal composites (IPMCs), are applied widely to make actuators. These actuators can sense and respond to environmental changes to achieve desired goals. However, backlash and hysteretic behaviors exist widely in the smart materials and affect the performance of actuators, even make the system with these actuators exhibit undesirable oscillations and instability. To compensate the effect of the backlash and hysteretic behavior, operator-based nonlinear control designs are developed in Sections 5.1 and 5.2. Experimental results to control these actuators are described in Section 5.3.

5.1 OPERATOR-BASED ROBUST NONLINEAR FEEDBACK CONTROL SYSTEMS DESIGN FOR NONSYMMETRIC BACKLASH

5.1.1 Introduction

Backlash is one of the most important nonlinearities that limit plant performance. Control of plants with backlash nonlinearity is an important area of control system research. As a result, control of backlash has attracted research efforts over several decades and is still an open source of both theoretical and practical problems because backlash has nondifferentiable nonlinearity which is often unknown [67]. For compensating backlash, there exist two approaches: adaptive and nonadaptive. In the adaptive-based approach, some researchers proposed identifying the backlash and constructing a backlash inverse. For continuous-time plants with symmetric backlash, in [68], an adaptive solution to identification and control was given. The work of [69] implemented a dynamic inversion using neural networks and backstepping. Using a smooth adaptive backlash inverse, [70] inverse designed output feedback backstepping adaptive controllers to compensate for backlash. Rather than constructing an inverse model, [71] defined a dynamic backlash model to mitigate the effect of

Operator-Based Nonlinear Control Systems: Design and Applications, First Edition. Mingcong Deng.

backlash and applied backstepping techniques in the control design. For a discrete-time linear plant with nonsymmetric backlash, [72] designed an adaptive backlash inverse controller and an adaptive scheme to guarantee global signal boundedness. The second approach to compensate for backlash is a nonadaptive one to design a robust controller. The objective of this approach is to make the controller robust enough to cope with the uncertain backlash effect. In [73], the stability of a linear plant with uncertain backlash was analyzed based on L_∞ and L_2. The considered slopes of uncertain backlash in [73] were symmetric. In practice, abrasion exists in gears and the slopes of backlash will be changed little by little. For the double-sided tooth impact case where the driver tooth collides with both the forward- and backward-driven teeth, using a nonsymmetric model to describe the effect of backlash is closer to the fact. For example, [74] analyzed self-oscillation in a thermal process with unsymmetrical backlash and phase leading phenomena. In addition, in the field of neurophysiology, nonsymmetric backlash was also considered in [75]. Nonsymmetric backlash is being applied gradually in different research fields. In this chapter, a nonlinear plant with uncertain nonsymmetric backlash is considered, and the desired control performance is achieved by the designed controller based on the this model.

Using the operator-theoretic approach of the above literature, nonlinear plants can be described by plant operator $P = ND^{-1}$, where D^{-1} and N are two factors of the right coprime factorization of these plants. The backlash which precedes the nonlinear plant can also be described as a Lipschitz operator. In practice, the signal between the backlash and the nonlinear plant is unavailable. So the backlash should be considered as one part of the nonlinear plant. Since, in this chapter, the considered backlash is uncertain nonsymmetric, the left the right slopes are unknown and are not same. We design a slope m_0 to substitute for the unknown slopes and to represent the backlash by a generalized Lipschitz operator term and a bounded parasitic term. In this way, a generalized Lipschitz operator term could be considered as one part of the nonlinear plant. We propose a new robust condition using robust right coprime factorization to design controllers that guarantee the stability of the whole system, as m_0 is used to substitute for the real slopes. To realize the output tracking performance of the considered plant, an operator-based exponential iteration theorem is applied to design a tracking controller to eliminate the effect of the parasitic term. The merit of this tracking controller is not related to any information of the uncertain backlash. Therefore, it is fit for eliminating the effect of the unknown parasitic term of the backlash.

5.1.2 Problem Statement

Let X and Y be linear spaces over the field of real numbers, and let X_s and Y_s be normed linear subspaces, called the stable subspaces of X and Y, respectively. Let $Q : X \to Y$ be an operator mapping from X to Y and denote by $\mathcal{D}(Q)$ and $\mathcal{R}(Q)$, respectively, the *domain* and *range* of Q. We always assume that $\mathcal{D}(Q) = X$ with $\mathcal{R}(Q) \subseteq Y$ and call operator Q a stable operator if $Q(X_s) \subseteq Y_s$.

In this chapter, we consider a nonlinear plant described by operator $P : U \to Y$, where U and Y are the input and output spaces, respectively, and P is described by the right coprime factorization

$$P = ND^{-1} \tag{5.1}$$

where $D : W \to U$ and $N : W \to Y$ are stable operators from the quasi-state space W to the input and output spaces. A feedback control system is said to be well-posed if every signal in the control system is uniquely determined for any input signal in U. For the above nonlinear plant (5.1), under the condition of well-posedness, since N and D are right coprime factorizations of P, there exist two stable operators $A : Y \to U$ and $B : U \to U$ satisfying the Bezout identity

$$AN + BD = M \qquad \text{for some } M \in \mathcal{S}(W, U) \tag{5.2}$$

where $\mathcal{S}(W, U)$ is the set of unimodular operator.

Conside the plant as either nonlinear or preceded by a nonsymmetric backlash (Figure 5.1) defined as [70]

$$u(t) = B_a(v(t)) = \begin{cases} m_r[v(t) - h], & \text{if } \dot{v}(t) > 0 \text{ and } u(t) = m_r[v(t) - h] \\ m_l[v(t) + h], & \text{if } \dot{v}(t) < 0 \text{ and } u(t) = m_l[v(t) + h] \\ u(t_-) & \text{otherwise} \end{cases} \tag{5.3}$$

where the parameters m_r and m_l stand for the right and left slopes of the backlash, $h > 0$ is the backlash distance and is unknown, and $u(t_-)$ means that no change occurs

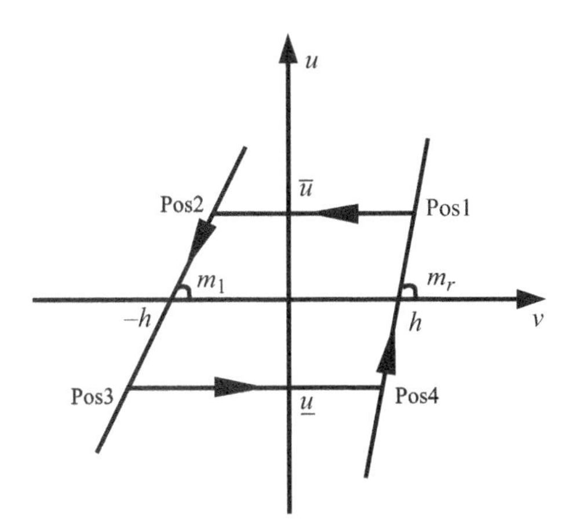

FIGURE 5.1 Nonsymmetric backlash.

in the output $u(t)$. We assume that (1) the coefficients m_r, m_l are strictly positive and unknown; (2) they satisfy this condition,

$$\frac{v_m - h}{v_m + h} m_r < m_l \leq m_r$$

where v_m $\max[v(t)]$; and (3) the maximum and the minimum values of the slopes of the nonsymmetric backlash are known, $\max\{m_l, m_r\} = \overline{m} < \infty$, $\min\{m_l, m_r\} = \underline{m}$.

To describe the backlash by operator, we describe it as

$$u(t) = B_a[v(t)] = m[v(t) + \Delta] \tag{5.4}$$

where $m = \{m_l, m_r, m_0\}$ and

$$\Delta = \left[h, -h, \frac{u(t_-)}{m_0} - v(t)\right]$$

The different conditions in (5.3) equal different values. Assume m_0 is the designed slope. We also assume that v is a piecewise monotone input and $v \in C_{pm}[0, t_E]$, where $[0, t_E]$ is a monotonicity partition for v. Let the relationship between input v and output u of nonsymmetric backlash be shown as Figure 5.1. Then, when v increases, u will vary along Pos3→Pos4→Pos1; when v decreases, u will vary along Pos1→Pos2→Pos3. Since Pos2 and Pos4 are the points of intersection, $m_l(v_{\text{Pos2}} + h) = m_0(v_{\text{Pos2}} + \overline{u}/m_0 - v_{\text{Pos2}}) = \overline{u}$ and $m_r(v_{\text{Pos4}} - h) = m_0(v_{\text{Pos4}} + \underline{u}/m_0 - v_{\text{Pos4}}) = \underline{u}$ are obtained, respectively.

Let v_1 and v_2 be in the same monotonicity partition and $v_1 \neq v_2$. If v_1 and v_2 are in the monotonic increasing partition, then there are three cases:

1. If $v_{\text{Pos3}} \leq v_{1,2} \leq v_{\text{Pos4}}$, then

$$\begin{aligned}\|B_a(v_1) - B_a(v_2)\| &= \left\| m_0\left(v_1 + \frac{\underline{u}}{m_0} - v_1\right) - m_0\left(v_2 + \frac{\underline{u}}{m_0} - v_2\right)\right\| \\ &= 0 \end{aligned} \tag{5.5}$$

In this case, the backlash operator B_a is a constant operator and a Lipschitz operator.

2. If $v_{\text{Pos4}} \leq v_{1,2} \leq v_{\text{Pos1}}$, then

$$\begin{aligned}\|B_a(v_1) - B_a(v_2)\| &= \|m_r(v_1 - h) - m_r(v_2 - h)\| \\ &= \|m_r(v_1 - v_2)\| \\ &\leq \|m_r\| \|v_1 - v_2\| \end{aligned} \tag{5.6}$$

Since $m_r \leq \overline{m} < \infty$, in this case, the backlash operator B_a is proved to be a Lipschitz operator.

3. If $v_{\mathrm{Pos4}} \leq v_1 \leq v_{\mathrm{Pos1}}$ and $v_{\mathrm{Pos3}} \leq v_2 \leq v_{\mathrm{Pos4}}$ (the same as $v_{\mathrm{Pos4}} \leq v_2 \leq v_{\mathrm{Pos1}}$ and $v_{\mathrm{Pos3}} \leq v_1 \leq v_{\mathrm{Pos4}}$), then

$$\begin{aligned}
\|B_a(v_1) - B_a(v_2)\| &= \left\| m_r(v_1 - h) - m_0 \left(v_2 + \frac{u}{m_0} - v_2 \right) \right\| \\
&= \left\| m_r(v_1 - h) - m_0 \left(v_{\mathrm{Pos4}} + \frac{u}{m_0} - v_{\mathrm{Pos4}} \right) \right\| \\
&= \| m_r(v_1 - v_{\mathrm{Pos4}}) \| \\
&\leq \| m_r \| \| v_1 - v_{\mathrm{Pos4}} \| \\
&\leq \| m_r \| \| v_1 - v_2 \|
\end{aligned} \tag{5.7}$$

Since $m_r \leq \overline{m} < \infty$, in this case, the backlash operator B_a is proved to be a Lipschitz operator.

In the same way, if v_1 and v_2 are in the monotonic decreasing partition, then the backlash operator B_a can also be proven to be a Lipschitz operator. As a result, this backlash operator is applied to describe our considered uncertain nonsymmetric backlash. The stability and output tracking performance of the nonlinear plant with this nonsymmetric backlash will be discussed in the following sections.

5.1.3 Nonsymmetric Backlash Control Design Scheme

With the above-mentioned backlash operator B_a, the control design scheme is shown as Figure 5.2, where each block denotes one operator. That is, P is the plant operator from space U^* to space Y, and A and B are two stable operators used to stabilize the considered plant with uncertain nonsymmetric backlash and are from space Y, U to space U, U, respectively. Let $v \in U$ be the input of uncertain nonsymmetric backlash, $u \in U^*$ be the input of the original nonlinear plant, and $U^* \subseteq U$. Generally, the signal of v is easy to obtain. However, the signal of u is unavailable. So the uncertain nonsymmetric backlash is considered as one part of the plant during controller design.

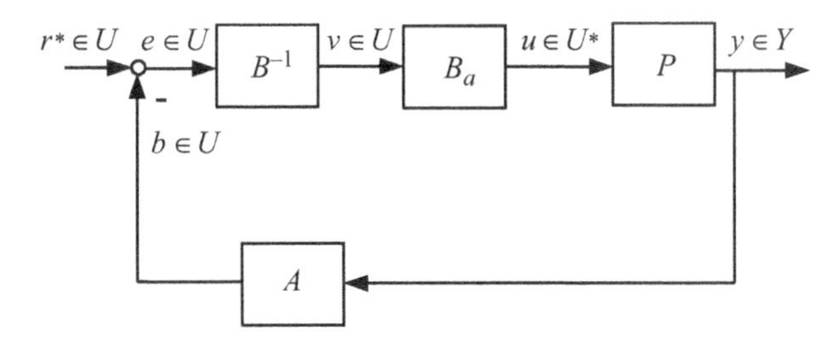

FIGURE 5.2 Control design scheme.

We represent the uncertain nonsymmetric backlash (5.4) as

$$u(t) = B_a[v(t)] = D_b[v(t)] + m\Delta = mv(t) + m\Delta \tag{5.8}$$

where D_b is a Lipschitz operator from space U to space U^* and $m\Delta$ is a bounded parasitic term.

Definition 5.1 Let U^e and Y^e be two extended linear spaces associated respectively with two given Banach spaces U and Y of measurable functions defined on the time domain $[0, \infty)$, where a Banach space is a complete vector space with a norm. Let D^e be a subset of U^e. A nonlinear operator $A : D^e \to Y^e$ is called a generalized Lipschitz operator on D^e if there exists a constant L such that $\|[A(x)]_T - [A(\tilde{x})]_T\|_Y \leq L\|x_T - \tilde{x}_T\|_U$ for all $x, \tilde{x} \in D^e$ and for all $T \in [0, \infty)$.

Let $v_1, v_2 \in D^e$ under the same slope $m = m_1 = m_2$. We can obtain that

$$\|[D_b(v_1)]_T - [D_b(v_2)]_T\|_{U^*} = \|[m_1 v_1]_T - [m_2 v_2]_T\|_{U^*} \leq \|m\|\|v_{1T} - v_{2T}\|_U$$

So the operator D_b is proven to be a generalized Lipschitz operator. Based on (5.8), the operator D_b could be fused in D, which is one part of the plant $P = ND^{-1}$, to be $\tilde{D} = D_b^{-1}D$. Considering that the slope m is unknown, we design a slope m_0 to substitute for the unknown slope m during control design. In this way, the uncertainty would be introduced into $\tilde{D}$ and would affect the stability of the considered plant. For this uncertainty, we have the following theorem to guarantee the stability of the nonlinear feedback control system with the uncertain nonsymmetric backlash. In this theorem, $\tilde{D}$ can be represented by $\tilde{D}_0$, $\tilde{D}_1$, and $\tilde{D}_2$ when m equals $m_0, \overline{m}$, and $\underline{m}$, respectively.

Theorem 5.1 Let U^e and Y^e be two extended linear spaces associated respectively with two given Banach spaces U and Y. Let D^e be a linear subspace of U^e and let $(B\tilde{D}_1 - B\tilde{D}_0)M^{-1} \in \text{Lip}(D^e)$, $(B\tilde{D}_2 - B\tilde{D}_0)M^{-1} \in \text{Lip}(D^e)$. Let the Bezout identity of the nominal plant be $AN + B\tilde{D}_0 = M \in S(W, U)$, $AN + B\tilde{D}_1 = \tilde{M}_1$ or $AN + B\tilde{D}_2 = \tilde{M}_2$, when the slope of the backlash is $\overline{m}$ or $\underline{m}$. Under the condition of controller A satisfying the bezout identity [$AN + B(D + \Delta D) = M$, where M is unimodular operator], if

$$\|[B\tilde{D}_1 - B\tilde{D}_0]M^{-1}\| < 1 \qquad \|[B\tilde{D}_2 - B\tilde{D}_0]M^{-1}\| < 1 \tag{5.9}$$

the system is stable, where $\|\cdot\|$ is defined as

$$\|F\| := \sup_{T\in[0,\infty)} \sup_{\substack{x,\,\tilde{x} \in D^e \\ x_T \neq \tilde{x}_T}} \frac{\|[Fx]_T - [F\tilde{x}]_T\|_Y}{\|x_T - \tilde{x}_T\|_U} \tag{5.10}$$

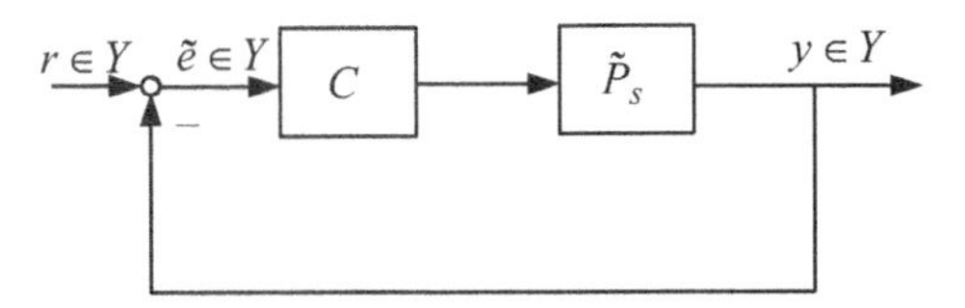

FIGURE 5.3 Tracking design scheme.

Proof Assume M is unimodular operator. Then M is invertible. From $AN + B\tilde{D}_0 = M$, $AN + B\tilde{D}_1 = \tilde{M}_1$, we have

$$\tilde{M}_1 = M + [B\tilde{D}_1 - B\tilde{D}_0]$$

Since $\tilde{M}_1 = M + [B\tilde{D}_1 - B\tilde{D}_0] = [I + (B\tilde{D}_1 - B\tilde{D}_0)M^{-1}]M$ and $(B\tilde{D}_1 - B\tilde{D}_0)M^{-1} \in \mathrm{Lip}(D^e)$, $I + (B\tilde{D}_1 - B\tilde{D}_0)M^{-1}$ is invertible based on the result in [2], where I is the identity operator. Consequently, we have $\tilde{M}_1^{-1} = M^{-1}[I + (B\tilde{D}_1 - B\tilde{D}_0)M^{-1}]^{-1}$. Meanwhile, since $\tilde{M}_1 = M + [B\tilde{D}_1 - B\tilde{D}_0]$, $(B\tilde{D}_1 - B\tilde{D}_0)M^{-1} \in \mathrm{Lip}(D^e)$, and $M \in \mathcal{U}(W, U)$, we have $\tilde{M}_1 \in \mathcal{U}(W, U)$. For $AN + B\tilde{D}_2 = \tilde{M}_2$, it can also be proven by the same method.

Then we define $Z = \max\{1/m_0 - 1/\overline{m}, 1/\underline{m} - 1/m_0\}$. Let the exact plant be $AN + B\tilde{D} = \tilde{M}$. Because $m \in [\underline{m}, \overline{m}]$, we obtain

$$\|[B\tilde{D} - B\tilde{D}_0]M^{-1}\| \leq \|[BD \cdot Z]M^{-1}\| < 1 \tag{5.11}$$

So $\tilde{M} \in \mathcal{U}(W, U)$. Thus the system shown in Figure 5.2 is well-posed. The output of the above system is obtained as

$$y(t) = N(AN + B\tilde{D})^{-1}\left[r^*(t) + BD_b^{-1}(m\Delta)\right] \tag{5.12}$$

As mentioned above, operators N, A, B, D_b^{-1}, and D_b are all stable and $r^*(t)$, Δ and m are bounded. Since $\mathcal{U}(W, U)$ is the set of unimodular operators, $(AN + B\tilde{D})^{-1}$ is stable. So the output y is bounded. The system is considered to be stable. ■

According to Theorem 5.1, the plant shown in Figure 5.2 has been stabilized by controllers A and B, but the output tracking performance has not been considered yet. In this section, based on the above stabilized plant, we design a tracking controller based on the exponential iteration theorem in Chapter 3.

The tracking design scheme is shown as Figure 5.3, where $\tilde{P}_S$ is the stabilized system (Figure 5.2) and C is the tracking controller. The controller C is designed so that the open loop $\tilde{P}_S C$ of a feedback system in Figure 5.3 consists of an integrator in cascade with a system P_T (Figure 5.4) and satisfies the following conditions:

1. For all t in $[0, T]$, $P_T(r) \geq K_1 > 0$ as $T \geq t \geq t_1 \geq 0, r > 0$.
2. $\tilde{P}_S C(0) = 0$.
3. $\|\tilde{P}_S C(x) - \tilde{P}_S C(y), t\| \leq h_P \int_0^t \|x - y, \tau\| \, d\tau$ and h_P is any constant and is the gain of P_T in the first norm.

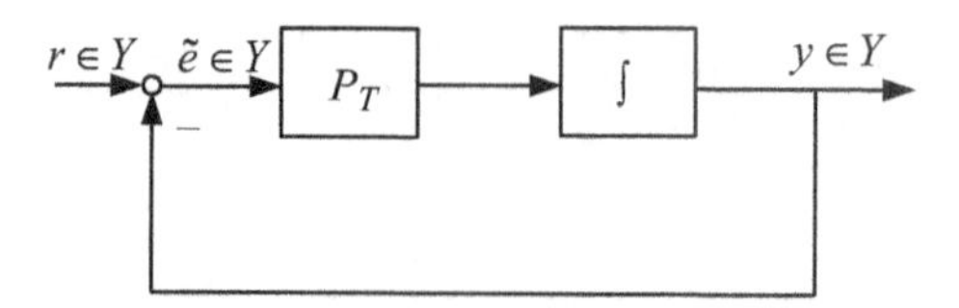

FIGURE 5.4 Equivalent block diagram of Figure 5.3.

Defining an operator $\tilde{G}$ from r to y, we have $\tilde{G} = \tilde{P}_S C(I - \tilde{G})$ as the feedback equation. Then the exponential iteration theorem is shown in Lemma 3.1. The feedback equation $\tilde{G} = \tilde{P}_S C(I - \tilde{G})$, in which all operators map the Banach space Y into itself, has a unique solution for $\tilde{G}$, which converges uniformly on $[0, T]$, provided that conditions 2 and 3 are satisfied. The plant Output is bounded.

Since $\tilde{e} = [I + \tilde{P}_S C)^{-1}(r)]$, $\tilde{P}_S C = \int_0^t P_T \, d\tau$, then

$$\begin{aligned} y(t) &= r(t) - (I + \tilde{P}_S C)^{-1}(r)(t) \\ &= r(t) - \{r(t) + \tilde{P}_S C[r(t)]\}^{-1} \\ &= r(t) - \left(r(t) + \int_0^t P_T \, d\tau \right)^{-1} \end{aligned} \tag{5.13}$$

Considering $P_T(r) \geq K_1 > 0$ as $T \geq t \geq t_1 \geq 0, r > 0$, we obtain

$$\int_0^t P_T[r(\tau)] \, d\tau \geq \int_0^{t_1} P_T[r(\tau)] \, d\tau + K_1 \int_{t_1}^t d\tau \tag{5.14}$$

where $K_1 \int_{t_1}^t d\tau$ can be made arbitrarily large by making $t \leq T$ large enough. Then $\{r(t) + \tilde{P}_S C[r(t)]\}^{-1}$ becomes arbitrarily small and leads $y(t)$ to $r(t)$.

According to the control design scheme, the stability of the nonlinear plant with uncertain nonsymmetric backlash is guaranteed with the robust condition. The main difference between the robust condition Lemma 2.7 and the condition in Theorem 5.1 is that the one in Theorem 5.1 can control systems with uncertain symmetric backlash. However, the method in Lemma 2.7 cannot be applied to systems with uncertain nonsymmetric backlash. The reason is that the slopes of the nonsymmetric backlash are not parallel. For this case, the systems can be controlled by the robust condition. Naturally, the robust condition can also be applied to systems with uncertain symmetric backlash. According to the control design scheme, the output tracking performance is also ensured. The merit of this tracking design method is that the tracking controller C can be designed without any information of uncertain nonsymmetric backlash as well as eliminate the effect of the parasitic term of uncertain nonsymmetric backlash in the feedback control system (Figure 5.3). The main reason is that the parasitic term is an internal term of robustly stable system P_S. In the following, simulation results are used to show the effectiveness of the control design scheme.

5.1.4 Simulation Results

In this section, we illustrate the effectiveness of the above control design scheme for an unstable nonlinear plant with uncertain nonsymmetric backlash. The simulation results show the stability and tracking performance of the unstable nonlinear plant.

The unstable nonlinear plant is from [3] and is described by the operator

$$\begin{aligned} P[u^*(t)] &= ND^{-1}[u^*(t)] \\ &= \int_0^t u^{*1/3}(\tau)\,d\tau + e^{t/3}u^{*1/3}(t) \end{aligned} \tag{5.15}$$

$$\begin{aligned} N[w(t)] &= \int_0^t e^{-\tau/3}w^{1/3}(\tau)\,d\tau + w^{1/3}(t) \\ D[w(t)] &= e^{-t}w(t) \end{aligned} \tag{5.16}$$

where $u^* \in U^*$, $P(u^*) \in Y^*$, and $w \in W$. We can observe that

$$ND^{-1}[u^*(t)] = N[e^t u^*(t)] = \int_0^t e^{-\tau/3}[e^\tau u^*(\tau)]^{1/3}\,d\tau + [e^t u^*(t)]^{1/3} = P$$

To verify that both N and D are stable, we pick any $x \in W$. There is a constant k such that $\|x\|_\infty < k$. Then for all $t \in [0, \infty]$, $|D[x(t)]| = e^{-t}|x(t)| < k$ and $|\int_0^t e^{-\tau/3}x^{1/3}(\tau)\,d\tau| < k^{1/3}|\int_0^t e^{-\tau/3}\,d\tau| \le 3k^{1/3}$, so that $|N[x(t)]| < 4k^{1/3}$. Thus both N and D are stable, where D^{-1} is unstable provided that the considered plant is unstable. In the following, the explanation on $P = ND^{-1}$ being right coprime factorized is given. That is, the proof of D and N satisfying the Bezout identity is shown. We choose the spaces $W = U^*$, $U^* = C[0, \infty)$, $Y = \{\int_0^t u^{*1/3}(\tau)\,d\tau + e^{t/3}u^{*1/3}(t)|u^* \in C[0, \infty)\} \subset U^*$ and the input space of uncertain nonsymmetric backlash $U = C[0, \infty)$, where $C[0, \infty)$ is the space of continuous functions. The nonsymmetric backlash parameters m_l, m_r, and h are unknown, but the bounds of slopes of the nonsymmetric backlash are $m_r \in [1, 1.2]$, $m_l \in [0.92, 1.1]$. The actual parameter values are chosen as $m_l = 1.08$, $m_r = 1.2$, $h = 0.5$. According to Theorem 5.1, we design two stable controllers A and B as

$$A[y(t)] = \begin{cases} (e^t - 1)g^3(t) & \text{if } y(t) = \int_0^t g(\tau)\,d\tau + e^{t/3}g(t) \\ 0 & \text{otherwise} \end{cases} \tag{5.17}$$

$$B[u(t)] = m_0 u(t) \tag{5.18}$$

where $g(t) = e^{-t/3}w^{1/3}(t)$, $m_0 = 1.05$. Since $\tilde{D} = D/m$ can be represented by $\tilde{D}_0$, $\tilde{D}_1$, and $\tilde{D}_2$ when $m = m_0, \overline{m}, \underline{m}$, if $m = m_0$, we obtain

$$AN[w(t)] + B\tilde{D}_0[w(t)] = (e^t - 1)[e^{-t/3}w^{1/3}(t)]^3 + m_0\frac{e^{-t}w(t)}{m_0} = I[w(t)]$$

Then, we have

$$\|[B\tilde{D}_1 - B\tilde{D}_0]I^{-1}\| = \left\| \left(m_0 \frac{e^{-t}w(t)}{\underline{m}} - m_0 \frac{e^{-t}w(t)}{m_0} \right) I^{-1}[w(t)] \right\| < 1$$

We can also get

$$\|[B\tilde{D}_2 - B\tilde{D}_0]I^{-1}\| < 1$$

and the robust condition in Theorem 5.1 is satisfied. To design tracking controller C, we should make it satisfy three conditions which are mentioned in Section 5.1.3. Then for the obtained robust stabilized system, we design tracking controller C as

$$C(\tilde{e}) = k[\int_0^t \tilde{e}(\tau)\,d\tau]^3 \tag{5.19}$$

where k is a constant. In this simulation, we set initialization of $C[\tilde{e}(0)] = 1$ and choose $k = 100$.

The objective is to stabilize the unstable nonlinear plant with the uncertain non-symmetric backlash and control the plant output $y(t)$ to follow a desired trajectory $r(t)$. We choose two kinds of desired trajectories to validate the tracking performance.

In the first example, we choose a squarelike wave as the desired trajectory described as

$$r(t) = \begin{cases} 1 - 0.5e^{-(t-100i)/3} & 100i \le t < 50 + 100i \\ 0.5 + 0.5e^{-(t-50-100i)/3} & 50 + 100i \le t < 100(i+1) \end{cases} \quad i = 0, 1, 2 \tag{5.20}$$

Without considering backlash compensation, we design the controllers by Lemmas 2.7 and 3.1 and have the simulation result shown in Figure 5.5. In Figure 5.5, since the output tends to infinity, the considered nonlinear system is not stabilized. With backlash compensation, we obtain a desired output (dash-dotted line) which tracks the reference input (solid line) completely in Figure 5.6. In Figure 5.7, the response of P_T with the effect of uncertain nonsymmetric backlash is shown, which satisfies condition 1 of tracking controller design. The tracking error and control input are shown in Figures 5.8 and 5.9, respectively. And the tracking error tends to zero. In the following, we apply the inverse backlash compensation of [72] in the considered nonlinear system, and unfortunately, an unstable output in Figure 5.10 is obtained. Since the considered plant (5.15) is unstable, the result in Figure 5.10 tends to infinity under the effect of the error of inverse backlash compensation, where the slopes of the nonsymmetric backlash are not parallel. For brevity, the results of control input and tracking error which correspond to Figures 5.5 and 5.10 are omitted. In this chapter, based on the robust condition using right coprime factorization and the three conditions for tracking controller design, the design parameters are selected by trial and error. Obtaining the best design parameters is left to future work.

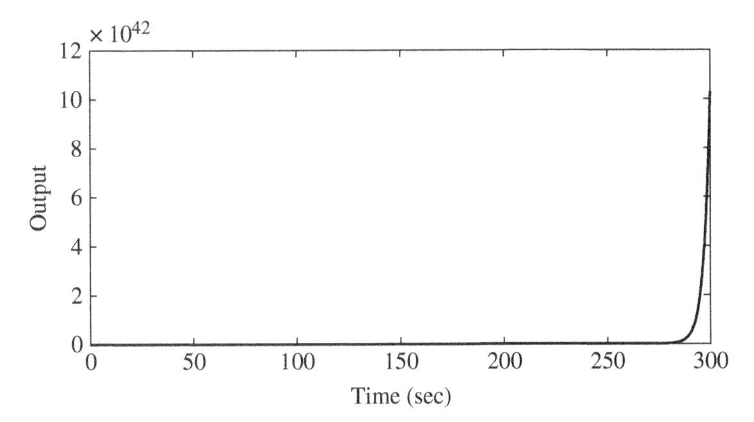

FIGURE 5.5 Plant output without considering backlash compensation.

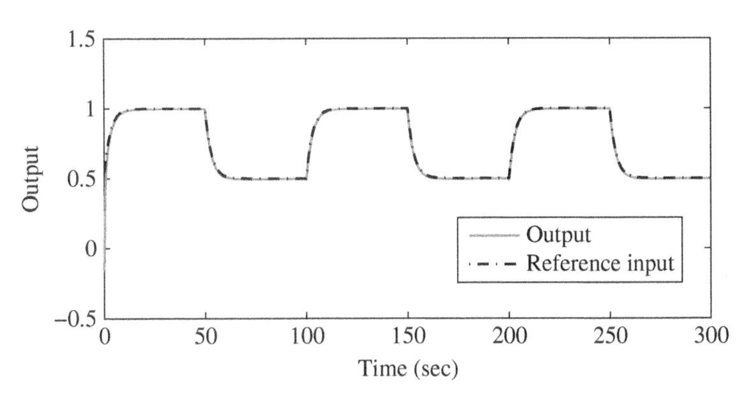

FIGURE 5.6 Plant output and reference input using method.

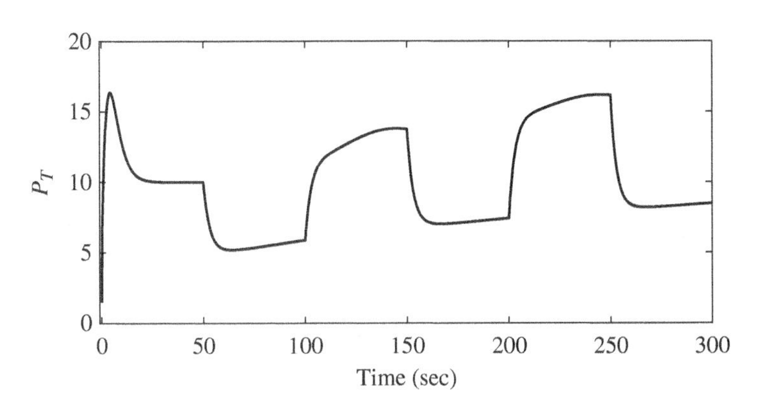

FIGURE 5.7 Response of P_T.

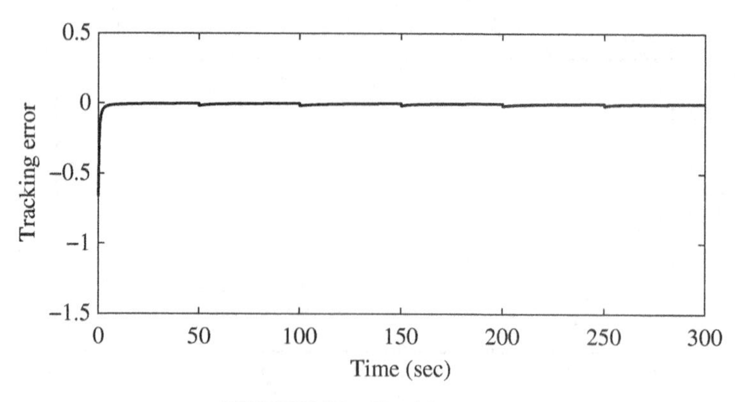

FIGURE 5.8 Tracking error.

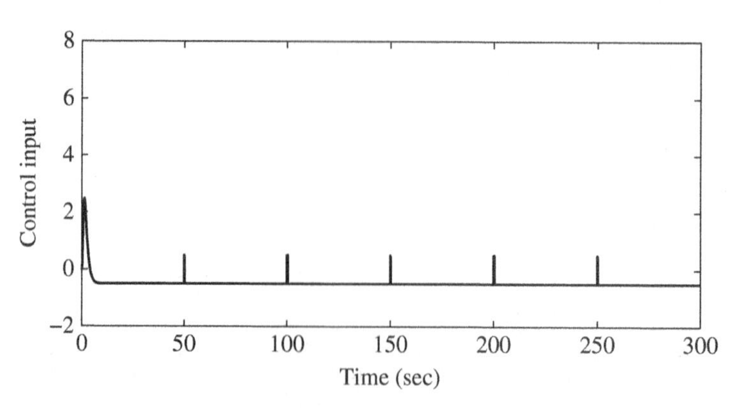

FIGURE 5.9 Control input.

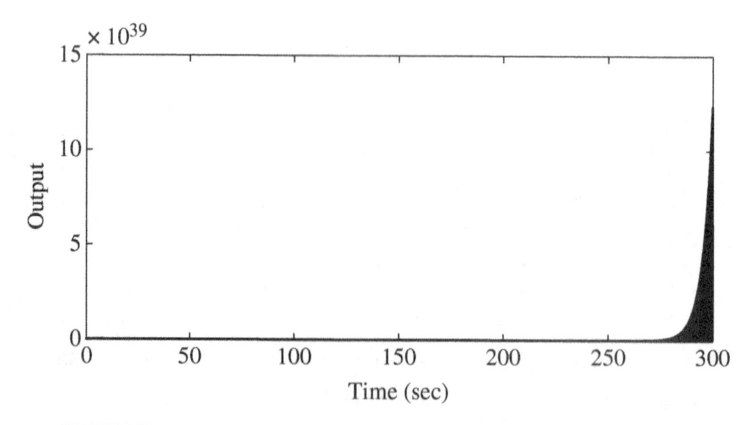

FIGURE 5.10 Plant output using inverse backlash compensation.

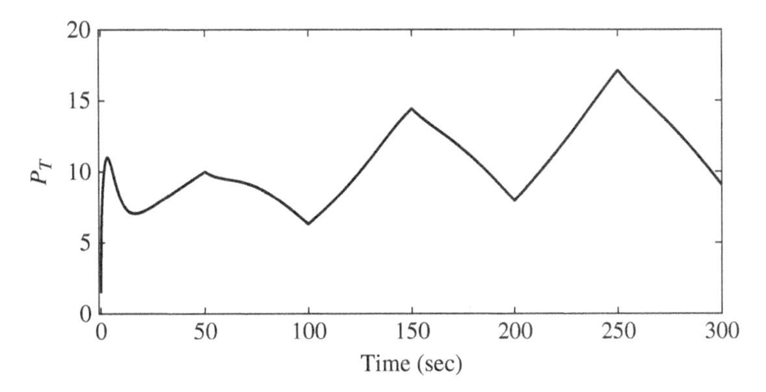

FIGURE 5.11 Response of P_T.

To validate the method further, in the second example, we choose a triangle wave to be the desired trajectory defined as

$$r(t) = \begin{cases} 0.5 + 0.01t & 100i \le t < 50 + 100i \\ 1 - 0.01t & 50 + 100i \le t < 100(i+1) \end{cases} \quad i = 0, 1, 2 \qquad (5.21)$$

Then, the response of P_T with the effect of uncertain nonsymmetric backlash is obtained (Figure 5.11) which satisfies condition 1 of tracking controller design. With this P_T, we obtain simulation results which are presented in Figures 5.12–5.14. In Figure 5.12, the plant output (dash-dotted line) tracks the reference input (solid line). The tracking error and control input are shown in Figures 5.13 and 5.14, respectively. And the tracking error tends to zero. To use this kind of reference input, the considered unstable nonlinear plant with uncertain nonsymmetric backlash is also stabilized.

Clearly, the simulation results verify our theoretical findings and show the effectiveness of our control scheme.

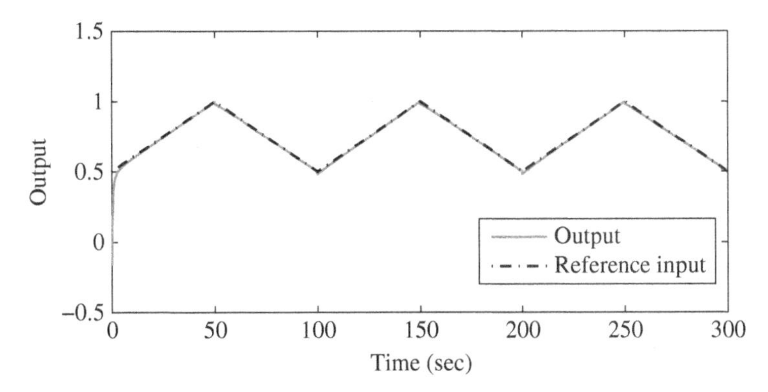

FIGURE 5.12 Plant output and reference input.

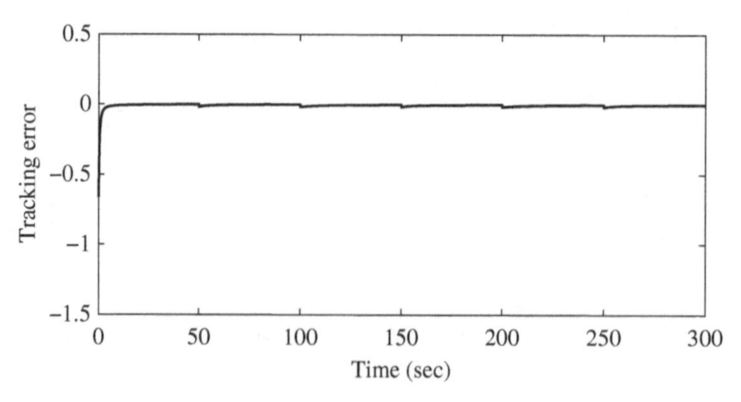

FIGURE 5.13 Tracking error.

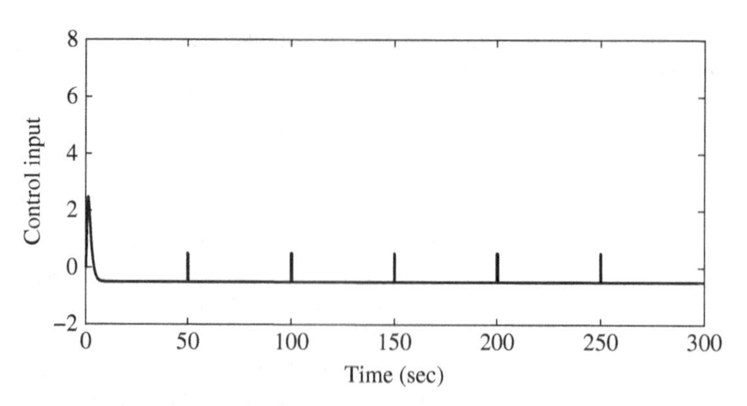

FIGURE 5.14 Control input.

5.1.5 Summary

The operator-based controller was designed based on the robust condition and the exponential iteration theorem. The stability and output tracking performance of the nonlinear plant with uncertain nonsymmetric backlash are guaranteed by the controller. The effectiveness of the control design method was confirmed through simulations.

5.2 OPERATOR-BASED ROBUST NONLINEAR FEEDBACK CONTROL SYSTEMS DESIGN FOR SYMMETRIC AND NONSYMMETRIC HYSTERESIS

5.2.1 Introduction

The application of actuators [76, 77] has attracted much attention from researchers. One of the important characterestics of actuators is hysteretic behavior. Many phenomenological models have been proposed to describe the hysteresis nonlinearity, such as the Preisach model, the Krasnosel'skii-Pokrovskii model, the and Prandtl–Ishlinskii (PI) model [78]. Moreover, these models have been widely applied to control the plants with hysteresis effects. In [79, 80] PI models are described with elementary hysteresis operators, which are noncomplex hysteresis nonlinearities with a simple mathematical structure, such as stop hysteresis operators and play hysteresis operators. These operators are symmetric, which means that the slopes of their upward and downward actions are the same.

However, the slopes may not be the same in practice. To describe input nonlinearity, [72] and [81] proposed a backlash model and a dead-zone model with non-symmetric slopes, respectively. Especially, [81] solved the unstable problems from the effect of dead-zone dynamics. To describe hysteresis and make the hysteresis model resemble more the real hysteretic behavior, [91] proposed a nonsymmetric PI hysteresis model using a nonsymmetric play hysteresis operator with unknown slopes. A rate-dependent asymmetric hysteresis model is proposed in [83]. So far, research work on nonsymmetric PI hysteresis has been in modeling. The control problem of the nonsymmetric PI hysteresis has not been studied yet. In this chapter, the modified nonsymmetric PI hysteresis model of [91] is given. To stabilize the nonlinear plants with hysteretic behavior, an operator-theoretic method is used. It leads to the BIBO stability of the feedback control system depending on robust right coprime factorization [2, 84]. A nonlinear plant can be described by plant operator $P = ND^{-1}$, where D^{-1} and N are two factors of right coprime factorization of the plant. The hysteresis which precedes the nonlinear plant can also be described as an operator. In practice, the signal between the hysteresis and the nonlinear plant is unavailable. So the hysteresis should be considered as part of the nonlinear plant. Since in this chapter the considered hysteresis is described by the nonsymmetric PI hysteresis model, the left and right slopes are unknown and not the same. We design a slope m_0 to substitute for the unknown slopes and modify the nonsymmetric hysteresis model of [91]. As a result, the obtained hysteresis model can be presented by a generalized Lipschitz operator term and a bounded parasitic term. In this way, the generalized Lipschitz operator term could be considered as part of the nonlinear plant. Then we propose a condition using robust right coprime factorization to design controllers for guaranteeing the stability of the whole system.

5.2.2 Problem Setup

Consider a nonlinear unstable plant $P : U \to Y$, where U and Y are the input and output spaces, respectively. The plant is described as

$$P = ND^{-1} \tag{5.22}$$

where $D : W \to U$ and $N : W \to Y$ are stable operators from the quasi-state space W to the input and output spaces, respectively. A feedback control system is said to be well-posed if every signal in the control system is uniquely determined for any input signal in U. For the above nonlinear plant (5.22), under the condition of well-posedness, N and D are right coprime factorized if there exist two stable operators $A : Y \to U$ and $B : U \to U$ satisfying the Bezout identity

$$AN + BD = M \quad \text{for some } M \in \Sigma(W, U) \tag{5.23}$$

where $\Sigma(W, U)$ is the set of unimodular operators. Let the overall plant with perturbation ΔP be

$$\tilde{P} = P + \Delta P \tag{5.24}$$

where $\tilde{P}$ and P are nonlinear and unstable operators. As mentioned above, the right coprime factorization of $\tilde{P}$ is

$$\tilde{P} = P + \Delta P = (N + \Delta N)D^{-1} \tag{5.25}$$

We assume that ΔN is unknown but the upper and lower bounds of ΔN are known. According to (5.23), we can obtain

$$A(N + \Delta N) + BD = M \quad \text{for some } M \in \Sigma(W, U) \tag{5.26}$$

if the range of ΔN is the included in the null set of A, where ΔP is the perturbation of the plant which can represent only ΔN as ΔP is an additive uncertainty. However, in some cases, (5.26) is not satisfied since ΔN is unknown. So the stability of the nonlinear feedback system with perturbation should be guaranteed by the following lemma. Proof of this lemma is given in Lemma 2.7.

Lemma 5.1 Let U^e and Y^e be two extended linear spaces which are associated respectively with two given Banach spaces U_B and Y_B. Let D^e be a linear subspace of U^e and let $[A(N + \Delta N) - AN]M^{-1} \in \text{Lip}(D^e)$. Let the Bezout identity of the nominal plant and the exact plant be $AN + BD = M \in \Sigma(W, U)$, $A(N + \Delta N) + BD = \tilde{M}$, respectively. If controller A satisfies (5.26) and

$$||[A(N + \Delta N) - AN]M^{-1}|| < 1 \tag{5.27}$$

the system is stable, where $||\cdot||$ is defined as

$$||F|| := \sup_{T \in [0,\infty)} \sup_{\substack{x, \tilde{x} \in D^e \\ x_T \neq \tilde{x}_T}} \frac{||[Fx]_T - [F\tilde{x}]_T||_{Y_B}}{||x_T - \tilde{x}_T||_{U_B}} \tag{5.28}$$

In this chapter, the controlled plant either is nonlinear or has hysteresis nonlinearity. So a phenomenological model should be used to describe this hysteresis. Further, robust stability of the controlled plant with the hysteresis needs to be considered. Therefore, in Section 5.2.3, a nonsymmetric PI hysteresis model is given, and in Section 5.2.4, the design of robust stable control systems is introduced.

5.2.3 Nonsymmetric Prandtl–Ishlinskii Hysteresis Model

In this chapter, the PI hysteresis model is used to describe the hysteresis. Generally, the PI hysteresis model is described with stop hysteresis operators or play hysteresis operators. We consider the play type where the play hysteresis operator $F_h(\cdot; u^*_{-1}) : C_m[0, t_E] \times u^*_{-1} \to C_m[0, t_E]$ is defined as

$$F_h[u(t)] = \begin{cases} u(t) + h & u(t) \leq F_h[u(t_i)] - h \\ F_h[u(t_i)] & -h < u(t) - F_h[u(t_i)] < h \\ u(t) - h & u(t) \geq F_h[u(t_i)] + h \end{cases} \tag{5.29}$$

for $t_i < t \leq t_{i+1}$ and $0 \leq i \leq N-1$, where $0 = t_0 < t_1 < \cdots < t_N = t_E$ is a partition of $[0, t_E]$ such that the function u is monotone on each of the subintervals $[t_i, t_{i+1}]$. For brevity $F_h(u)$ is used instead of $F_h(u; u^*_{-1})$ and $F_h[u(0)] = \max(u(0) - h, \min(u(0) + h, u^*_{-1}))$. Assume $u^*_{-1} \in R$ is an initial value and h is the threshold value of the play hysteresis operator. The play hysteresis operator is shown in Figure 5.15 and it is symmetric because $m_r = m_l = 1$, which is obtained from (5.29). The play PI model is a weighted superposition of play hysteresis operators,

$$u^*(t) = \int_{h_0}^{H} p(h) F_h[u(t)]\, dh \tag{5.30}$$

where $p(h)$ is a given continuous density function for satisfying $p(h) \geq 0$ with $\int_{h_0}^{\infty} hp(h) < \infty$. Since the density function $p(h)$ vanishes for large values of h, we assume that there exists a constant H such that $p(h) = 0$ for $h > H$.

In fact, m_r and m_l, the slopes of the play hysteresis operator, are never the same in practice, as shown in Figure 5.16. So we consider a nonsymmetric play hysteresis

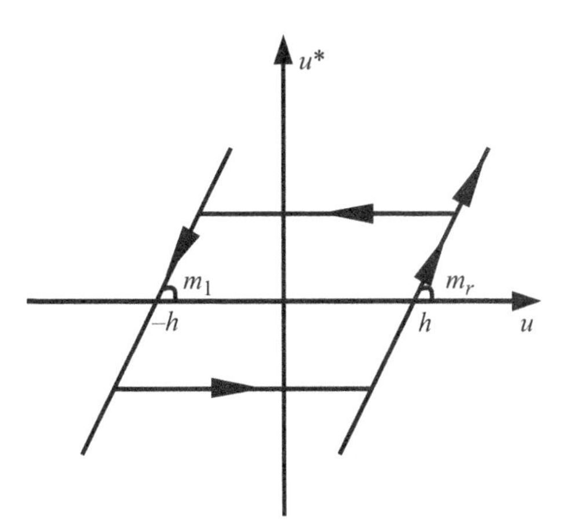

FIGURE 5.15 Play hysteresis operator.

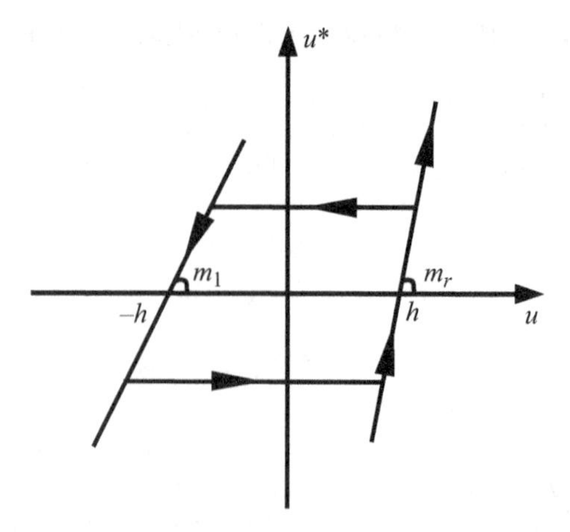

FIGURE 5.16 Nonsymmetric play hysteresis operator.

operator instead of a symmetric play hysteresis operator to describe the hysteresis nonlinearity. The nonsymmetric play hysteresis operator is defined as [72, 81].

$$
\begin{aligned}
v(t) &= B_a[u(t)] \\
&= \begin{cases} m_r[u(t) - h] & \text{if } \dot{u}(t) > 0 \text{ and } v(t) = m_r[u(t) - h] \\ m_l[u(t) + h] & \text{if } \dot{u}(t) < 0 \text{ and } v(t) = m_l[u(t) + h] \\ v(t_i) & \text{otherwise} \end{cases}
\end{aligned}
\tag{5.31}
$$

where the parameters m_r and m_l stand for the right and left slopes of the play hysteresis operator, and $h > 0$ is the threshold value. We assume that (1) the coefficients m_r and m_l are strictly positive and unknown; (2) m_r and m_l satisfy this condition,

$$
\frac{u_m - h}{u_m + h} m_r < m_l < m_r
$$

where $u_m = \max[u(t)]$; and (3) the maximum and the minimum values of the slopes of the nonsymmetric play hysteresis operator are known, $\overline{m} = \max(m_r)$ and $\underline{m} = \min(m_l)$. A nonsymmetric play PI hysteresis model [91] is described by the weight superposition of the above nonsymmetric play hysteresis operator as

$$
P_I[u(t)] = \int_{h_0}^{H} p(h) B_a[u(t)]\, dh \tag{5.32}
$$

where $p(h)$ is a continuous density function.

The controlled plant is nonlinear and with the hysteresis described by a nonsymmetric PI hysteresis model. For this kind of nonlinear plant, the operator-theoretic approach is used based on nonlinear Lipschitz operators from one normed linear space to another normed linear space. Based on this theory, the nonlinear plant can be described by robust right coprime factorization and can be stabilized by the designed controllers which satisfy the Bezout identity. In the given hysteresis model, the slopes are different and unknown. That is, there is an uncertainty factor in the hysteresis model. Depending on the robust condition using robust right coprime factorization, the nonlinear plant with the hysteresis model can be stabilized.

To facilitate the utilization of robust right coprime factorization, the model of [8] needs to be modified. First we represent (5.31) as

$$v(t) = B_a[u(t)] = m[u(t) + \Delta] \tag{5.33}$$

where

$$m = \{m_l, m_r\} \qquad \Delta = \left\{h, -h, \frac{v(t_i)}{m} - u(t)\right\}$$

Based on (5.33) we again look at the hysteresis model (5.32). In $[h_0, H]$, we can find h_x to satisfy $h \leq |u(t) - v(t_i)|$ when $h \in [h_0, h_x]$. Then (5.32) can be represented as [92]

$$u^*(t) = P_I[u(t)] = D_{\mathrm{PI}}[u(t)] + \Delta_{\mathrm{PI}} \tag{5.34}$$

where

$$D_{\mathrm{PI}}[u(t)] = Ku(t)$$

$$K = \int_{h_0}^{H} p(h) m \, dh \qquad m = m_l, m_r$$

$$\Delta_{\mathrm{PI}} = \begin{cases} -\displaystyle\int_{h_0}^{h_x} m_r h p(h)\, dh + \int_{h_x}^{H} m_r p(h) \left(\frac{v(t_i)}{m_r} - u(t)\right) dh & \text{if } u(t) > v(t_i) \\ \displaystyle\int_{h_0}^{h_x} m_l h p(h)\, dh + \int_{h_x}^{H} m_l p(h) \left(\frac{v(t_i)}{m_l} - u(t)\right) dh, & \text{if } u(t) \leq v(t_i) \end{cases}$$

Since $\int_{h_0}^{\infty} hp(h)\, dh < \infty$ and $h_x \in [h_0, H]$, we get that $\pm \int_{h_0}^{h_x} mhp(h)\, dh$ is bounded. As input $u(t)$ is bounded, $v(t)$ is bounded according to (5.33) for $h \in [h_x, H]$. Hence $v(t_i)/m - u(t)$ is proven to be bounded. Then we have that

$$\int_{h_x}^{H} mhp(h) \left(\frac{v(t_i)}{m} - u(t)\right) dh$$

is bounded, so Δ_{PI} is bounded. As an illustration, responses of the nonsymmetric play hysteresis operator and the obtained hysteresis model are shown in Figures 5.17–5.19. Assume $u(t) = 6\sin(3t)/(1+t)$ is the input of the obtained hysteresis model, where $t \in [0, 2\pi]$ and the initial value of the symmetric/nonsymmetric

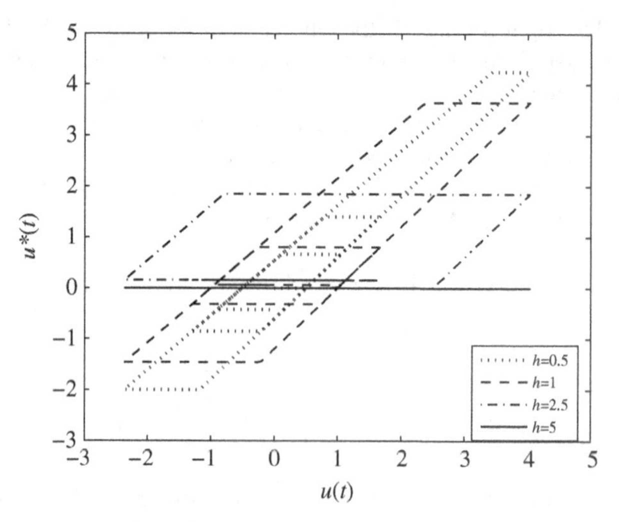

FIGURE 5.17 Response of nonsymmetric play hysteresis operator.

play hysteresis operator output is set to be zero. The density function is given as $p(h) = e^{-0.067(h-1)^2}/3$, $h \in [0.2, 10]$. We set the slopes of the play hysteresis operator as $m_r = 1.2$, $m_l = 1.09$, respectively.

Figure 5.20 shows the response of the symmetric play hysteresis operator under different thresholds of h. The result clearly shows that the output of the play hysteresis operator decreases with increasing value of h.

Figure 5.17 shows the response of the nonsymmetric play hysteresis operator under different thresholds h. The outputs of the nonsymmetric play hysteresis operator are

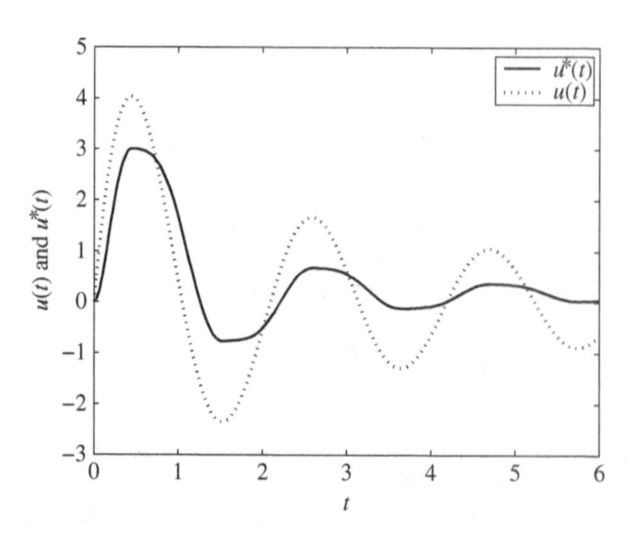

FIGURE 5.18 Input and output of obtained PI hysteresis model.

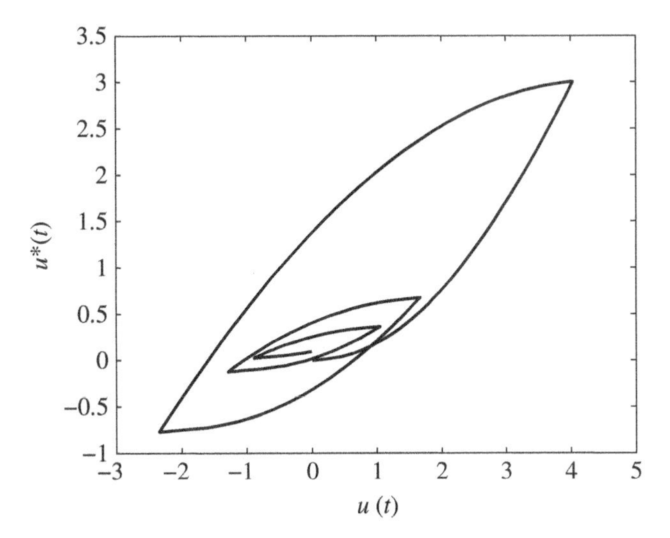

FIGURE 5.19 Response of obtained PI hysteresis model.

similar to Figure 5.20 and differ only due to the difference in slopes. Figure 5.18 shows the input and output of the obtained PI hysteresis model. The response of the obtained PI hysteresis model described by the nonsymmetric play hysteresis operator is shown in Figure 5.19. It illustrates that the obtained nonsymmetric PI hysteresis model can describe the hysteretic behavior. In the next section, the robust stability of a controlled plant with the obtained hysteresis is discussed.

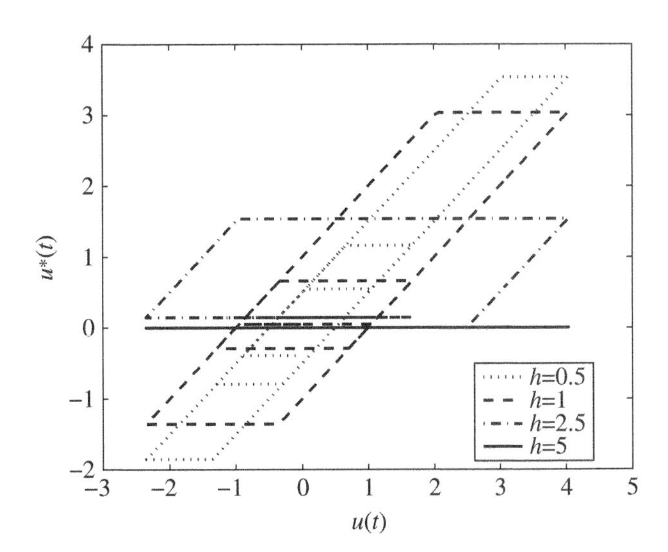

FIGURE 5.20 Response of symmetric play hysteresis operator.

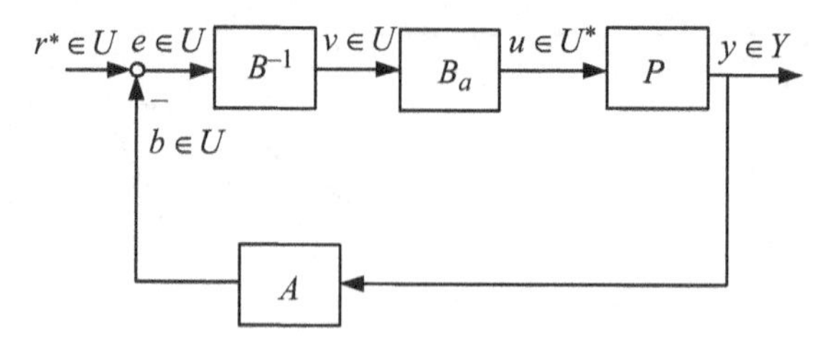

FIGURE 5.21 Robust control system.

5.2.4 Design of Robust Stable Control System

In this section, an operator-based design scheme of a nonlinear plant with nonsymmetric PI hysteresis is introduced based on robust right coprime factorization.

The original nonlinear plant is preceded by the hysteresis. Namely, the output of the nonsymmetric PI hysteresis model is the input of the original nonlinear plant. Two stable controllers A and B which satisfy the Bezout identity are designed to stabilize the system. The framework of the controlled system is shown as Figure 5.21.

Let $u \in U$ be the input of the nonsymmetric PI hysteresis model, $u^* \in U^*$ be the input of the original nonlinear plant, and $U \cap U^* \neq \emptyset$. Generally, the signal of u is easy to obtain. However, the signal u^*, which is the output of the actuator from the practical system, is unavailable. So the hysteresis of the actuator is considered part of the plant during the design of controllers. From (5.34), we regard $D_{\text{PI}}[u(t)]$ of the nonsymmetric PI hysteresis model as a part of $\tilde{D} = D_{\text{PI}}^{-1} D$, where D is a part of the original plant $P = ND^{-1}$. To make the controlled plant stable, two controllers A and B are designed to satisfy $AN + B\tilde{D} = M$, where $M \in \Sigma(W, U)$. Since the slopes m_r and m_l are unknown, we design a slope m_0 to substitute for the unknown slopes during control design. In this way, $D_{\text{PI}}[u(t)] = Ku(t)$, where $K = \int_{h_0}^{H} p(h) m_0 \, dh$.

Definition 5.2 Let U^e and Y^e be two extended linear spaces associated respectively with two given Banach spaces U and Y of measurable functions defined on the time domain $[0, \infty)$, where a Banach space is a complete vector space with a norm. Let D^e be a subset of U^e. A nonlinear operator $F : D^e \to Y^e$ is called a generalized Lipschitz operator on D^e if there exists a constant L such that $\|[F(x)]_T - [F(\tilde{x})]_T\|_V \leq L\|x_T - \tilde{x}_T\|_U$, for all $x, \tilde{x} \in D^e$ and for all $T \in [0, \infty)$.

According to Definition 5.2, let $u_1, u_2 \in D^e$. We can obtain that

$$\begin{aligned} \|[D_{\text{PI}}(u_1)]_T - [D_{\text{PI}}(u_2)]_T\|_{U^*} &= \|[Ku_1]_T - [Ku_2]_T\|_{U^*} \\ &\leq \| \int_{h_0}^{H} m_0 p(h) \, dh \| \|[u_1]_T - [u_2]_T\|_U \end{aligned} \tag{5.35}$$

Since $\| \int_{h_0}^{H} m_0 p(h) \, dh \| < \infty$, the operator D_{PI} is proven to be a generalized Lipschitz operator. Then we have the following theorem to guarantee the stability of the nonlinear feedback control system based on Lemma 5.1 as m_0 is substituted for the real slopes.

Before presenting the theorem, some notation should be introduced. From (5.34), $D_{\mathrm{PI}}[u(t)] = Ku(t)$, so $D_{\mathrm{PI}}^{-1} = 1/K$; $\tilde{D}$ can be represented by $\tilde{D}_0$, $\tilde{D}_1$, and $\tilde{D}_2$ when $m = m_0, \overline{m}, \underline{m}$, respectively; and m_0 is the designed slope, $m_0 \in [\underline{m}, \overline{m}]$.

Theorem 5.2 Let U^e and Y^e be two extended linear spaces associated respectively with two given Banach spaces U_B and Y_B. Let D^e be a linear subspace of U^e and $(B\tilde{D}_1 - B\tilde{D}_0)M^{-1} \in \mathrm{Lip}(D^e)$, $(B\tilde{D}_2 - B\tilde{D}_0)M^{-1} \in \mathrm{Lip}(D^e)$. Let the Bezout identity of the nominal plant be $AN + B\tilde{D}_0 = M \in \Sigma(W, U)$, $AN + B\tilde{D}_1 = \tilde{M}_1$, and $AN + B\tilde{D}_2 = \tilde{M}_2$ when the slopes of the nonsymmetric play hysteresis operator are $m_0, \overline{m}$, and $\underline{m}$, respectively. Under the condition of controller A to satisfy the Bezout identity, if

$$\|[B\tilde{D}_1 - B\tilde{D}_0]M^{-1}\| < 1 \tag{5.36}$$

$$\|[B\tilde{D}_2 - B\tilde{D}_0]M^{-1}\| < 1 \tag{5.37}$$

the system is stable, where $\|\cdot\|$ is defined as

$$\|F\| := \sup_{T\in[0,\infty)} \sup_{\substack{x,\tilde{x} \in D^e \\ x_T \neq \tilde{x}_T}} \frac{\|[Fx]_T - [F\tilde{x}]_T\|_{Y_B}}{\|x_T - \tilde{x}_T\|_{U_B}} \tag{5.38}$$

Proof Since M is a unimodular operator, M is invertible. From $AN + B\tilde{D}_0 = M$, $AN + B\tilde{D}_1 = \tilde{M}_1$, we have that $\tilde{M}_1 = M + [B\tilde{D}_1 - B\tilde{D}_0]$. Since $\tilde{M}_1 = M + [B\tilde{D}_1 - B\tilde{D}_0] = [I + (B\tilde{D}_1 - B\tilde{D}_0)M^{-1}]M$ and $(B\tilde{D}_1 - B\tilde{D}_0)M^{-1} \in \mathrm{Lip}(D^e)$, $I + (B\tilde{D}_1 - B\tilde{D}_0)M^{-1}$ is invertible based on the result in (5.36) [2], where I is the identity operator. Consequently, we obtain $\tilde{M}_1^{-1} = M^{-1}[I + (B\tilde{D}_1 - B\tilde{D}_0)M^{-1}]^{-1}$. Meanwhile, since $\tilde{M}_1 = M + [B\tilde{D}_1 - B\tilde{D}_0]$, $(B\tilde{D}_1 - B\tilde{D}_0)M^{-1} \in \mathrm{Lip}(D^e)$, and $M \in \Sigma(W, U)$, we have that $\tilde{M}_1 \in \Sigma(W, U)$. For $AN + B\tilde{D}_2 = \tilde{M}_2$, $\tilde{M}_2 \in \Sigma(W, U)$ can be obtained by the same method.

Then we define $Z = \max\{1/K_0 - 1/\overline{K}, 1/\underline{K} - 1/K_0\}$, where K_0, $\overline{K}$, and $\underline{K}$ are the values of K when $m = m_0, \overline{m}, \underline{m}$, respectively. Let the exact plant be $AN + B\tilde{D} = \tilde{M}$. Because $m \in [\underline{m}, \overline{m}]$, we obtain

$$\begin{aligned} \|[B\tilde{D} - B\tilde{D}_0]M^{-1}\| &= \|\left[\frac{1}{K}BD - \frac{1}{K_0}BD\right]M^{-1}\| \\ &\leq \|[Z \cdot BD]M^{-1}\| < 1 \end{aligned} \tag{5.39}$$

So $\tilde{M} \in S(W, U)$. Thus the system shown in Figure 5.21 is well-posed. The output of the above system is obtained as

$$y(t) = N(AN + B\tilde{D})^{-1}\left[r^*(t) + BD_{\text{PI}}^{-1}(\Delta_{\text{PI}})\right] \tag{5.40}$$

As mentioned above, operators N, A, B, and D_{PI}^{-1} are all stable, and $r^*(t)$ and Δ_{PI} are bounded. Since $\Sigma(W, U)$ is the set of unimodular operators, $(AN + B\tilde{D})^{-1}$ is stable. So output y is bounded. Hence the considered system is stable. ■

The main difference between the above condition and Lemma 2.7 lies in that the condition is fit for an uncertain $\tilde{D}$. That is, it is used to guarantee the robust stability of the controlled plant with hysteresis when $\tilde{D}$ varies between the known boundary. The above condition is extended from Lemma 2.7, which is just fit for a certain $\tilde{D}$. In the following, a numerical example is considered.

5.2.5 Numerical Example

In this section, a numerical example from [3] is given to show the effectiveness of the condition of robust stability.

Let the given plant operator P be defined by

$$\begin{aligned} P[u^*(t)] &= \int_0^t u^{*1/3}(\tau)\,d\tau + e^{t/3}u^{*1/3}(t) \\ &= ND^{-1}[u^*(t)] \\ N[w(t)] &= \int_0^t e^{-\tau/3}w^{1/3}(\tau)\,d\tau + w^{1/3}(t) \\ D[w(t)] &= e^{-t}w(t) \end{aligned} \tag{5.41}$$

where $u^* \in U^*$, $P(u^*) \in Y$. We choose the spaces $W = U, U = C_{[0,\infty)}$, $Y = \{\overline{u} + e^{t/3}\overline{u}' \,|\, \overline{u} \in C^1_{[0,\infty)}\} \subset U$, and $U^* = C_{[0,\infty)}$, where $C_{[0,\infty)}$ is the space of continuous functions and $C^1_{[0,\infty)}$ is the subspace of $C_{[0,\infty)}$ that consists of all the functions having a continuous first derivative. From (5.41), operator D^{-1} is unstable obviously.

Let the density function of the obtained modified PI hysteresis model be $p(h) = e^{-0.067(h-1)^2}/3$, $h \in [0.2, 10]$ and the bounds of the slopes be $m_r \in [1, 1.2]$, $m_l \in [0.92, 1.1]$. That is, $\overline{m} = \max(m_r) = 1.2$, $\underline{m} = \min(m_l) = 0.92$. The input of the considered system is $r^* = 6 - e^{-20t}$. According to the designed framework, we design the controllers A and B as

$$A[y(t)] = \begin{cases} \left(e^t - \dfrac{1}{K_0}\right)[g(t)]^3 & \text{if } y(t) = \int_0^t g(\tau)\,d\tau + e^{t/3}g(t) \\ 0 & \text{otherwise} \end{cases} \tag{5.42}$$

$$B[u(t)] = I[u(t)] \tag{5.43}$$

where $g(t) = e^{-t/3}w^{1/3}(t)$, $K_0 = \int_{h_0}^H m_0 p(h)\,dh$, $m_0 = 1.05$.

Then, the output of the considered system is obtained as

$$y(t) = N(AN + B\tilde{D})^{-1}[r^*(t) + BD_{\mathrm{PI}}^{-1}(\Delta_{\mathrm{PI}})] \tag{5.44}$$

From (5.34), Δ_{PI} is bounded and D_{PI} and D_{PI}^{-1} are stable operators. To prove that the operators D and N are stable [3], we pick any $x \in W$. There is a constant k such that $\|x\|_\infty < k$. Then, for all $t \in [0, \infty)$,

$$|D[x(t)]| = |e^{-t}x(t)| = e^{-t}|x(t)| < k$$

and

$$\left|\int_0^t e^{-\tau/3}x^{1/3}(\tau)\,d\tau\right| < k^{1/3}\left|\int_0^t e^{-\tau/3}\,d\tau\right| \leq 3k^{1/3}$$

so that $|N[x(t)]| < 4k^{1/3}$. Thus both D and N are stable. Since input r^* is bounded, output $y(t)$ is bounded if $(AN + B\tilde{D})^{-1}$ is stable.

According to known parameters,

$$\begin{aligned}\|[B\tilde{D}_1 - B\tilde{D}_0]M^{-1}\| &= \left\|\left(\frac{1}{\overline{K}} - \frac{1}{K_0}\right)e^{-t}M^{-1}\right\| \\ &= \left\|\left(\frac{1}{\int_{h_0}^H p(h)\overline{m}\,dh} - \frac{1}{\int_{h_0}^H p(h)m_0\,dh}\right)e^{-t}M^{-1}\right\| \\ &< 1\end{aligned}$$

is obtained. We also have that $\|[B\tilde{D}_2 - B\tilde{D}_0]M^{-1}\| < 1$, and the robust condition in Theorem 5.2 is satisfied (see Figure 5.22: dotted line for $\tilde{D}_1$, solid line for $\tilde{D}_2$). In a real calculation, the definition of the Lipschitz seminorm in Section 2.1.2.5 is usually employed.

Let $AN + B\tilde{D} = \tilde{M}$. Because $m \in [\underline{m}, \overline{m}]$, we obtain that $\tilde{M} \in \Sigma(W, U)$. Since $\Sigma(W, U)$ is the set of unimodular operators, $(AN + B\tilde{D})^{-1}$ is stable. Thus, output $y(t)$ is bounded. The unstable nonlinear plant with the nonsymmetric PI hysteresis model is stabilized depending on the robust condition.

5.2.6 Summary

A nonsymmetric PI hysteresis model described using a nonsymmetric play hysteresis operator with unknown slopes was given to express hysteretic behavior. Nonlinear plants with the obtained hysteresis model are stabilized by a robust controller designed using the robust stability condition. The effectiveness of the method was demonstrated by a numerical example.

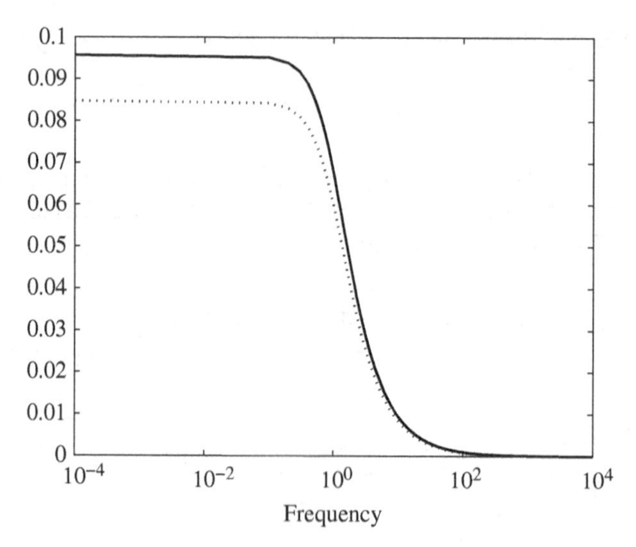

FIGURE 5.22 Verification of robust condition.

5.3 OPERATOR-BASED NONLINEAR FEEDBACK SYSTEMS APPLICATION FOR SMART ACTUATORS

5.3.1 Nonlinear Control of Piezoelectric Actuator

A piezoelectric actuator is lightweight and has high operational speed and can be bounded or embedded along a robot arm easily and so has been proved to be an effective control device of flexible arm vibration. In general, the piezoelectric actuator has hysteresis, and thus the system using the actuator usually exhibits undesirable oscillations and even instability. So far, for the vibration control of a flexible arm experimental system using the piezoelectric actuator, an operator-based nonlinear system control technique has been given in [66]. However, in [66], the hysteresis of the piezoelectric actuator is not considered. On the other hand, [92] proposed a new controller for the actuator having hysteresis characteristic. It considered the first vibration mode of the flexible arm, and based on the concept of Lipshitz operators and robust right coprime factorization, a vibration control system was designed and a tracking operator was considered [21]. The PI model was adopted to describe hysteresis [79]. Using the PI model, a nonlinear compensator was introduced to compensate for the hysteretic effects. To make the nonlinear system be BIBO stable, the PI model is decomposed into two parts: an invertible part and a disturbance part [92, 93]. But the effectiveness of the controller was only confirmed by numerical simulation. The problems of controller in realtime and confirming the effectiveness experimentally are still not solved.

This section shows how the designed controller is modified in the experimental system, the controller parameter if the experimental system is obtained through identification of the experimental system. Finally, experimental results are shown

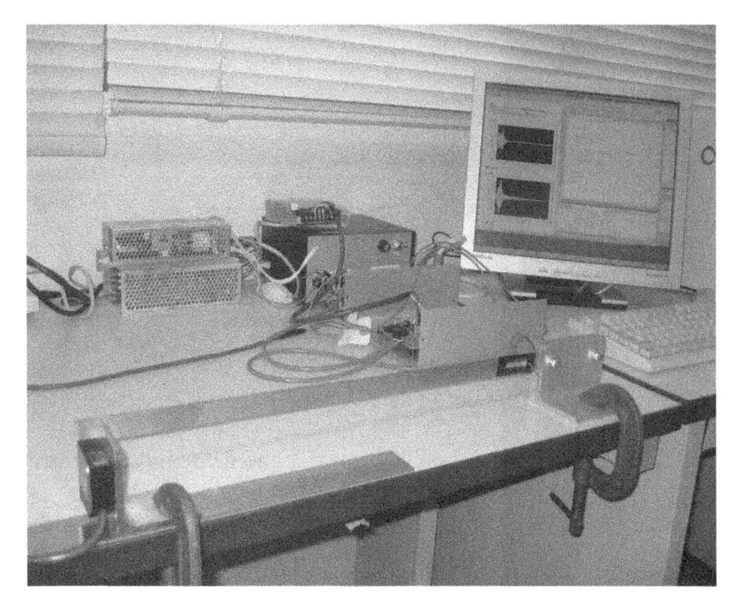

FIGURE 5.23 Experimental setup of vibration control system.

to support the effectiveness of the design method. In Section 5.2, the symmetric PI hysteresis model is used to design a controller that guarantees the robust right coprime factorization stability and ensures a desired tracking performance.

The experimental setup of the vibration control system is shown in Figure 5.23. The system consists of a flexible arm, a piezoelectric actuator (FUJI CERAMICS, PM50*20*0.5), a laser sensor (OMRON, Z4M-W40), and a PC-based controller. The system is a SISO feedback control system [66]. The control input signal generated by the PC-based controller is fed to an interface board of 12 bits resolution (CONTEC, AD12-16(PCI)E). Then the signal is amplified and applied to the piezoelectric actuator which is mounted close to the root of the flexible arm. The supplied voltage is limited as −100–100 V and the sampling period of the controller is 1 msec. The feedback signal is the displacement of the top of the flexible arm which is generated by the laser sensor.

Note that the control program of the PC-based controller is written in the C programming language on the Windows operating system, and the wide voltage limit is what was used in the experiment in [66]

The configuration of the flexible arm with a piezoelectric actuator is shown in Figure 5.24. Using the bending moment $M_p(t)$, which is the input of the plant, the dynamics of the flexible arm is described by the partial differential equation

$$\rho S \frac{\partial^2 y}{\partial t^2} + \frac{\partial^2}{\partial x^2}\left[EI\left(1 + C\frac{\partial}{\partial t}\right)\frac{\partial^2 y}{\partial x^2}\right] = \frac{\partial^2}{\partial x^2}\{M_p[H(x-l_1) - H(x-l_2)]\} \tag{5.45}$$

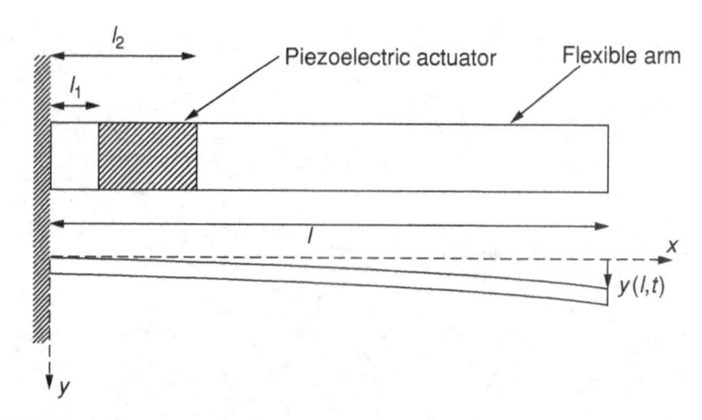

FIGURE 5.24 Configuration of flexible arm with piezoelectric actuator.

where x is the distance along the flexible arm, y is the flexual displacement of the flexible arm, $H(\cdot)$ is the Heaviside function, l_1, l_2 are the attachment positions of the piezoelectric actuator, and ρ, S, E, I, and C are the density of the arm, the cross-sectional area of the arm, the Young's modulus of the arm, the moment of inertia of the area, and the damping modulus.

The bending moment $M_p(t)$ of the piezoelectric actuator generated by voltage $u(t)$ is described by

$$M_p(t) = u^*(t) = P_I[u(t)] \tag{5.46}$$

where $u^*(t)$ is the output of the PI model. The flexible arm is actuated by the piezoelectric actuator, so the input of the flexible arm $u^*(t)$ is the output of the PI model. Figure 5.25 shows the relationship between the PI model and the plant.

Then the elastic deformation of the flexible arm is modeled using modal analysis, and the displacement $y(x, t)$ is represented as an infinite sum of the form using convolution on two functions about the plant and the PI model:

$$y(x,t) = \sum_{m=1}^{\infty} \left(A_m \int_0^t e^{-(\alpha_m/2)(t-\tau)} \sin \frac{\beta_m}{2}(t-\tau) \cdot u^*(\tau)\, d\tau \right) \tag{5.47}$$

where

$$A_m = \frac{2\omega_m(x)}{\rho S \phi_m \beta_m} \left[\dot{\omega}_m(l_2) - \dot{\omega}_m(l_1)\right], \quad \alpha_m = k_m^2 C_m$$

$$\beta_m = \sqrt{4k_m^2 - k_m^4 C_m^2} \qquad k_m = \sqrt{\frac{\lambda_m^4 EI}{\rho S}}$$

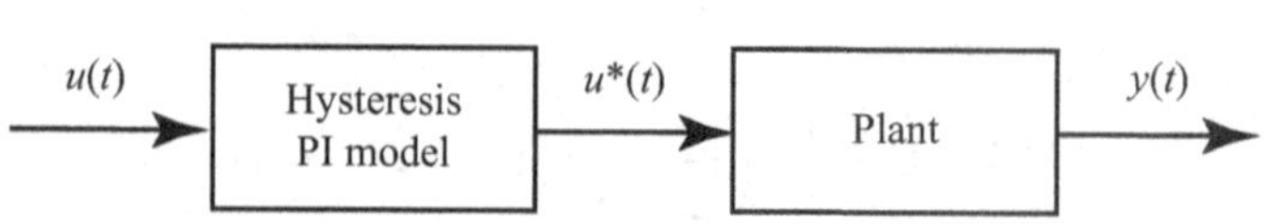

u(t)
Hysteresis PI model
u*(t)
Plant
y(t)

FIGURE 5.25 Plant with hysteresis.

where ϕ_m, k_m, λ_m, and C_m, are the constant, the natural frequency, the solution of $1+\cos(\lambda_m l)\cosh(\lambda_m l)=0$, with l the length of the arm, and the damping modulus of the mth mode, respectively. Then ω_m, the mth mode function, can be expressed by the following formula using an arbitrary constant B_m:

$$\omega_m(x) = B_m[(\sinh\lambda_m l + \sin\lambda_m l)(\cosh\lambda_m x - \cos\lambda_m x) \\ -(\cosh\lambda_m l + \cos\lambda_m l)(\sinh\lambda_m x - \sin\lambda_m x)] \quad (5.48)$$

In this section we consider the first vibration mode of the flexible arm, and the unconsidered vibration modes are regarded as perturbation ΔP. Then the operator representation of the experimental system can be described as

$$[P+\Delta P](u^*)(t) = (1+\Delta)A_1\int_0^t e^{-(\alpha_1/2)(t-\tau)}\sin\frac{\beta_1}{2}(t-\tau)\cdot u^*(\tau)\,d\tau \quad (5.49)$$

$$[N+\Delta N](\omega)(t) = (1+\Delta)\,A_1 e^{-(\alpha_1/2)t}\int_0^t \sin\frac{\beta_1}{2}(t-\tau)\cdot\omega(\tau)\,d\tau \quad (5.50)$$

where

$$P(u^*)(t) = A_1\int_0^t e^{-(\alpha_1/2)(t-\tau)}\sin\frac{\beta_1}{2}(t-\tau)\cdot u^*(\tau)\,d\tau \quad (5.51)$$

$$N(\omega)(t) = A_1 e^{-(\alpha_1/2)t}\int_0^t \sin\frac{\beta_1}{2}(t-\tau).\omega(\tau)\,d\tau \quad (5.52)$$

$$D(\omega)(t) = e^{-(\alpha_1/2)t}w(t) \quad (5.53)$$

where ΔN is the numerator of the coprime factorization presentation of ΔP, N, ΔN, and D are the stable operators, respectively, and D is invertible. In the following, the robust right coprime factorization condition is described. The right coprime factorization of the nominal plant P and the overall plant $P+\Delta P$ are

$$P = ND^{-1} \qquad P+\Delta P = (N+\Delta N)D^{-1} \quad (5.54)$$

where we assume that ΔN is unknown but the upper and lower bounds of ΔN are known.

In the simulation, we used the parameters shown in Table 5.1. The moment of inertia of area I is calculated using the width w and the thickness t of the flexible arm.

TABLE 5.1 Parameters for Simulations

l	0.5 m	ρ	8030 kg/m^3
l_1	10×10^{-3} m	l_2	60×10^{-3} m
C_1	0.0018797	λ_1	2.641
E	1.97×10^{11} N/m^2	S	10×10^{-6} m^2
w	0.5×10^{-3} m	t	20×10^{-3} m

To describe the hysteresis of the piezoelectric actuator, the PI model, one of the popular phenomenological models, is adopted. Generally, the PI model is described with stop operators or play operators [79]. Here, the PI model defined by play operators with threshold value $h > 0$ is used. Play operators are designed as follows [92]. Analytically, suppose that $C_m[0, t_E]$, where $0 = t_0 < t_1 < \cdots < t_N = t_E$ is a partition of $[0, t_E]$, is the space of the piecewise monotone continuous function. The input function $u(t) \in C_m[0, t_E]$ is monotone on each of the subintervals $[t_i, t_{i+1}]$. The play operator $F_h(\cdot; u^*_{-1}) : C_m[0, t_E] \times u^*_{-1} \rightarrow C_m[0, t_E]$ for an initial vaule $u^*_{-1} \in R$ is defined as

$$F_h(u(0); u^*_{-1}) = f_h(u(0), u^*_{-1}) \tag{5.55}$$

$$F_h(u(t); u^*_{-1}) = f_h(u(t), F_h(u(t_i); u^*_{-1})) \tag{5.56}$$

for $t_i < t \leq t_{i+1}$ and $0 \leq i \leq N - 1$ with

$$f_h(u, q) = \max(u - h, \min(u + h, q)) \tag{5.57}$$

The spaces of input and output can be extended to the space $C[0, t_E]$ of continuous functions. Here the play operator is represented in another form [92]:

$$F_h[u(t)] = \begin{cases} u(t) + h & u(t) \leq F_h[u(t_i)] - h \\ F_h[u(t_i)] & -h < u(t) - F_h[u(t_i)] < h \\ u(t) - h & u(t) \geq F_h[u(t_i)] + h \end{cases} \tag{5.58}$$

where $F_h(u)$ shows $F_h(u; u^*_{-1})$ in brief. The initial condition of (5.58) is given by $F_h[u(0)] = \max(u(0) - h, \min(u(0) + h, u^*_{-1}))$. The PI model can be described by a weighted superposition of the above play operator with threshold value H:

$$u^*(t) = P_I[u(t)] = D_{\mathrm{PI}}[u(t)] + \Delta[u(t)] \tag{5.59}$$

where

$$D_{\mathrm{PI}}[u(t)] = Ku(t),\ K = \int_0^{h_x} p(h)\, dh$$

$$\Delta[u(t)] = -\int_0^{h_x} S_n h p(h)\, dh + \int_{h_x}^{H} p(h) F_h[u(t_i)]\, dh$$

$$S_n = \mathrm{Sgn}\,\{u(t) - F_h[u(t_i)]\}$$

where the choice of $H = \infty$ as the upper limit of integration in the literature is just a matter of convenience [78]. On $[0, H]$ we can find h_x to make $h \leq |u(t) - F_h[u(t_i)]|$, where $h \in [0, h_x]$. The hysteresis operator is $P_I(\cdot)$ and $p(h)$ is a given continuous

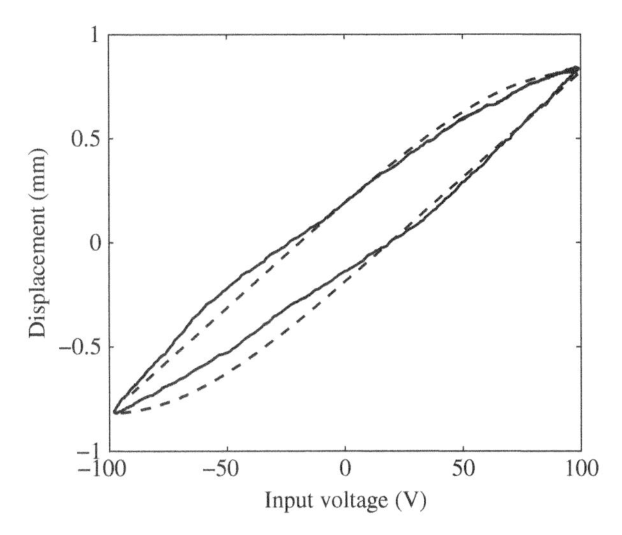

FIGURE 5.26 Hysteresis curve of (—) experimental value and (- -) PI model.

density function satisfying $p(h) \geq 0$ with $\int_0^{h_x} hp(h)\,dh < \infty$ and is expected to be identified from experimental data. To identity the density function parameters, for example, for $p(h) = a \times e^{b(h-1)^2}$, we set the threshold value h. The weight parameters a and b are found by performing nonlinear least squares fit. With the defined density function, the P_I operator maps $C[t_0, \infty)$ into $C[t_0, \infty)$. Further, the operators Δ and D_{PI} also map $C[t_0, \infty)$ into $C[t_0, \infty)$. The case of $D_{\mathrm{PI}} = 0$ is considered in [92], and for brevity the case of $D_{\mathrm{PI}} \neq 0$ is shown in this chapter. In the following, the two cases of $D_{\mathrm{PI}} = 0$ and $D_{\mathrm{PI}} \neq 0$ are considered in simulation and experiment. Since the density function $p(h)$ vanishes for large values of h, we assume that there exists a constant H such that $p(h) = 0$ for $h > H$. In an experiment to evaluate the above PI model, input voltage $u(t) = 100\sin(2\pi f_1 t)$ is applied to operate the piezoelectric actuator. In the equation of $u(t)$, $f_1 = 1.83$ Hz, which is the first mode frequency of the flexible arm used. A laser sensor is used to measure the displacement response of the piezoelectric actuator. Figure 5.26 shows the measured data obtained from the actuator and the data derived from the PI model given in (5.59), where parameters $p(h) = 0.00032 \times e^{-0.00086(h-1)^2}$ and $h \in [0, 100]$ of the PI model were found by performing a nonliear least-squares fit using a Levenberg–Marquardt method. In Figure 5.26, the hysteresis curve of the PI model closely approximates the hysteresis of the piezoelectric actuator, so the PI model generates the hysteresis.

In the following, we design robust stable controllers of a nonlinear system with hysteresis based on robust right coprime factorization. The framework of our control system is shown in Figure 5.27. Assume U is the input space of the PI model and U^* is the output space of the PI model. Let the output space of the original nonlinear plant and quasi-state space be Y and W and N, ΔN, D are $N : W \to Y$, $\Delta N : W \to Y$, $D : W \to U^*$, respectively; A, B are the controllers and the stable operators and

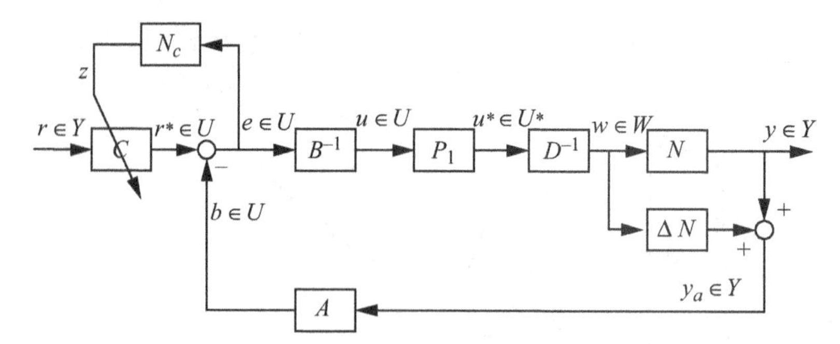

FIGURE 5.27 Control system.

B is invertible. We can choose $W = U$. In Figure 5.27, the output u^* of the PI model is the input of the original nonlinear plant. Generally, it is difficult to observe the signal of u^* which is the output of the actuator from the practical system. So the hysteresis of the actuator is considered as one part of controller design. We consider $D_{\mathrm{PI}}[u(t)]$ in the hysteresis model given in (5.59) and D of the above plant as a new invertible factor of right coprime factorization $\tilde{D}[\omega(t)] = D_{\mathrm{PI}}^{-1} D[\omega(t)]$ and let $\Delta_{\mathrm{PI}} = \Delta[u(t)]$. The corresponding controllers A and B which satisfy the Bezout identity $AN + B\tilde{D} = M \in S(W, U)$ are also designed to make the nonlinear system BIBO stable, where $S(W, U)$ is the set of unimodular operators [92]. Then we obtain

$$y(t) = (N + \Delta N)[A(N + \Delta N) + B\tilde{D}]^{-1}[r^*(t) + BD_{\mathrm{PI}}^{-1}\Delta_{\mathrm{PI}}] \tag{5.60}$$

Let the operator $N + \Delta N$ be stable and the Bezout identity of the exact plant be $A(N + \Delta N) + B\tilde{D} = \bar{M}$. If

$$\|[A(N + \Delta N) - AN]M^{-1}\| < 1 \tag{5.61}$$

the system retains a robust coprime factorization and is robustly stable based on Lemma 2.7.

Since A and B satisfy the Bezout identity, (5.60) can be represented as

$$y(t) = (N + \Delta N)\bar{M}^{-1}[r^*(t) + BD_{\mathrm{PI}}^{-1}\Delta_{\mathrm{PI}}] \tag{5.62}$$

Accorging to the above system framework, the signal between D^{-1} and $N + \Delta N$ can be obtained. Depending on this signal, the stable tracking operator C is designed to satisfy the condition

$$(N + \Delta N)\bar{M}^{-1}[C(r(t), z) + BD_{\mathrm{PI}}^{-1}\Delta_{\mathrm{PI}}] = r(t) \tag{5.63}$$

where z is adjusted by nonlinear compensator N_c based on an error signal e. Then, we can have $y(t) = r(t)$. To compensate for $\Delta_{\rm PI}$, the designed tracking operators should include information of them. In this chapter, a nonlinear compensator is designed to realize this purpose. Using the error signal e, the nonlinear compensator could adjust the tracking operator to realize the tracking function. The adjusted variable z includes z_1 and z_2 and is described as

$$z = (z_1, z_2) = (\bar{D}_{\rm PI}^{-1}, B\bar{\Delta}_{\rm PI}) \tag{5.64}$$

where

$$\begin{aligned}
\bar{D}_{\rm PI}^{-1} &= \bar{K}^{-1} = \left(\int_0^{h_x} p(h)\,dh\right)^{-1} \\
\bar{\Delta}_{\rm PI} &= -\int_0^{h_x} S_n h p(h)\,dh + \int_{h_x}^{H} p(h) F_h\{B^{-1}[e(t_i)]\}\,dh \\
S_n &= \mathrm{Sgn}(B^{-1}[e(t)] - F_h\{B^{-1}[e(t_i)]\})
\end{aligned}$$

When the nonlinear compensator N_c is installed in a real-time practical system according to the control system in Figure 5.27, variable z is adjusted by the nonlinear compensator N_c using a former error signal e. So we modify the controller to compensate for the hysteresis corresponding to the current error signal e in the experiment.

Based on the design scheme, we design controllers as follows:

$$A[y_a(t)] = \left((e^{t/2} + \Delta)^2 - \frac{e^{(1-\alpha_1/2)t} + \Delta}{K}\right) g^2(t) \tag{5.65}$$

$$B[u(t)] = I[u(t)] \tag{5.66}$$

where $g(t) = e^{-\frac{t}{2}} \omega^{\frac{1}{2}}(t)$, $K = \int_0^{h_x} p(h)\,dh$, and $I(\cdot)$ is the identity operator. If $\Delta = 0$, we have

$$[AN + B\tilde{D}](\omega)(t) = \left\{\left(\omega(t) - \frac{e^{-(\alpha_1/2)t}}{K}\omega(t)\right) + e^{-(\alpha_1/2)t} D_{\rm PI}^{-1}\omega(t)\right. \tag{5.67}$$

$$= I(\omega)(t) \tag{5.68}$$

for all $w \in W$. We design the tracking operator

$$C(r(t), z) = \frac{1}{1 - e^{(-\alpha_1/2)t} z_1} A[r(t)] - z_1 z_2 \tag{5.69}$$

where $t > 0$ and z is obtained by the nonlinear compensator. Here, we provide the following condition to estimate the stability of the considered system. If $\Delta \neq 0$, for any $\omega \in W$, we have

$$\Delta_s = [A(N + \Delta N) - AN](\omega)(t) = \left[2\Delta e^{-t/2} + \Delta^2 e^{-t} - \frac{\Delta e^{-t}}{K}\right]\omega(t) \quad (5.70)$$

If Δ_s satisfies the following condition, the considered system is BIBO stable:

$$\|\Delta_s \omega^{-1}(t)\| < 1 \quad (5.71)$$

To demonstrate the effectiveness of the method, two kinds of experiments, without and with a PI model, are conducted. The experimental results are presented in this section. The parameters and input voltage $V_d(t)$ ($t < 5$ sec) used in the two experiments are $V_d(t) = 20\sin(2\pi f_1 t)$, $p(h) = 0.13 \times e^{-0.01(h-1)^2}$, $h \in [0, 100]$. The input voltage is applied to the piezoelectric actuator, and the piezoelectric actuator is used as a control actuator for $t \geq 5$ sec. The mode frequency and the damping modulus of the first-order mode are obtained by experimental modal analysis, and the effect of unconsidered vibration modes which are more than the second vibration modes is confirmed to define the disturbance. Figure 5.28 shows the control input for the case with and without the PI model. In Figure 5.28, we can see these control input voltages are less than the limit of the piezoelectric actuator (−100–100 V). Figure 5.29 shows the output of the system without control. The control output shown in Figure 5.30 is for the cases with and without the PI model. In Figure 5.30, the control performance with the PI model is better than the performance without the PI model, namely, when $t > 10$ sec, the displacement with the PI model is reduced about 50% more than the case without PI model.

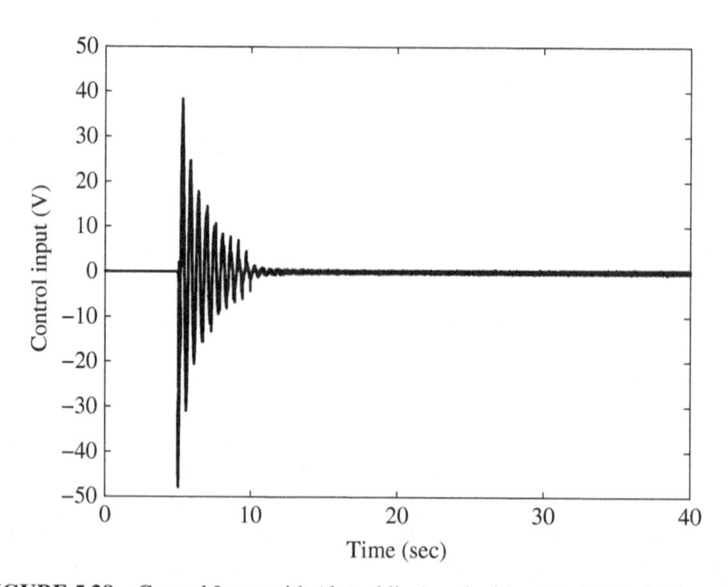

FIGURE 5.28 Control Input with (dotted line) and without (solid line) PI model.

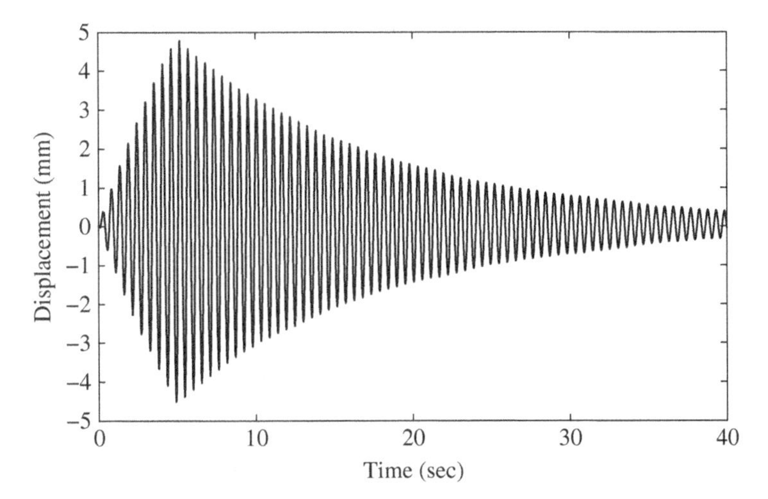

FIGURE 5.29 Output of system without control.

5.3.2 Nonlinear Control of Shape Memory Alloy Actuator

The SMA actuators have several advantages, such as large ratio of force to mass, light weight, low driving voltage, and large displacement, when compared with other actuators. Therefore, SMA actuators have been applied in various applications. Hashimoto et al. [94] used SMA strings as a mobile robot actuator. Stevens and Buckner [95] confirmed that SMA actuators can be effective for miniature surgical robotic systems. Grant and Hayward [96] developed a two-stage controller for a simplified

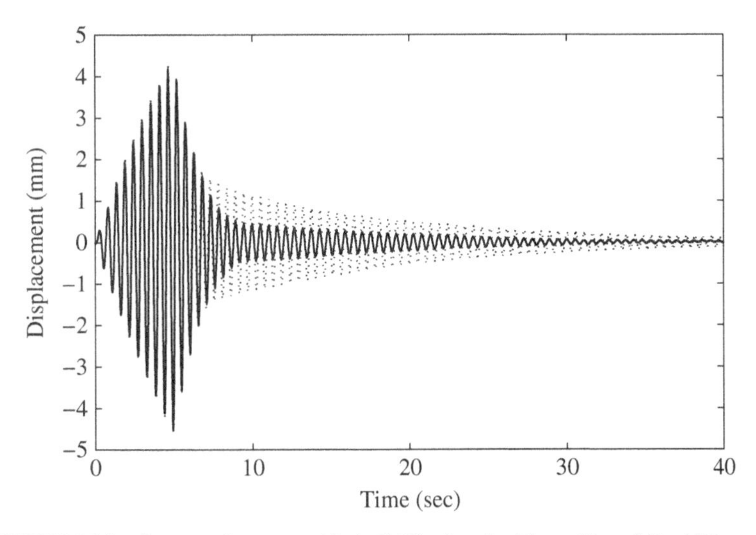

FIGURE 5.30 Output of system with (solid line) and without (dotted line) PI model.

SMA model with improved accuracy and speed. While SMA actuators possess interesting properties, they have such undesirable characteristics as hysteresis and slow response speed. In particular, due to the hysteretic behavior of SMA actuators, their performance deteriorates over time.

The complex hysteretic nonlinearity of SMA actuators is a challenging topic in the control field, and the design of these control systems has been considered by many researchers. Brinson [97] proposed a constitutive model that included the SMA hysteretic behavior. Madill and Wang [98] used a hysteresis model to prove the L_2-stability of position control. Choi et al. [99] considered a linear model with uncertainty and presented a robust force tracking control method. Majima et al. [100] used the Preisach hysteresis model and proposed a tracking control system. Moallem and Tabrizi [101] considered the nonlinear behavior and thermal characteristics of SMA actuators and presented a scheme for tracking control. Some researchers have proposed vibration control of flexible arms using SMA actuators [102, 103]. These reports, however, did not consider the hysteretic nonlinearity of SMA actuators. Compensating for the hysteresis of SMA actuators is an important issue for realizing desired control.

This section considers the design of a nonlinear vibration control system design of a flexible arm using an SMA actuator with hysteresis. To eliminate the effect of hysteresis, an operator-based PI model is used where the hysteresis is compensated for by the operator-based controller of Chapter 2. The operator-based method has been proved effective for nonlinear system control design, which is easier than other theories such as the Lyapunov method to guarantee robust stabilization. Based on the concept of the Lipschitz operator and the robust right coprime factorization condition, vibration controllers and a tracking controller are designed to guarantee BIBO stability and to ensure system output tracking performance [21].

SMA wire actuators have often been used because the wire is easy to cut, connect, and activate electrically. Besides the advantages mentioned above, a SMA wire is used to simplify the evaluation of the characteristic of the SMA actuator. In this section, the SMA actuator is modeled, and the model is used to design the vibration controller of a flexible arm and a hysteresis compensator. In general, SMA actuators are able to get large strain (4–6%) and actuation force under a thermal input [104]. The heating process is based on the Joule effect, which causes current to flow through a SMA wire. On the other hand, SMA actuators are well known to have a slow response speed and thus lose effectiveness at high frequencies of vibration. In this section, to improve response, the size of the SMA wire is reduced to increase the rate of heat transfer. In addition, the offset electric current, which is selected to have stability in a certain temperature in the R phase region [103], is added. In vibration control, because SMA material cannot reverse completely with load repeatedly for the whole area between the martensite phase and austenite phase, only a transition within the R phase is used. The deformation of the R phase is limited by only 1–2%. However, it is bigger than that of other materials.

Figure 5.31 shows an SMA actuator model composed of a thermal model which represents the temperature–electric power relationship T_h and the hysteresis model which represents strain–temperature relationship PI.

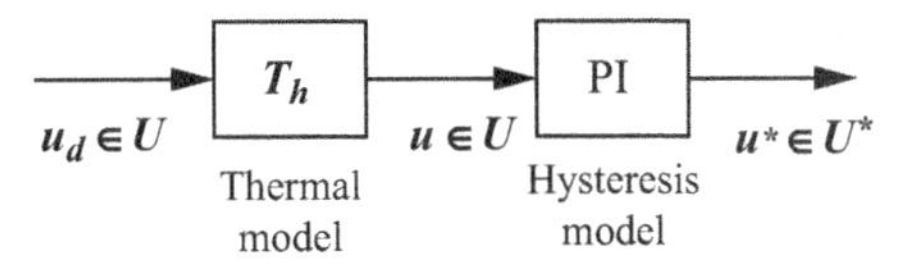

FIGURE 5.31 SMA actuator model.

The heat transfer equation for a SMA wire consists of natural convection and electrical heating:

$$mc_p \frac{d(T - T_a)}{dt} = i^2 R - h_c A_c (T - T_a) \tag{5.72}$$

where i is the electric current passing through the SMA wire and R is the resistance per unit length which can be considered as a constant to simplify the analysis, A_c is the surface area, c_p is the specific heat, m is the mass, h_c is the heat convection coefficient, T_a is the ambient temperature, and T is the the temperature of the SMA wire. Note that the surface area of the SMA wire may change during the operation, but it is assumed that the effect is negligible. For convenience, the electric power $i^2 R$ and the difference in temperature $T - T_a$ are defined as follows:

$$u_d(t) = i^2 R \tag{5.73}$$

$$u(t) = T - T_a \tag{5.74}$$

where the electric power u_d represents the input of the thermal model. Using (5.73) and (5.74), equation (5.72) is transformed as follows:

$$u(t) = T_h(u_d) = \frac{1}{mc_p} \int e^{-\gamma(t-\tau)} u_d(\tau)\, d\tau \tag{5.75}$$

where

$$\gamma = \frac{h_c A_c}{mc_p}$$

The PI model, one of the popular phenomenological models, is used to compensate for the hysteretic behavior of a SMA wire. It is formulated through a weighted superposition of elementary hysteresis operators, which are stop hysteresis operators or play hysteresis operators [105]. In this chapter, the PI model defined by the play hysteresis operator with threshold value $h > 0$ is used. Analytically, suppose that $C_m[0, t_E]$, where $0 = t_0 < t_1 < \cdots < t_N = t_E$ is a partition of $[0, t_E]$, is the space of piecewise monotone continuous functions. The play hysteresis operator

$F_h(\cdot; u_{-1}^*) : C_m[0, t_E] \times u_{-1}^* \to C_m[0, t_E]$ for an initial value $u_{-1}^* \in R$ is generally defined as [79]

$$\begin{aligned} F_h(u(0); u_{-1}^*) &= f_h(u(0), u_{-1}^*) \\ F_h(u(t); u_{-1}^*) &= f_h(u(t), F_h(u(t_i); u_{-1}^*)) \end{aligned} \tag{5.76}$$

for $t_i < t \leq t_{i+1}$ and $0 \leq i \leq N-1$ with

$$f_h(u, q) = \max(u - h, \min(u + h, q)) \tag{5.77}$$

where h is the threshold value of the play hysteresis operator. Based on (5.76), the play hysteresis operator can be represented as

$$F_h[u(t)] = \begin{cases} u(t) + h & u(t) \leq F_h[u(t_i)] - h \\ F_h[u(t_i)] & -h < u(t) - F_h[u(t_i)] < h \\ u(t) - h & u(t) \geq F_h[u(t_i)] + h \end{cases} \tag{5.78}$$

The initial condition of (5.78) is given by $F_h[u(0)] = \max(u(0) - h, \min(u(0) + h, u_{-1}^*))$. The Lipschitz operator-based PI model can be described by using the weighted superposition of the above play hysteresis operator [92] , and it is decomposed into an invertible term and a bounded parasitic term as disturbance, represented as

$$u^*(t) = D_{\mathrm{PI}}[u(t)] + \Delta_{\mathrm{PI}}[u(t)] \tag{5.79}$$

where

$$\begin{aligned} &D_{\mathrm{PI}}[u(t)] = Ku(t) \qquad K = \int_{h_0}^{h_x} p(h)\,dh \\ &\Delta_{\mathrm{PI}}[u(t)] = -\int_{h_0}^{h_x} S_n h p(h)\,dh + \int_{h_x}^{H} p(h) F_h[u(t_i)]\,dh \\ &S_n = \begin{cases} 1 & \text{if } u(t) - F_h[u(t_i)] \geq 0 \\ -1 & \text{if } u(t) - F_h[u(t_i)] < 0 \end{cases} \end{aligned}$$

In $[h_0, H]$, h_x can be found to make $h \leq |u(t) - F_h[u(t_i)]|$ where $h \in [h_0, h_x]$. $P_I(\cdot)$ represents the hysteresis operator and $p(h)$ is the given continuous density function, satisfying $p(h) \geq 0$ with $\int_{h_0}^{\infty} hp(h)\,dh < \infty$, and $\int_{h_0}^{h_x} hp(h)\,dh$ is bounded. With the defined density function, the P_I operator maps $C[t_0, \infty)$ into $C[t_0, \infty)$ [106]. Moreover, the operators D_{PI} and Δ_{PI} also map $C[t_0, \infty)$ into $C[t_0, \infty)$. Since the density function $p(h)$ vanishes for large values of h, it is assumed that there exists a constant H such that $p(h) = 0$ for $h > H$. For example, we set the threshold value H in a density function $p(h) = a \times e^{b(h-1)^2}$ to identity the density function

TABLE 5.2 Physical Parameters of SMA Wire

Description	Symbol (unit)	Value
Length	l_a (mm)	100
Diameter	d (mm)	0.1
Surface area per unit length	A_c (mm^2)	0.314
Resistance per unit length	R (Ω/m)	135
Weight per unit length	m (kg)	5×10^{-6}

parameters. Then the weight parameters a and b can be found by performing a nonlinear least-squares fitting [107]. Parameter identification results To evaluate the above PI model for describing the hysteresis of the SMA actuator and to identify some parameters, experiments were conducted. In the experiments, a SMA wire (BMF100, Toki Corporation) is used for which the physical parameters are described in Tables 5.2 and 5.3. The ambient temperature T_a is 25°C. The strain is derived by measuring the displacement of the SMA wire using a laser sensor (Z4M-W40, Omron). The surface temperature of the SMA wire is measured by a thermocouple (CHAL-002, Omega), where the output of the thermocouple is amplified by an operational amplifier (AD595) with gain of 100 V/V, and then the amplified value is fed to a PC through an interface board (AD12-16(PCI)E, Contec). The interface board is used to measure not only some voltages but also the output of the control voltage.

In the real SMA wire, the resistance is changed according to the stretch of the SMA wire, so both the voltage of the SMA wire $v(t)$ and the electric current passing through the SMA wire $i(t)$ are measured in real time, the real electric power $vi(t)$ is computed, the real electric power is adjusted to track the reference electric power using a proportional integral control scheme with a control period of 1 msec. Thermal parameters c_p and h_c were identified. When the input is the electric power and the output is the surface temperature of the SMA wire, the step response was measured with an input electric power 0.6 W. The following parameters were obtained using MATLAB's curve fitting toolbox: $c_p = 7349$ J/kg °C and $h_c = 689$ W/m^{2}°C [104]. The experiment to evaluate the PI model was conducted. Figure 5.32 shows the

TABLE 5.3 Parameters of Flexible Arm

Description	Symbol (unit)	Value
Length	l (m)	0.8
Attachment position of SMA wire	l_1 (m)	0.1
Density	ρ (kg/m^3)	2700
Cross-sectional area	S (m^2)	10×10^{-6}
Young's modulus	E (N/m^2)	6.9×10^{10}
Area moment of inertia	I (m^2)	1.67×10^{-12}
Damping modulus of first-order mode	C_1	0.0015
Thickness	t (m)	1.0×10^{-3}
Width	w_1 (m)	20×10^{-3}

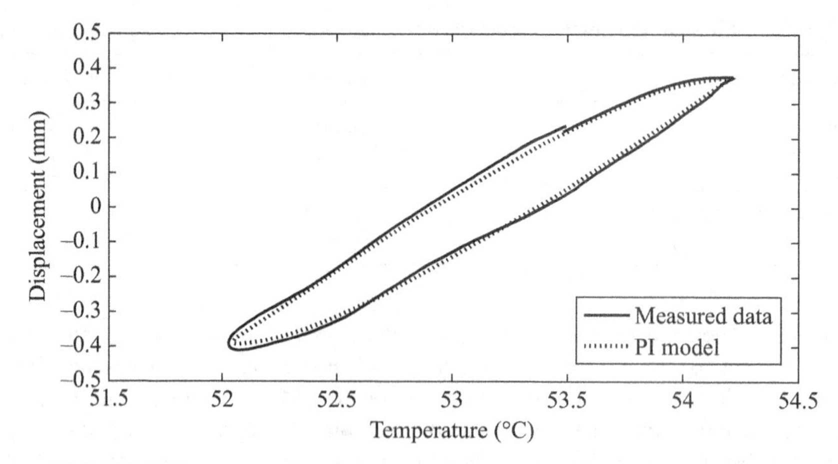

FIGURE 5.32 Hysteresis curves of (–) experimental value and (- -) PI model.

measured displacement and the data derived from the PI model given in (5.77). The density function $p(h) = 0.0012 \times e^{2.8(h-1)^2}$, $h \in [0.01, 1]$, of the PI hysteresis model was found by performing a nonlinear least-squares fit using the Levenberg–Marquardt method. In Figure 5.32, the hysteresis curve of the PI model closely approximates the hysteresis of the SMA actuator, so the operator-based PI hysteresis model generates the hysteresis of the SMA actuator.

The configuration of a flexible arm with a SMA wire is shown in Figure 5.33. In this section, the relationship between the bending moment $M_a(t)$, which is the input of the plant and the strain of the SMA wire is considered to be linear based on Mohr's theorem. The moment $M_a(t)$ is defined as

$$M_a(t) = M_{a0} \cdot P_I(u)(t) \tag{5.80}$$

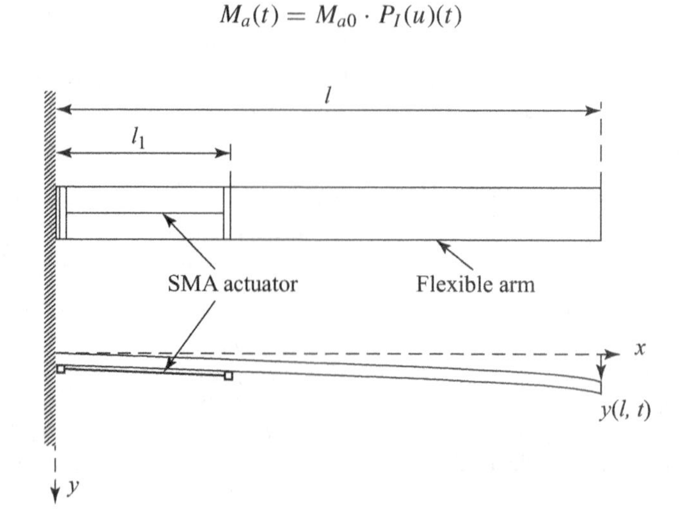

FIGURE 5.33 Configuration of flexible arm with SMA actuator.

where M_{a0} is given by

$$M_{a0} = \frac{3EI}{l_1^2} \tag{5.81}$$

where E is Young's modulus, I is the area moment of inertia, and l_1 is the attachment position of the SMA wire. According to the Euler–Bernoulli beam equation, the dynamics of the flexible arm can be described by the partial differential equation

$$\rho S \frac{\partial^2 y}{\partial t^2} + \frac{\partial^2}{\partial x^2}\left[EI\left(1 + C\frac{\partial}{\partial t}\right)\frac{\partial^2 y}{\partial x^2}\right] = \frac{\partial^2}{\partial x^2}[M_a \delta(x - l_1)] \tag{5.82}$$

where x is the distance along the arm, y is the flexural displacement of the arm, and $\delta(\cdot)$ is the delta function. Assume ρ, S, and C are the density of the arm, the cross-sectional area of the arm, and the damping modulus, respectively. Equation (5.82) can be represented using the convolution on two functions about the plant and the PI model:

$$y(x,t) = \sum_{m=1}^{\infty}\left(A_m \int_0^t e^{-(\alpha_m/2)(t-\tau)} \sin\frac{\beta_m}{2}(t-\tau)\cdot u^*(\tau)\,d\tau\right) \tag{5.83}$$

where

$$A_m = \frac{2\omega_m(x)\omega_{l_1}}{\rho S \phi_m \beta_m}, \qquad \alpha_m = k_m^2 C_m$$

$$\beta_m = \sqrt{4k_m^2 - k_m^4 C_m^2} \quad k_m = \sqrt{\frac{\lambda_m^4 EI}{\rho S}}$$

where ϕ_m, k_m, λ_m, and C_m are a constant, the natural frequency, the roots of $1 + \cos(\lambda_m l)\cosh(\lambda_m l) = 0$, where l is the length of arm, and the damping modulus of the mth mode, respectively. Then ω_m is the mth vibration mode function and can be expressed by the following formula using an arbitrary constant B_m:

$$\begin{aligned}\omega_m(x) = B_m[&(\sinh\lambda_m l + \sin\lambda_m l)(\cosh\lambda_m x - \cos\lambda_m x)\\ &-(\cosh\lambda_m l + \cos\lambda_m l)(\sinh\lambda_m x - \sin\lambda_m x)]\end{aligned} \tag{5.84}$$

When ω_m is the first vibration mode function, $\lambda_1 l = 1.875$. Consider the first vibration mode of the flexible arm as a nominal plant. Then the operator representation of the nominal plant can be described as

$$P(u^*)(t) = A_1 \int_0^t e^{-(\alpha_1/2)(t-\tau)} \sin\frac{\beta_1}{2}(t-\tau)\cdot u^*(\tau)\,d\tau \tag{5.85}$$

Let the input space, output space, and quasi-state space be U, Y, and W, respectively. The right factorization of $P = ND^{-1}$ with $N : W \to Y$ and $D : W \to U^*$, where U^* is the output space of the PI model and $U^* \subseteq U$, can be described by

$$N(w)(t) = A_1 e^{-(\alpha_1/2)t} \int_0^t \sin \frac{\beta_1}{2}(t-\tau) \cdot w(\tau)\, d\tau \tag{5.86}$$

$$\begin{aligned} &D(w)(t) = e^{-(\alpha_1/2)t} w(t) \qquad D^{-1}(u^*)(t) = e^{(\alpha_1/2)t} u^*(t) \\ &DD^{-1}(u^*)(t) = D^{-1}D(w)(t) = I \end{aligned} \tag{5.87}$$

The right factorization can be confirmed by the following:

$$\begin{aligned} ND^{-1}(u^*)(t) &= N[e^{(\alpha_1/2)t} u^*(t)] \\ &= A_1 \int_0^t e^{-(\alpha_1/2)(t-\tau)} \sin \frac{\beta_1}{2}(t-\tau) \cdot u^*(\tau)\, d\tau \\ &= P(u^*)(t) \end{aligned}$$

The unconsidered vibration modes are regarded as perturbation ΔP, and the overall plant $\tilde{P}$ can be described as [66]

$$\begin{aligned} \tilde{P} &= [P + \Delta P](u^*)(t) \\ &= (1+\Delta) A_1 \int_0^t e^{-(\alpha_1/2)(t-\tau)} \sin \frac{\beta_1}{2}(t-\tau) \cdot u^*(\tau)\, d\tau \end{aligned} \tag{5.88}$$

The right factorization of the overall plant $\tilde{P}$ can be redescribed using ΔN, which is the numerator of the coprime factorization presentation of ΔP:

$$P + \Delta P = (N + \Delta N) D^{-1} \tag{5.89}$$

where

$$\begin{aligned} &[N + \Delta N](w)(t) = (1+\Delta)\, A_1 e^{-(\alpha_1/2)t} \int_0^t \sin \frac{\beta_1}{2}(t-\tau) \cdot w(\tau)\, d\tau \\ &D^{-1}(u^*)(t) = e^{(\alpha_1/2)t} u^*(t) \end{aligned}$$

where N, ΔN, and D are BIBO stable. We assume that ΔN is unknown but the upper and lower bounds of ΔN are known.

The vibration controllers of a nonlinear system with hysteresis based on the robust right coprime factorization condition are designed. First, an operator-based feedback control system without hysteresis compensator is considered. Figure 5.34 shows the

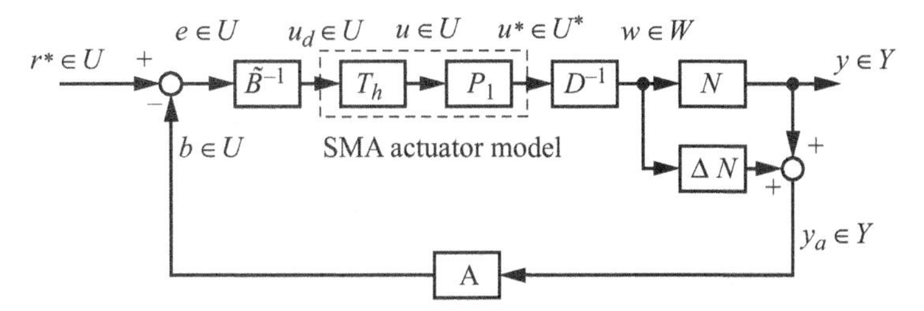

FIGURE 5.34 Operator-based control system without hysteresis compensator.

framework of the control system without hysteresis compensator. The controller $\tilde{B}$, which is included in the SMA model T_h, is defined as

$$\tilde{B} = T_h B \tag{5.90}$$

For simplicity, B is designed as I, and then the controller of (5.90) is $\tilde{B} = T_h$. From equation (5.72), the inverse of T_h is described as

$$T_h^{-1}[u(t)] = mc_p \left(\gamma u(t) + \frac{du(t)}{dt} \right) \tag{5.91}$$

Since T_h^{-1} is BIBO stable, we obtain

$$\tilde{B}^{-1} = T_h^{-1} \tag{5.92}$$

In this chapter, the thermal model, which is a part of the SMA model, is compensated by the controller $\tilde{B}^{-1}$. The output u^* of the PI model is the input of the overall plant $\tilde{P}$. Generally, it is difficult to observe the signal of u^*, which is the output of the actuator from the practical system, so the hysteresis of the SMA actuator is considered as one part of the plant. We consider $D_{\mathrm{PI}}[u(t)]$ given in (5.77) and D of the above plant as a new invertible factor of the right coprime factorization $\tilde{D}(w) = D_{\mathrm{PI}}^{-1} D(w)$ and $\Delta_{\mathrm{PI}}[u(t)]$ is regarded as the bounded disturbance part, where $\tilde{D} : W \to U$. Here, we consider the system after having eliminated the disturbance $\Delta_{\mathrm{PI}}[u(t)]$. The controllers A and B which satisfy the Bezout identity $AN + B\tilde{D} = M \in \mathcal{U}(W, U)$ are designed to make the nonlinear system BIBO stable. For $N \to N + \Delta N$, the satisfying design scheme for robust stabilization is given in [2]. That is, the controllers satisfy the perturbed Bezout identity:

$$A(N + \Delta N) + B\tilde{D} = \tilde{M} \in \mathcal{U}(W, U) \tag{5.93}$$

Under this condition if

$$\|[A(N + \Delta N) - AN]M^{-1}\| < 1 \tag{5.94}$$

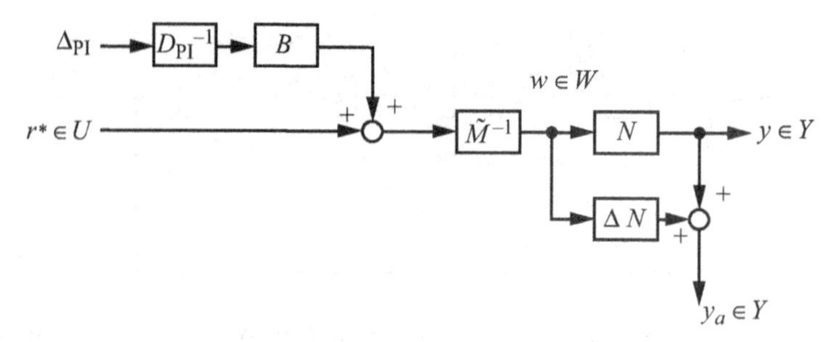

FIGURE 5.35 Equivalent diagram of Figure 5.34.

the overall system $\tilde{P}$ is BIBO stable, and if $\|[A(N+\Delta N) - AN]M^{-1}\|$ of (5.94) can be obtained using the bounded information of ΔN, the detailed ΔN is not necessary. When the perturbed Bezout identity is satisfied, the outputs of the overall plant and the nominal plant can be represented as

$$y_a(t) = (N+\Delta N)[A(N+\Delta N) + B\tilde{D}]^{-1}\left[r^*(t) + BD_{PI}^{-1}\Delta_{PI}\right] \quad (5.95)$$

$$y(t) = N(AN + B\tilde{D})^{-1}\left[r^*(t) + BD_{PI}^{-1}\Delta_{PI}\right] \quad (5.96)$$

Since A and B satisfy the Bezout identity, (5.95) can be represented as

$$y_a(t) = (N+\Delta N)\tilde{M}^{-1}[r^*(t) + BD_{PI}^{-1}\Delta_{PI}] \quad (5.97)$$

which is shown in Figure 5.35. Equation (5.96) also can be represented as

$$y(t) = NM^{-1}\left[r^*(t) + BD_{PI}^{-1}\Delta_{PI}\right] \quad (5.98)$$

To eliminate Δ_{PI}, which represents the effect of hysteresis, the designed operator should include information on OPI. In this chapter, a nonlinear compensator is designed to realize this purpose. The adjusted variable z includes z_1 and z_2 and is described as

$$z = (z_1, z_2) = \left(\bar{D}_{PI}^{-1}, B\bar{\Delta}_{PI}\right) \quad (5.99)$$

where

$$\bar{D}_{PI}^{-1} = \bar{K}^{-1} = \left(\int_{h_0}^{h_x} p(h)\,dh\right)^{-1}$$

$$\bar{\Delta}_{PI} = -\int_{h_0}^{h_x} S_n h p(h)\,dh + \int_{h_x}^{H} p(h)F_h\{B^{-1}[e(t_i)]\}\,dh$$

$$S_n = \begin{cases} 1 & \text{if } B^{-1}[e(t)] - F_h(B^{-1}[e(t_i)] \geq 0 \\ -1 & \text{if } B^{-1}[e(t)] - F_h(B^{-1}[e(t_i)] < 0 \end{cases}$$

In $[h_0, H], h_x$ can be found to make $h \leq |B^{-1}[e(t)] - F_h(B^{-1}[e(t_i)]|$ when $h \in [h_0, h_x]$.

We consider a way to design the tracking controller after compensating for the hysteresis by using the above compensator. Suppose that the following operator design condition is satisfied for the tracking operator C:

$$(N + \Delta N)\tilde{M}^{-1}C(r)(t) = I(r)(t) \tag{5.100}$$

where $r(t)$ is a given reference input. Then, the output of the overall plant $y_a(t)$ tracks to the reference input $r(t)$, that is, we can obtain $y_a(t) = I(r)(t) = r(t)$. According to the same argument, if $NM^{-1}C(r)(t) = I(r)(t)$, we can obtain $y(t) = I(r)(t) = r(t)$. The tracking controller C can be designed by compensating for the hysteresis, which is represented in equation (5.99), to satisfy the condition

$$(N + \Delta N)\tilde{M}^{-1}\left[C(r(t), z) + BD_{\mathrm{PI}}^{-1}\Delta_{\mathrm{PI}}\right] = r(t) \tag{5.101}$$

where z is adjusted by a nonlinear compensator N_c based on error signal e in real time. The framework of our system with a hysteresis compensator is shown as Figure 5.36. The output of the overall plant $y_a(t)$ tracks to the reference input $r(t)$. Also, if

$$NM^{-1}\left[C(r(t), z) + BD_{\mathrm{PI}}^{-1}\Delta_{\mathrm{PI}}\right] = r(t) \tag{5.102}$$

then the output of the nominal plant $y(t)$ tracks to the reference input $r(t)$.

Next, we design two controllers A and B such that the nominal plant P has a right coprime factorization. The controller B is designed as I, so we design the controller A as

$$A[y_a(t)] = \frac{2}{A_1\beta_1}[(e^{(\alpha_1/4)t} + \Delta)^2 - (1 + \Delta)^2]\frac{1}{K}\eta(t) \tag{5.103}$$

$$\eta(t) = \frac{d^2y(t)}{dt^2} + \alpha_1\frac{dy(t)}{dt} + \frac{\alpha_1^2 + \beta_1^2}{4}y(t)$$

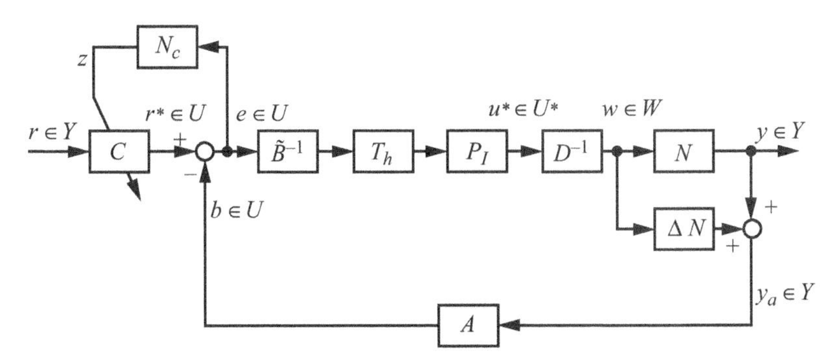

FIGURE 5.36 Operator-based control system with hysteresis compensator.

where $K = \int_0^{h_x} p(h)\,dh$ and Δ represents the effect of ΔN to the output of controller A. To verify that the controller A is BIBO stable (note that in this case $\Delta = 0$), we have

$$A[y(t)] = (1 - e^{-(\alpha_1/2)t})\frac{1}{K}\frac{2}{A_1\beta_1}e^{(\alpha_1/2)t}\eta(t) \tag{5.104}$$

where, from (5.86),

$$w(t) = \frac{2}{A_1\beta_1}e^{(\alpha_1/2)t}\eta(t) \tag{5.105}$$

Then (5.104) can be represented as

$$A[y(t)] = AN(w) = (1 - e^{-(\alpha_1/2)t})\frac{1}{K}w(t) \tag{5.106}$$

The signal w of a quasi-state space is bounded, so the controller $A[y(t)]$ is BIBO stable. The Bezout identity of the nominal plant is confirmed to be the unimodular operator M. When $\Delta = 0$, the following Bezout identity is obtained:

$$[AN + B\tilde{D}](w)(t) = \frac{1}{K}w(t) \quad \text{for all } w \in W \tag{5.107}$$

where $K = \int_0^{h_x} p(h)\,dh \neq 0$, so the above Bezout identity is satisfied. Here, we provide the following condition to estimate the stability of the overall system. If $\Delta \neq 0$, for any $w \in W$, we have

$$\begin{aligned} \Delta_s &= [A(N + \Delta N) - AN](w)(t) \\ &= \frac{2\Delta e^{-(\alpha_1/4)t}}{K}(w)(t) \end{aligned} \tag{5.108}$$

Based on the equation (5.94), if the following condition is satisfied, the overall system is BIBO stable:

$$\|\Delta_s M^{-1}(t)\| = \|2\Delta e^{-(\alpha_1/4)t}\| < 1 \tag{5.109}$$

According to (5.102), the tracking controller of the nominal plant is designed as

$$C(r(t), z) = \frac{1}{1 - e^{-(\alpha_1/2)t}}A[r(t)] - z_1 z_2 \tag{5.110}$$

where, too $t > 0$, $z_1 z_2$ is obtained by the nonlinear compensator based on (5.99). To confirm the validity of the compensator, we consider the case where only the signal

$z_1 z_2$ is added to the feedback control system to compensate for hysteresis. Let the input signal of the plant be u_1^*. Then the signal u_1^* is represented as

$$\begin{aligned} u_1^* &= D_{\mathrm{PI}} B^{-1}(-z_1 z_2) + \Delta_{\mathrm{PI}} \\ &= -\bar{\Delta}_{\mathrm{PI}} + \Delta_{\mathrm{PI}} \\ &\approx 0 \end{aligned} \tag{5.111}$$

where $D_{\mathrm{PI}} \approx \bar{D}_{\mathrm{PI}}$. So, the hysteresis can be eliminated using the compensator. Next, we confirm the tracking controller:

$$\begin{aligned} r(t) &= NM^{-1}\left[C(r(t), z) + BD_{\mathrm{PI}}^{-1}\Delta_{\mathrm{PI}}\right] \\ &= N(\omega)(t) \\ &= y(t) \end{aligned} \tag{5.112}$$

Therefore, the output $y(t)$ tracks to the reference input $r(t)$ [104].

Two kinds of experiments, namely without and with a hysteresis compensator, were conducted. Figures 5.37 and 5.38 show the experimental setup and the SMA wire installed in the flexible arm. To obtain the maximum moment, the SMA actuator is attached near the root of the flexible arm. The parameters of the SMA actuator and the flexible arm are the same as the parameters used in the simulation. The ambient temperature is 25°C, and the offset power is 0.27 W, seen as a fast response given the small change in the temperature of the SMA wire. The experiments were performed with a sampling time of 1 msec, and the input electric power was controlled

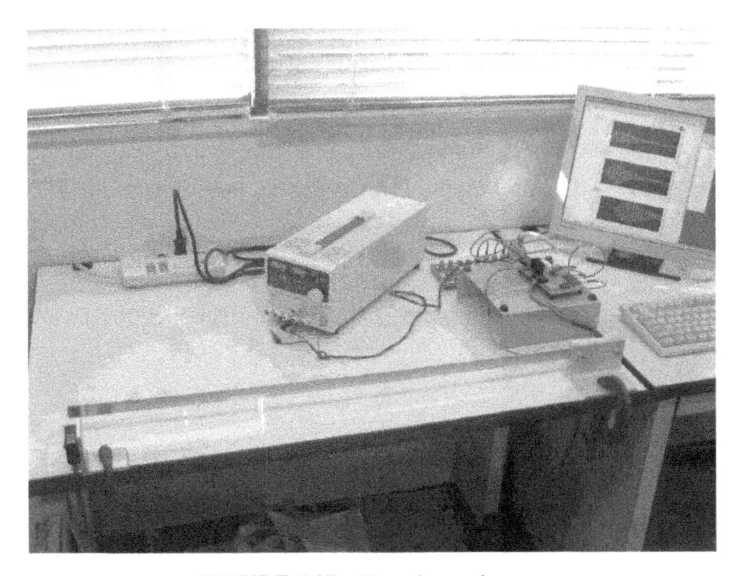

FIGURE 5.37 Experimental setup.

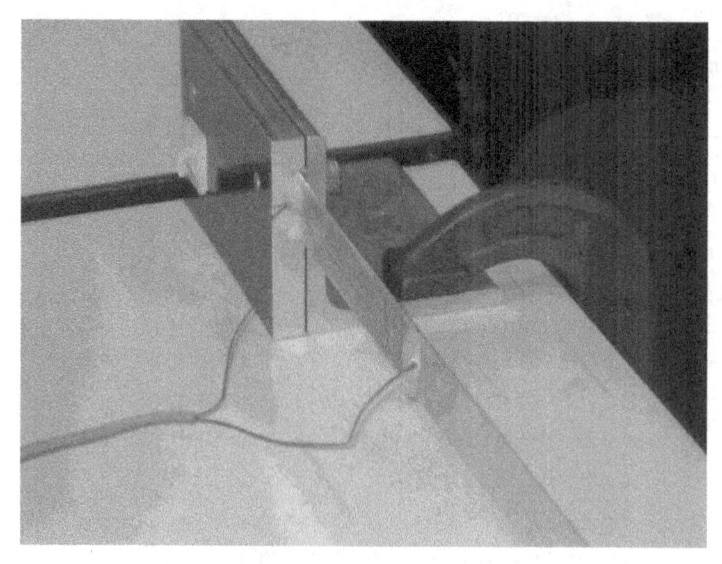

FIGURE 5.38 SMA wire on flexible arm.

using traditional proportional-integral similar to SMA parameter identification. The displacement of the top of the flexible arm was measured by the laser sensor, which the flexure of flexible arm can be measured. In the vibration control experiments, we selected the following input electric power $u_d(t)$ to make the arm vibrate until $t = 10$ sec:

$$u_d(t) = 0.15 \sin(2\pi f_1 t) \tag{5.113}$$

Figure 5.39 shows the result of the free vibration. After 10 sec, the displacement of the flexible arm decreases slowly without control, similar to the result of simulation. Figures 5.40 and 5.41 show input electric power with and without the hysteresis compensator, where the control input power is greater than zero and is affected by the vibration mode. The result of Figure 5.42 trends to small value compared with the result of Figure 5.43 which is similar to the simulation and the simulator is omitted for brevity. As a result, considering the effect of the hysteresis, the vibration suppression of the flexible arm is improved, and the experimental results are also in good agreement with the simulation results.

5.3.3 Nonlinear Control of IPMC

The IPMC, one of the most promising electroactive polymers (EAPs), also called an artificial muscle, is being developed to produce effective, miniature, light and low-power actuators. Because IPMCs have large strain and stress induced electrically, are light weight, are small and simple mechanisms, and have small electric consumption

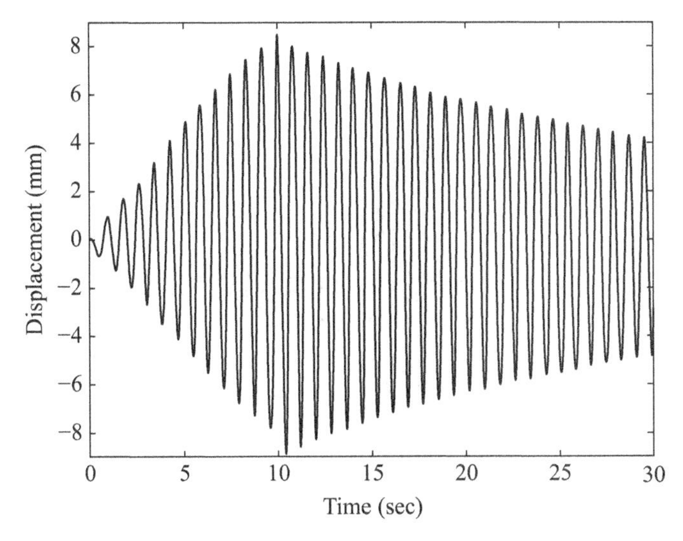

FIGURE 5.39 Experimental result: Output of system after being vibrated without control.

and low drive voltage, they have many potential applications in the developments of miniature robots and biomedical devices [108, 109].

The IPMC model is of three types: black-box, gray-box, and white-box [110]. The black-box models have no prior knowledge of the system. The gray-box models have some knowledge of the system or structure. The white-box models are obtained by physical system derivation and have a comprehensive knowledge of the physical system. Most black-box and gray-box models were developed to study certain response

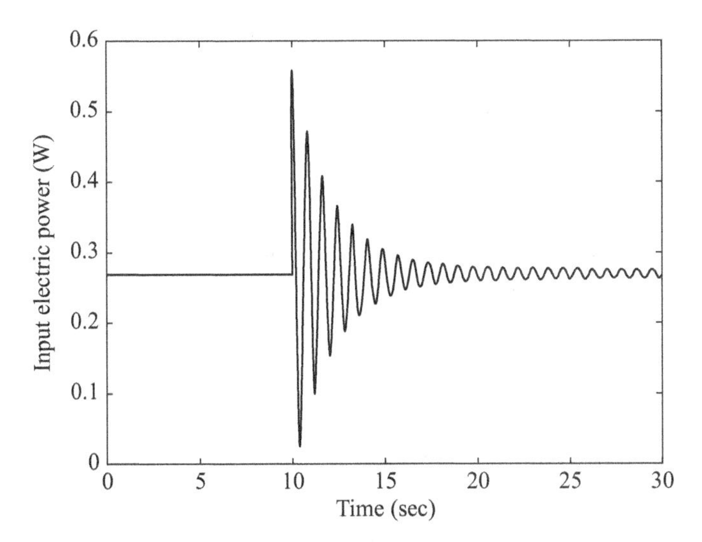

FIGURE 5.40 Input electric power.

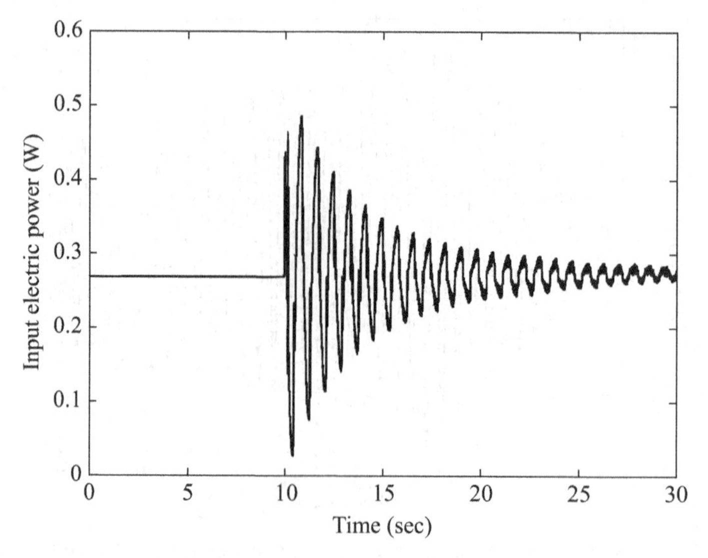

FIGURE 5.41 Input electric power.

characteristics or phenomena in the material and are mainly linear. The white-box versions, on the other hand, attempt to model physical processes taking place within the actuator, and are usually nonlinear. For the linear models, a linear quadratic regulator (LQR), proportional integral and derivative (PID), and adaptive fuzzy algorithm and impedance control scheme have been designed in precise position control [111]. Moreover, the IPMC shows mainly nonlinear behavior for high strain and stress,

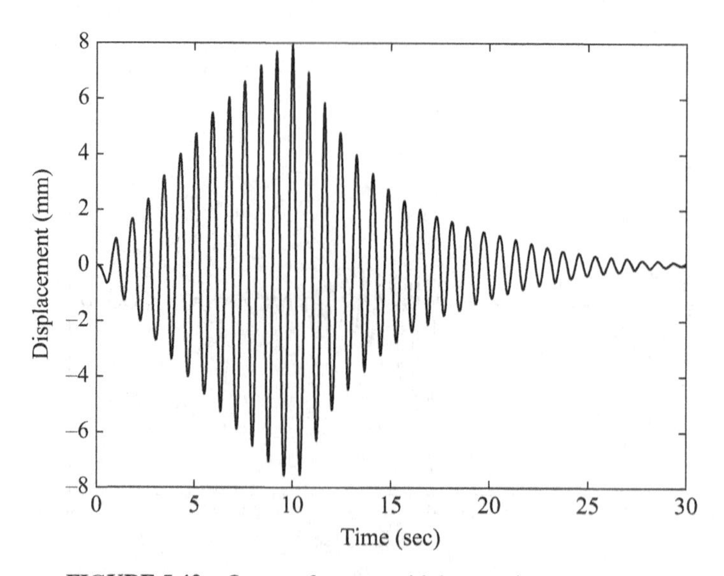

FIGURE 5.42 Output of system with hysteresis compensator.

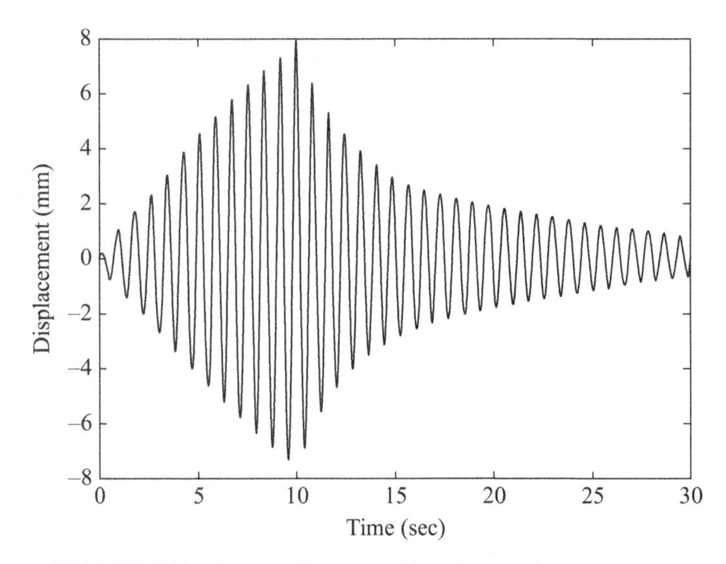

FIGURE 5.43 Output of system without hysteresis compensator.

and a practical mathematical model and an effective control method are desirable in precise position control.

Preciseon position control is critical for the safe operation of IPMC actuators in bio/micromanipulation. In a robotic manipulator, for exampler, the IPMC has to move arbitrarily from one specified position to another. Based on his or her experience, it needs a skilful operator to manually stop the swing at the right position. It is well known that right coprime factorization is a promising approach for the analysis, design, stabilization, and control of a nonlinear system [112]. Especially, robust right coprime factorization has attracted much attention due to its convenience in researching input–output stability problems of nonlinear systems [1, 21, 113, 114]. On the whole, this approach has been proved effective in theoretical studies and practical applications in nonlinear systems (see Figure 5.44). However, for nonlinear systems with uncertainties and input constraints, realizing output tracking performance is still

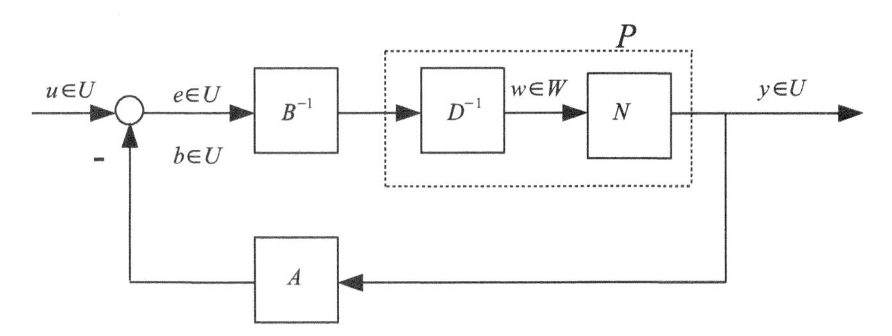

FIGURE 5.44 Nonlinear feedback control system.

a challenging issue. As a result, in this section, the robust nonlinear precision position control design of an IPMC with uncertainties and input constraints is studied. First, considering the measurement error of parameters and the model error of IPMC, an improved practical nonlinear model with uncertainties of IPMC is obtained. Second, an operator-based robust nonlinear control design for IPMC with uncertainties and input constraints is presented. Finally, some simulation and experimental results are shown to confirm the effectiveness of the control method based on the obtained nonlinear model.

This section is organized as follows. A nonlinear model with IPMC uncertainties and the problem statement are described. An operator theorem is introduced as well as robust stable control design using the operator-based approach. The simulation and experimental results are shown.

Nonlinear IPMC Model The dynamic IPMC models fall into two categories: linear and nonlinear. Linear models have no prior knowledge or some knowledge of the system. Nonlinear models have a comprehensive knowledge of the physical system. The nonlinear IPMC can be modeled by the equations [111]

$$\begin{aligned} \dot{v} &= -\frac{v + Y(v)(R_a + R_c) - u}{[C_1(v) + C_a(v)](R_a + R_c)} \\ y &= \frac{3\alpha_0 \kappa_e(\sqrt{2\Gamma(v)} - v)}{Y_e H^2} \end{aligned} \tag{5.114}$$

where v is the state variable, u is the control input voltage, y is the curvature output, R_a is the electrode resistance, R_c is the ion diffusion resistance, α_0 is the coupling constant, Y_e is the equivalent Young's modulus of IPMC, and κ_e is the effective dielectric constant of the polymer. Assume $\Gamma(v)$, $C_1(v)$, and $C_a(v)$ are functions of the state variable and some parameters:

$$\Gamma(v) = \frac{b}{a^2}\left[\frac{ave^{-av}}{1 - e^{-av}} - ln\left(\frac{ave^{-av}}{1 - e^{-av}}\right) - 1\right] \tag{5.115}$$

where

$$a = \frac{F(1 - C^- \Delta V)}{RT} \qquad b = \frac{F^2 C^-(1 - C^- \Delta V)}{RT\kappa_e} \tag{5.116}$$

and F is Faraday's constant, C^- is the anion concentrations, ΔV is the volumetric change, R is gas constant, and T is absolute temperature. Then

$$C_1(v) = S\kappa_e \frac{\dot{\Gamma}(v)}{\sqrt{2\Gamma(v)}} \tag{5.117}$$

Here, $S = WL$ is the surface area of the IPMC and L, W, and H denote the length, width and thickness of the IPMC, respectively. Assume that

$$C_a(v) = \frac{q_1 SF}{RT} \frac{K_1 C^{H^+} e^{-vF/(RT)}}{(K_1 C^{H^+} + e^{-vF/(RT)})^2} \tag{5.118}$$

where $K_1 = k_1/k_{-1}$, k_1 and k_{-1} are the chemical rate constants for the forward and reverse directions of the electrochemical surface process, q_1 is some constant, and C^{H+} is the concentration of the hydrion H^+. Also assume that

$$Y(v) = Y_1 v + Y_2 v^2 + Y_3 v^3 \tag{5.119}$$

where Y_1, Y_2, and Y_3 are the coefficients of the polynomial.

Problem Statement The IPMC dynamic model described above has a comprehensive knowledge of the physical system derivation and is accurate mathematically. However, it is difficult to be adopted in practice because it is difficult to accurately identify some of the physical parameters. Moreover, some physical parameters are small enough and without the influence for the dynamic model. As a result, in this section, a practical nonlinear model is obtained. In the following we will explain how to obtain the practical nonlinear model based on the above dynamic model.

In general, ΔV is small enough in (5.116) and C^- is a bound constant. Then $|C^- \Delta V| \to 0$, and the parameters a and b in (5.116) can be calculated approximately by the equations

$$a \approx \frac{F}{RT} \qquad b \approx \frac{F^2 C^-}{RT\kappa_e} \tag{5.120}$$

The IPMC can operate in a humid environment or a dry environment. In this section, the IPMC setup is investigated in a dry environment. Then $C^{H+} \to 0$, so

$$C_a(v) \approx 0 \tag{5.121}$$

In (5.119), Y_1, Y_2, and Y_3 are small enough, and $|Y(v)| \ll |v|$. So, in (5.114), because R_a and R_c are bounded, $Y(v)$ can be ignored and considered model error. In addition, physical constants such as T, L, W, H, R_a, and R_c must be measured or identified by experiment and will also create error. Therefore, the nonlinear model

$$\dot{v} = -\frac{v - u}{C_1(v)(R_a + R_c)} \qquad y = \frac{3\alpha_0 \kappa_e \sqrt{2\Gamma(v)}}{Y_e H^2} + \Delta P \tag{5.122}$$

is obtained, where ΔP is the uncertainty consisting of the parameter error and model error of the IPMC.

Substituting (5.115), (5.117), and (5.120) into (5.122), the following nonlinear dynamic IPMC model is obtained:

$$\dot{v} = -\frac{(v-u)\sqrt{2b\left[\dfrac{ave^{-av}}{1-e^{-av}} - \ln\left(\dfrac{ave^{-av}}{1-e^{-av}}\right) - 1\right]}}{S\kappa_e b(R_a+R_c)\left(1-\dfrac{1-e^{-av}}{ave^{-av}}\right)\dfrac{e^{-av}(1-e^{-av}-av)}{(1-e^{-av})^2}}$$
$$y = \frac{3\alpha_0\kappa_e\sqrt{2b\left[\dfrac{ave^{-av}}{1-e^{-av}} - \ln\left(\dfrac{ave^{-av}}{1-e^{-av}}\right) - 1\right]}}{aY_eH^2} + \Delta P \tag{5.123}$$

Defining a new state variable $x = av$, the above nonlinear dynamic model can also be described by the equations

$$\dot{x} = -\frac{(x-au)\sqrt{2b\left[\dfrac{xe^{-x}}{1-e^{-x}} - \ln\left(\dfrac{xe^{-x}}{1-e^{-x}}\right) - 1\right]}}{S\kappa_e b(R_a+R_c)\left(1-\dfrac{1-e^{-x}}{xe^{-x}}\right)\dfrac{e^{-x}(1-e^{-x}-x)}{(1-e^{-x})^2}}$$
$$y = \frac{3\alpha_0\kappa_e\sqrt{2b\left[\dfrac{xe^{-x}}{1-e^{-x}} - \ln\left(\dfrac{xe^{-x}}{1-e^{-x}}\right) - 1\right]}}{aY_eH^2} + \Delta P \tag{5.124}$$

For IPMC actuators, hysteresis is one of the most important properties. With regard to the existence of hysteretic nonlinearities, the system usually exhibits undesirable oscillations and even instability. Moreover, to ensure safety and longer service life of IPMC, the process input is subject to a constraint on its magnitude. Considering hysteresis, uncertainties, and input constraints, a nonlinear robust control design using operator-based robust right coprime factorization is studied, so that the validity of the obtained nonlinear model and the effectiveness of the control method can be confirmed.

The PI hysteresis model is used to describe the hysteresis of IPMC. The PI model is formulated through a weighted superposition of elementary hysteresis operators, that is, stop hysteresis operators and play hysteresis operators [78, 93, 107, 115]. Here, a PI model defined by play operators with threshold value $h > 0$ is used. Analytically, suppose that $C_m[0; t_E]$, where $0 = t_0 < t_1 < \cdots < t_N = t_E$ is a partition of $[0; t_E]$, is the space of piecewise monotone continuous functions. The input function $u(t) \in C_m[0; t_E]$ is monotone on each of the subintervals $[t_i; t_{i+1}]$. The play hysteresis operator $F_h(\cdot; u^*_{-1}) : C_m[0; t_E] \times u^*_{-1} \mapsto Cm[0; t_E]$ for an initial value $u^*_{-1} \in R$ is defined as

$$\begin{aligned} F_h[u(0); u^*_{-1}] &= f_h(u(0), u^*_{-1}) \\ F_h[u(t); u^*_{-1}] &= f_h(u(t), F_h[u(t_i); u^*_{-1}]) \end{aligned} \tag{5.125}$$

for $t_i < t \leq t_{i+1}$ and $0 \leq i \leq N-1$ with

$$f_h(u, q) = \max(u - h, \min(u + h, q)) \tag{5.126}$$

The spaces of input and output can be extended to the space $C[0; t_E]$ of continuous functions. Here the play operator is represented in another format [115]:

$$F_h(u)(t) = \begin{cases} u(t) + h & u(t) \leq F_h(u)(t_i) - h \\ F_h(u)(t_i) & -h < u(t) - F_h(u)(t_i) < h \\ u(t) - h & u(t) \geq F_h(u)(t_i) + h \end{cases} \tag{5.127}$$

where $F_h(u)$ shows $F_h(u; u^*_{-1})$ in brief. The initial condition of (5.127) is given by $F_h(u)(0) = \max(u(0) - h; \min(u(0) + h; u^*_{-1}))$. In the chapter, the robust nonlinear control design to an IPMC with hysteresis is investigated using the operator-based approach. The operator-based PI model can be described by a weighted superposition of the above play operator with threshold value Ψ:

$$u^*(t) = \Upsilon(u)(t) = D_{\text{PI}}(u)(t) + \Delta(u)(t) \tag{5.128}$$

where

$$\begin{aligned} D_{\text{PI}}(u)(t) &= Ku(t) \\ K &= \int_0^{h_x} p(h)\,dh \qquad p(h) = a \times e^{b(h-1)^2} \\ \Delta(u)(t) &= -\int_0^{h_x} S_n h p(h)\,dh + \int_{h_x}^{\Psi} p(h) F_h(u)(t_i)\,dh \\ S_n &= \text{Sgn}[u(t) - F_h(u)(t_i)] \end{aligned}$$

and on $[0, \Psi]$ we can find h_x to make $h_x \leq | u(t) - F_n(u(t_i)) |$, where $h \in [0, h_x]$. If $h_x = 0$, $D_{\text{PI}}[u(t)] = 0$. Assume PI$(\cdot)$ is the hysteresis operator and $p(h)$ is a given continuous density function satisfying $p(h) \geq 0$ with $\int_0^\infty hp(h)\,dh < \infty$. With the defined density function, P_I maps $C[t_0, \infty)$ into $C[t_0, \infty)$ [115]. Further, the operators Δ and D_{PI} also map $C[t_0, \infty)$ into $C[t_0, \infty)$. Since the density function $p(h)$ vanishes for large values of h, we assume that there exists a constant Ψ such that $p(h) = 0$ for $h > \Psi$. Since the original plant is preceded by hysteresis, the plant can be represented as $P = ND^{-1}(u^*)(t) = ND^{-1}[P_I(u)(t)]$, where $u(t)$ is the control input. It should be noted that the hysteresis behavior is decomposed into two terms by (5.128). The first term describes the invertible part, and the second term describes the hysteretic behavior as a disturbance. This decomposition is crucial since it facilitates the utilization of the robust right coprime factorisation for the controller design. So the control objective is to design operator-based controllers that stabilize the above system with hysteresis and realize the desired output tracking performance. How to identify the density function $p(h)$ will be given in the experimental results.

Considering the nonlinear system with bounded uncertainties, the robust control problem using robust right coprime factorization has been researched. Assume that

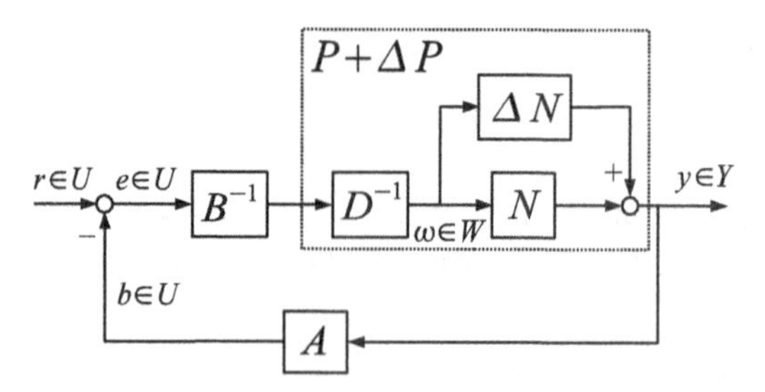

FIGURE 5.45 Nonlinear system with uncertainties based on right coprime factorization.

the uncertainties are given as ΔP, where ΔP is unknown but bounded. The right factorization of the nonlinear system is in the form

$$\tilde{P} = P + \Delta P = (N + \Delta N)D^{-1} \tag{5.129}$$

From [2], we can see that if the conditions

$$\begin{aligned} AN + BD &= L \\ A(N + \Delta N) + BD &= \tilde{L} \\ \|(A(N + \Delta N) - AN)L^{-1}\| &< 1 \end{aligned} \tag{5.130}$$

are satisfied, then the stability of the uncertain system $\tilde{P}$ is guaranteed, where L and $\tilde{L}$ are unimodular operators and $\|\cdot\|$ is the Lipschitz operator norm, shown in Figure 5.45.

Then, we consider the mentioned nonlinear IPMC control model using robust right coprime factorization. For the model described by equation (5.124), there exist some uncertainties ΔP in the IPMC. The uncertainties are unknown but bounded. In Figure 5.45, the uncertainties can be transformed into the uncertain operator ΔN. That is, the uncertain operator ΔN denotes the uncertainties caused by the approximate calculation.

Denote N, D, and ΔN as

$$N(\omega)(t) = \frac{3\alpha_0\kappa_e\sqrt{2b\left[\dfrac{\omega(t)e^{-\omega(t)}}{1-e^{-\omega(t)}} - \ln\left(\dfrac{\omega(t)e^{-\omega(t)}}{1-e^{-\omega(t)}}\right) - 1\right]}}{aY_eH^2} \tag{5.131}$$

$$D(\omega)(t) = \frac{S\kappa_e b(R_a + R_c)\dot{\omega}(t)\left(1 - \dfrac{1-e^{-\omega(t)}}{\omega(t)e^{-\omega(t)}}\right)\dfrac{e^{-\omega(t)}[1 - e^{-\omega(t)} - \omega(t)]}{(1-e^{-\omega(t)})^2}}{a\sqrt{2b\left[\dfrac{\omega(t)e^{-\omega(t)}}{1-e^{-\omega(t)}} - \ln\left(\dfrac{\omega(t)e^{-\omega(t)}}{1-e^{-\omega(t)}}\right) - 1\right]}} + \frac{\omega(t)}{a} \tag{5.132}$$

$$\Delta N(\omega)(t) = \Delta\sqrt{2b\left(\frac{\omega(t)e^{-\omega(t)}}{1-e^{-\omega(t)}} - \ln\left(\frac{\omega(t)e^{-\omega(t)}}{1-e^{-\omega(t)}}\right) - 1\right)} \tag{5.133}$$

To ensure the safety and longer service life of the IPMC, the process input $u_d(t)$ is subject to the following constraint on its magnitude:

$$u_d(t) = \sigma(u_1(t))$$
$$\sigma(v) = \begin{cases} u_{\max} & v > u_{\max} \\ v & u_{\min} \leq v \leq u_{\max} \\ u_{\min} & v < u_{\min} \end{cases} \tag{5.134}$$

where $u_1(t)$ is the control input before the constraint. Assume $u_{\max} = 3$ V and $u_{\min} = -3$ V are the maximum and minimum voltages to ensure safe operation of the IPMC, respectively. When the input is limited in (5.134), the limited part can be equivalent to the uncertainty of the system. Then, the entire uncertainty of the system is expressed in the form

$$\Delta\tilde{N} : W \rightarrow Y$$

Then, we can design operators A and B to satisfy the following Bezout equations. If $-u_{\max} \leq u_1 \leq u_{\max}$,

$$A_1 N + BD = I \qquad \|A_1(N + \Delta N) - A_1 N)\| < 1 \tag{5.135}$$

Otherwise

$$A_2 N + BD = I \qquad \|(A_2(N + \Delta\tilde{N}) - A_2 N)\| < 1 \tag{5.136}$$

where operators A_1 and A_2 are stable and B is invertible. Therefore, for the case of the IPMC control system with constraint inputs, we suppose that

$$B(u_d)(t) = a u_d(t) \tag{5.137}$$

According to the robust stable conditions, if $-u_{\max} \leq u_1 \leq u_{\max}$,

$$A_1(y)(t) = -\frac{aSY_e H^2(R_a + R_c)}{3\alpha_0}\dot{y}(t) \tag{5.138}$$

Otherwise

$$A_2(y)(t) = -\frac{aSY_e H^2(R_a + R_c)}{3\alpha_0}\dot{y}(t)\phi[\sigma(u_1)] \tag{5.139}$$

where $\phi(\cdot)$ is a constraint function related to $\sigma(u_1)$.

The research here is concerned with nonlinear control design with hysteretic nonlinearities. That is, the considered nonlinear plant is preceded by hysteresis (see

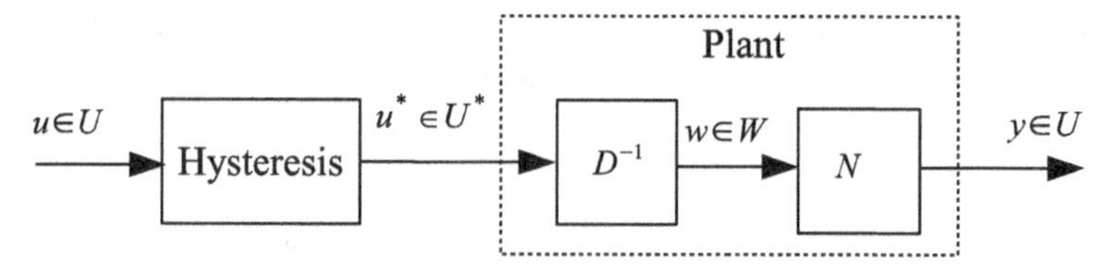

FIGURE 5.46 Nonlinear plant preceded by hysteresis nonlinearities.

Figure 5.46). Since the original plant is preceded by hysteresis, the plant can be represented as

$$P = ND^{-1}(u^*)(t) = ND^{-1}[P_I(u)(t)] \tag{5.140}$$

where $u(t)$ is the control input. It should be noted that the hysteresis behavior is decomposed into two terms by (5.128). The first term describes the invertible part, and the second term describes the hysteretic behavior as a disturbance. This decomposition is crucial since it facilitates the utilization of robust right coprime factorization for the controller design. So the objective is to design operator-based controllers that stabilize the above system with hysteresis.

Considering the nonlinear system with bounded uncertainties and hysteresis, based on robust right coprime factorization, robust stable control for a nonlinear model is designed. The framework of the robust control system with uncertainties and hysteresis is shown in Figure 5.47. Assume U is the input space of the PI model and U^* is the output spaces of the PI model. Let the output space of the original nonlinear plant and quasi-state space be Y and W; N, ΔN, and D are $N : W \rightarrow Y$, $\Delta N : W \rightarrow Y$, and $D : W \rightarrow U^*$, respectively; and A, B are the stable operator controllers and B is invertible. We can choose $W = U$. In Figure 5.47, the output u^* of the PI model is the input of the original nonlinear plant. In practice, the signal of u^* between the hysteresis and the nonlinear plant is unavailable. So the hysteresis of the actuator should be considered as a part of the nonlinear plant during controller design. We consider $D_{\text{PI}}(u)(t)$ in the hysteresis model given in (5.128) and D of the above plant

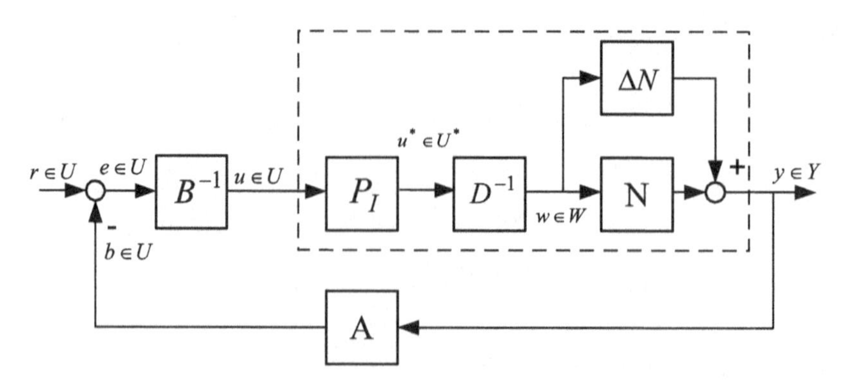

FIGURE 5.47 Robust control system with uncertainties and hysteresis.

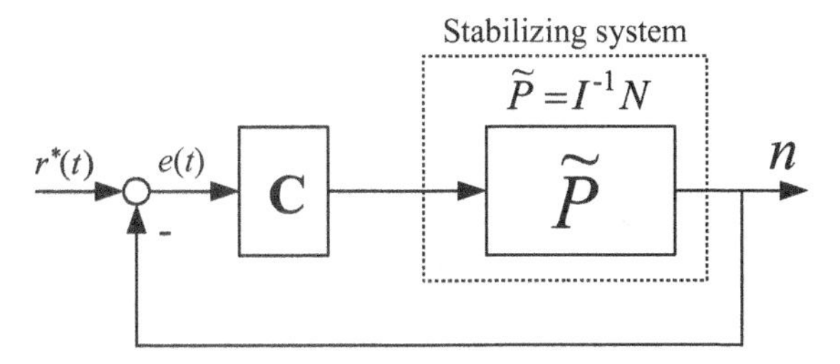

FIGURE 5.48 Tracking control system.

as a new invertible factor of right coprime factorization, $\tilde{D}(\omega)(t) = D_{\rm PI}^{-1}D(\omega)(t)$. The corresponding controllers A and B which satisfy the Bezout identity $AN + B\tilde{D} = M \in S(W, U)$ are also designed to make the nonlinear system BIBO stable, where $S(W, U)$ is the set of unimodular operator.

Based on the design scheme, D, N, and ΔN are denoted as equations (5.131)–(5.133), and the controllers are designed as follows:

$$A_1(y)(t) = -\frac{aSY_eH^2(R_a + R_c)}{3\alpha_0}\dot{y}(t) \tag{5.141}$$

$$A_2(y)(t) = -\frac{aSY_eH^2(R_a + R_c)}{3\alpha_0}\dot{y}(t)\phi[\sigma(u_1)] \tag{5.142}$$

$$B(u)(t) = aKu_d(t) \tag{5.143}$$

where $K = \int_0^{h_x} p(h)\,dh$.

In addition to guaranteeing the robust stability of the IPMC system, the tracking performance of the system must also be considered. Here, the tracking condition is difficult to obtain as the operator N is a complex nonlinear function such that we design the tracking system given in Figure 5.48, where the stabilizing system regarded as the plant is equal to the system in Figure 5.45 or 5.47. The tracking controller C is designed satisfying the conditions in [2] and is shown as

$$u(t) = k_pe(t) + k_i\int e(\tau)\,d\tau \tag{5.144}$$

The hysteretic properties of IPMC using experimental data are introduced, and some simulation and experimental results are given to illustrate the effectiveness of the nonlinear robust control design method. Figure 5.49 is a photograph of the experimental setup an shows an IPMC sample of dimensions 50 mm 10 mm 0.2 mm clamped at one end and subject to voltage excitation generated from the computer and board (PCI-3521). A laser displacement sensor (ZX-LD40: 40±10 mm) is used to measure the bending displacement d.

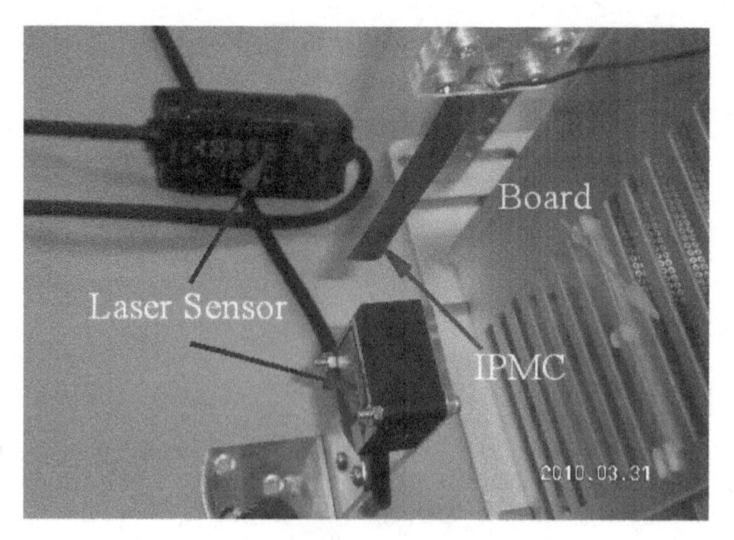

FIGURE 5.49 Photograph of experimental setup.

The displacement response of an IPMC is measured under a sequence of voltage values which are monotonically from −2.0 to 2.0 V and then back to −2.0 V (see Figure 5.50). Each voltage value v is held for 100 sec to guarantee that the IPMC reaches the steadystate, which ensures that the effects of other dynamics are eliminated or minimized and any measured output–input loop would indeed come from hysteresis. It is evident from the measured steady-state displacement versus the voltage input (see Figure 5.51) that the output–input relationship shows hysteresis.

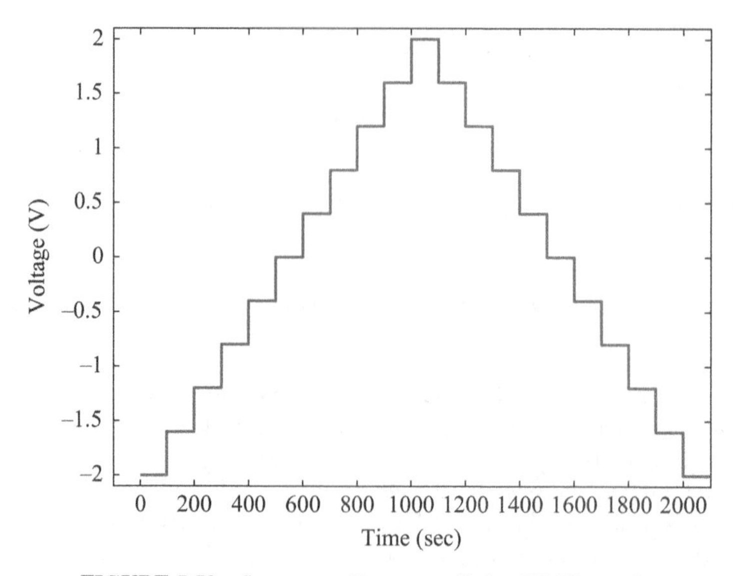

FIGURE 5.50 Sequence of inputs applied to IPMC sample.

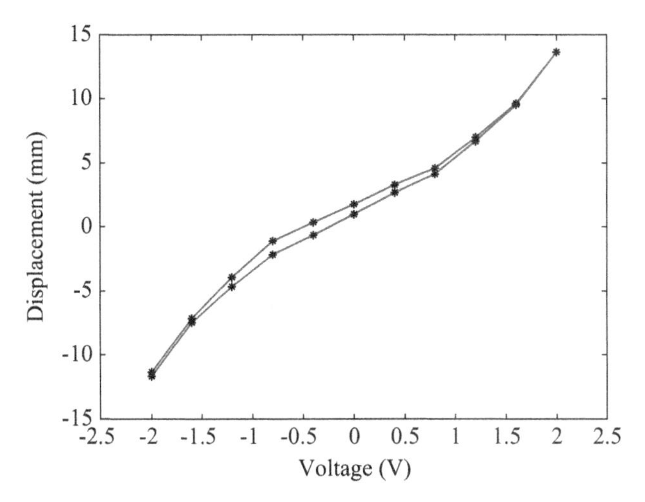

FIGURE 5.51 Steady-state displacement *vs* voltage input.

To identify the density function parameters in a density function $p(h) = a \times e^{b(h-1)^2}$, an experiment is conducted. In the experiment, input voltage $u(t) = 3\,\sin(2\pi f t)$ $(f = 0.1, 0.2, 0.4, 0.8)$ is applied to the IPMC actuator. Using the method in [107], the density function $p(h) = 0.00024 \times e^{-0.00075(h-1)^2}$ and $h \in [0, 3]$ of the PI model are identified. Figure 5.52 shows the input and output of the identified PI hysteresis model. The response of the identified PI hysteresis model is shown in Figure 5.53.

Some identified physical parameters are shown in Table 5.4. In the simulation, the uncertain factor in (5.133) is modeled as $\Delta = [3\alpha_0\kappa_e\sqrt{2b}/(aY_e h^2)]\times 5\%$. In

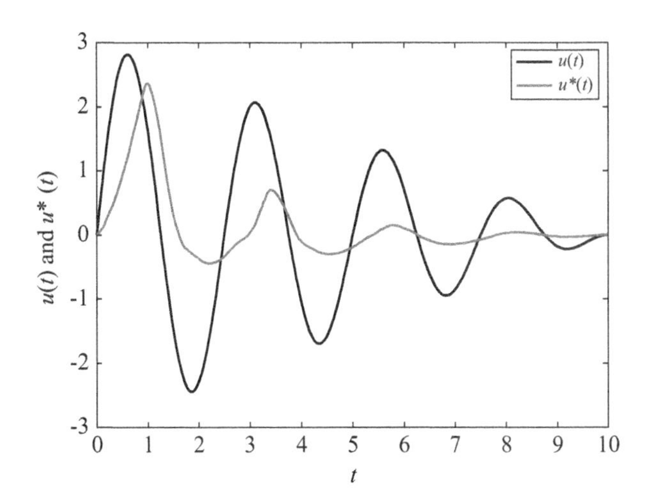

FIGURE 5.52 Input and output of identified PI hysteresis model.

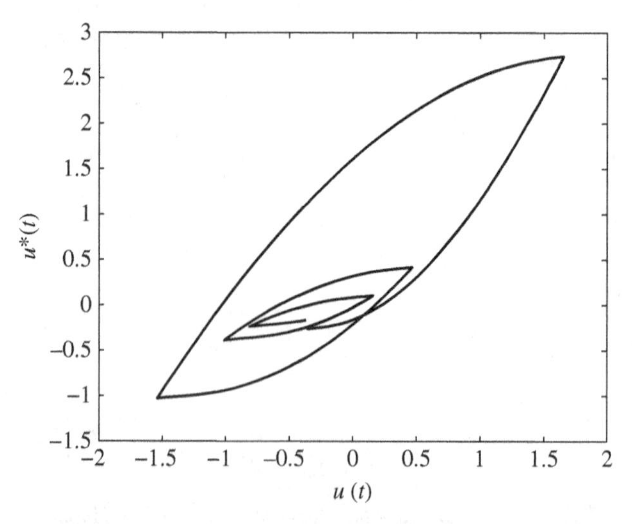

FIGURE 5.53 Response of identified PI hysteresis model.

fact, the uncertainty of the model is smaller than ΔN, so the robust stability of the system can be guaranteed. The curvature control simulation results of the IPMC based on right coprime factorization with and without uncertainties are shown in Figure 5.54, respectively. From Figure 5.54, we can see that the nonlinear IPMC with uncertainties using right coprime factorization is robust stable. Figures 5.55 and 5.56 show the simulation results without and with input constraints, respectively, where the reference input of the curvature is $r_f = 1\ \text{m}^{-1}$ and the tracking controller is given as

$$u(t) = 50e(t) + 0.000015 \int e(\tau)\,d\tau \tag{5.145}$$

From Figure 5.54, we can find that the IPMC control output can track the reference input using the tracking controller. In this book, a quantitative tracking performance analysis is omitted. It should be taken using Theorem 3.1.

TABLE 5.4 Parameters in IPMC

T	F	κ_e
290 K	96487 C mol^{-1}	1.12×10^{-6}Fm^{-1}
R_a	R	Y_e
18 Ω	8.3143 J mol^{-1}K^{-1}	0.56 GPa
R_c	°C^{-}	α_0
60 Ω	980 mol	0.12 J °C^{-1}
L	W	H
50 mm	10 mm	200 μm

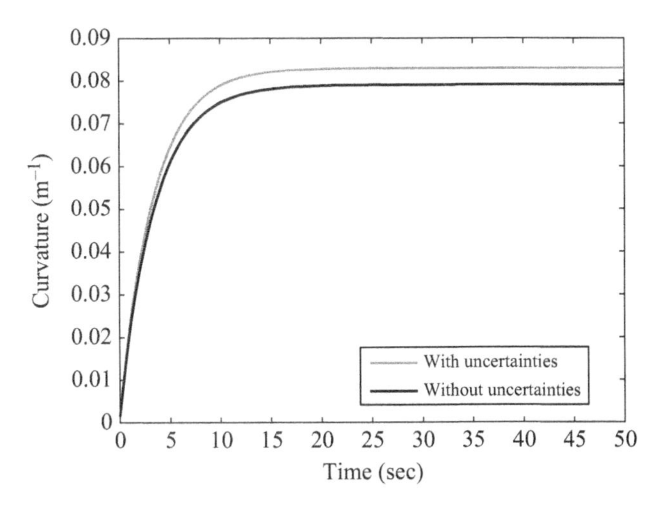

FIGURE 5.54 Curvature control simulation results based on robust coprime factorization.

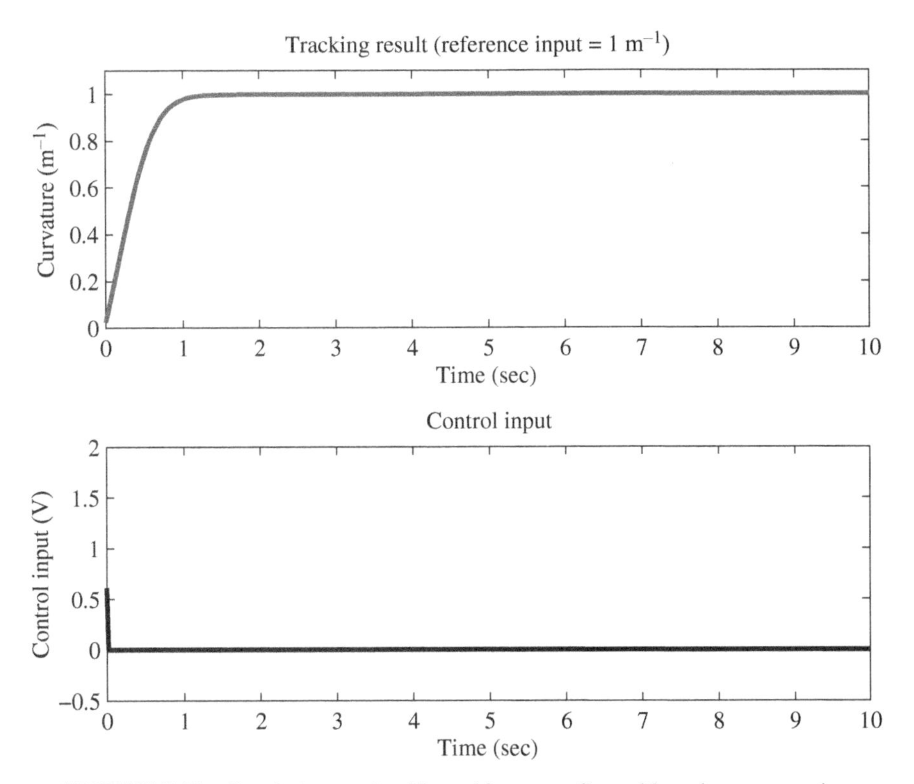

FIGURE 5.55 Simulation result with tracking controller, without input constraint.

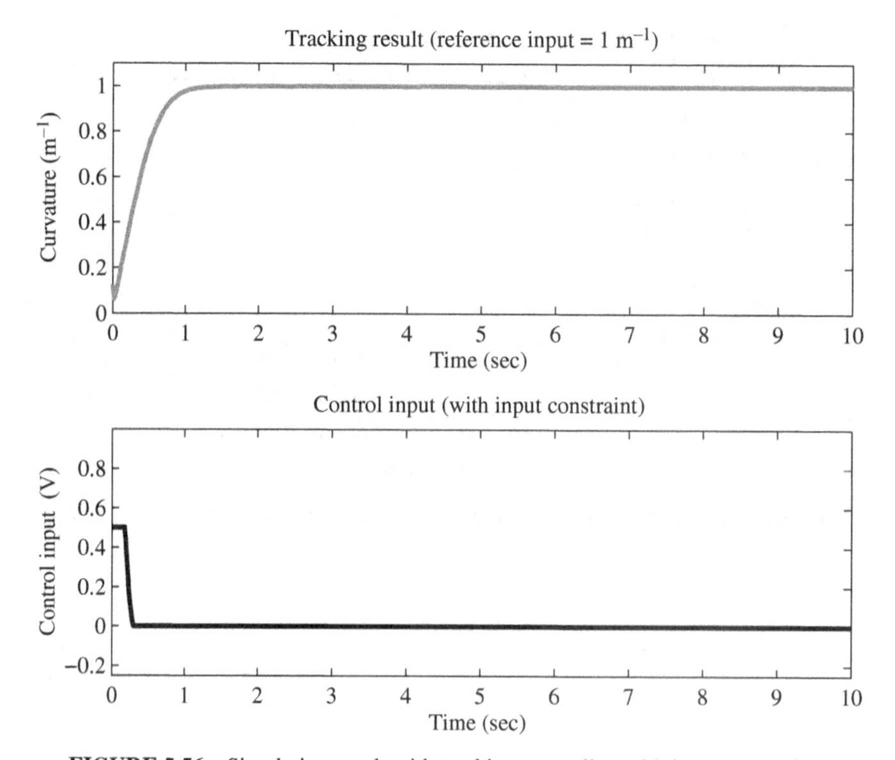

FIGURE 5.56 Simulation result with tracking controller, with input constraint.

Based on curvature output, the displacement can be calculated. The relationship between the bending curvature $1/\rho$ and the displacement d can be described by the equation (see Figure 5.57)

$$\frac{1}{\rho} = \frac{2d}{d^2 + h^2} \tag{5.146}$$

where h is the vertical distance.

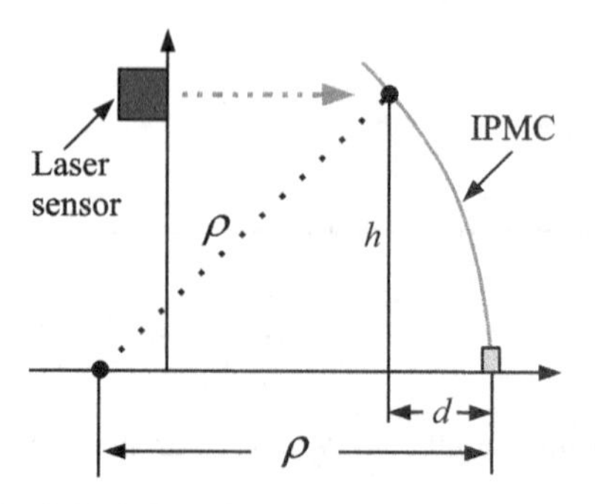

FIGURE 5.57 Calculation of displacement d based on curvature $1/\rho$.

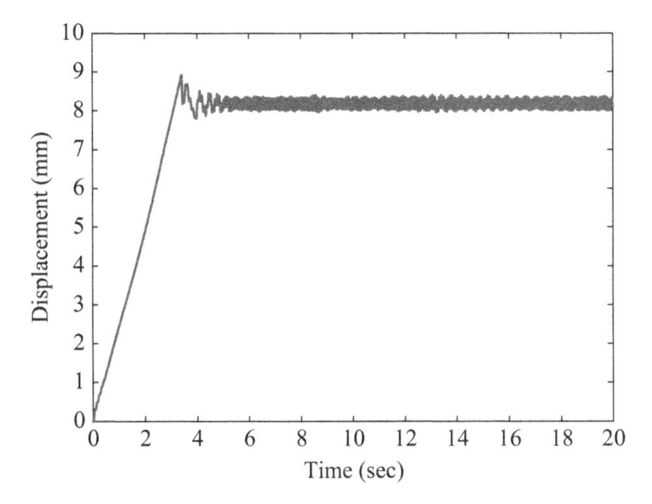

FIGURE 5.58 Experimental result on output response without considering hysteretic properties.

Figure 5.58 shows the displacement response without considering hysteretic properties in IPMC, and Figure 5.59 shows the displacement response considering hysteretic properties in IPMC, where the desired output of displacement d is 8 mm. The results show that the robust stability of the IPMC displacement control system is guaranteed and tracking performance is satisfied using the method. The oscillations in the system responses can be avoided by considering the hysteresis of the IPMC actuator.

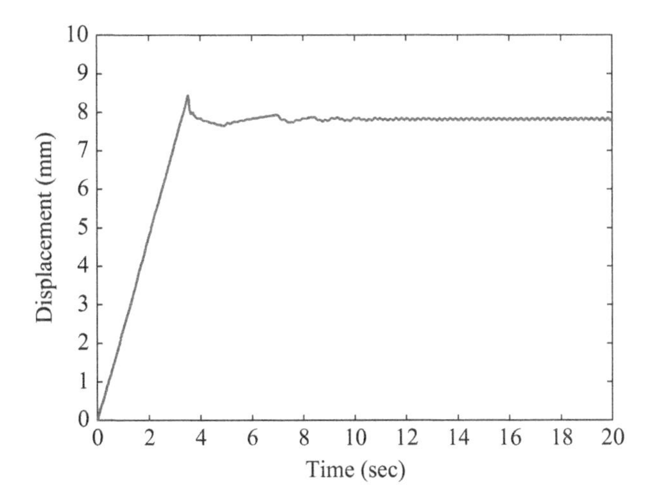

FIGURE 5.59 Experimental result on output response considering hysteretic properties.

5.3.4 Summary

Three kinds of smart actuators were introduced. The simplified, symmetric PI hysteresis model was employed instead of the model in Section 5.2. However, for the symmetric PI hysteresis model with input limitations, a design scheme for IPMC was shown.

CHAPTER 6

Application of Operator-Based Nonlinear Feedback Control to Large-Scale Systems Using Distributed Control System Device

In this chapter, an operator-based nonlinear distributed control system (DCS) device used to control large-scale industries is developed. To control water level, temperature, and water flow, robust nonlinear control systems using robust right coprime factorization are realized using the DCS device. Simulation and experimental studies are also given.

6.1 INTRODUCTION

In large-scale industrial process control systems, controlled variables are often liquid level, flow rate, temperature, pressure, and so on, most of which are nonlinear functions of the control inputs. That is, most industrial processes are nonlinear dynamic processes. For nonlinear processes, the application of existing linear methods to solve control problems has been studied by some researchers because there exist many effective analysis and design techniques for linear time-invariant processes. However, to use the linear control techniques, in general, nonlinear processes need to be approximated to obtain the linearization model, such as the transfer function [116, 117]. Hence, it may lead to inaccurate, complex implementation, consumed excessive energy, and thereby increased cost in an effort to maintain robust stability. On the other hand, nonlinear control design methodologies have attracted much attention due to, for example their simpler implementation, increased speed, and decreased control energy [118–120]. Until now, there exist some well-developed nonlinear control approaches for nonlinear processes, such as gain scheduling, feedback linearization, Lyapunov redesign, backstepping, and sliding-mode control [116–120]. Concerning the existing methods, each nonlinear control technique has its own advantages and limitations. Most of the techniques for nonlinear design assume

Operator-Based Nonlinear Control Systems: Design and Applications, First Edition. Mingcong Deng.

that measurements of all states are available [116–118]. Namely, all states have to be acquired in the modeling process such that the mathematical models are obtained to represent the nonlinear processes and some states are required to feed back to the designed nonlinear controllers. However, in practical control problems, all states are usually not measurable due to economical or technical reasons. As a result, the existing control design methods often do not generate satisfactory performance. To solve the problem, operator-based robust right coprime factorization is considered for nonlinear processes, where the input–output time function model given by basic physical rules from the real process is adopted [8, 17, 45, 121–126]. That is, approximation of the real process is avoided in operator-based robust right coprime factorization. Furthermore, it is easy to ensure the stability of nonlinear control systems using a Bezout identity. Especially, robust stability against uncertainties can be guaranteed under the inequality condition of Lemma 2.7.

In practical applications, robust stabilization design of nonlinear control processes has been investigated using robust right coprime factorization. For example, a water level experimental system with uncertainties was considered in [121]. Networked nonlinear control design was studied for an aluminum plate thermal process with time delays in Section 3.5. Fault diagnosis for a nonlinear process was discussed using operator-based control in Section 4.3. Applications of smart actuators are shown in Chapter 5. The above researches show operator-based robust right coprime factorization has been proved to be effective in theoretical studies and practical applications for nonlinear processes. However, to the best of our knowledge, almost no research of the method applied to an industrial process with unknown uncertainties using DCS device has been reported in the literature. Namely, the technique has not been investigated from a large-scale industrial application point of view. From the industrial viewpoint, only some simple calculation blocks, such as addition blocks, multiplication blocks, and division blocks, are included in the DCS device. However, there are often some nonlinear functions in designed controllers using operator-based robust right coprime factorization, such as the logarithm function and differential function, which cannot be implemented in a straightforward manner. As a result, we must realize approximately the designed right coprime factorization controllers with nonlinear functions, which will lead to some parasitic terms in the controllers. Therefore, the parasitic terms and the process unknown uncertainties should be considered simultaneously. Under the parasitic terms and the uncertainties, the robust stability of the nonlinear feedback control system may not be ensured or there may be a nonzero static error [121]. Namely, how to properly realize the designed controllers in the DCS device to guarantee the robust stability of nonlinear feedback control system and obtain satisfactory tracking performance is a significant issue. As a result, a robust stable condition is derived to guarantee the robust stability of the nonlinear feedback control system with the parasitic terms and uncertainties.

In general, an industrial process can be imitated by a multitank process. So, a multitank process consisting of water level and water flow techniques is set up in this chapter, which makes it nonlinear. Then, to guarantee the robust stability and obtain the satisfactory control performance for the multitank process, a robust nonlinear design method using operator-based robust right coprime factorization is given.

Further, from an industrial application viewpoint, a DCS (CENTUM CS3000) device is considered to realize the controllers using operator-based robust right coprime factorization. For the parasitic terms caused by the approximated realization of the controllers and the process unknown uncertainties, the robust stability of the control system is guaranteed by satisfying the derived robust stable condition. Robust stability of the system is analysed and the effectiveness of the design scheme is shown by experimental results.

6.2 MULTITANK PROCESS MODELING

The DCS-based multitank experimental system shown in Figure 6.1 consists of a water level process and a water flow process. The experimental system includes three parts: a supervision station, a field control station, and controlled processes. The supervision station is a human–machine interface (HMI) containing functions of the engineering station, the work station, and the remote monitor. The field control station

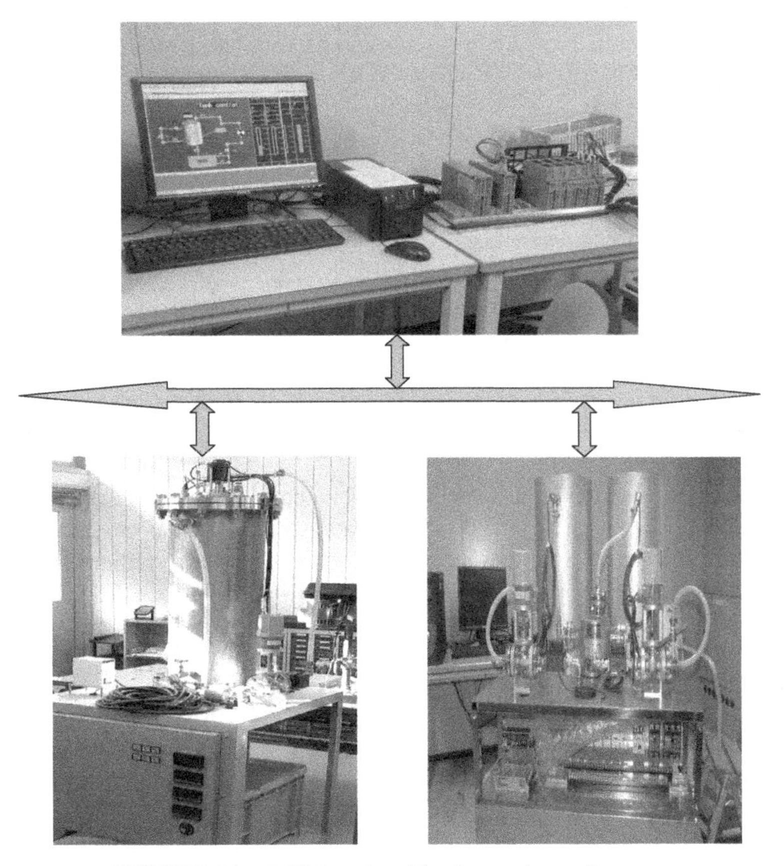

FIGURE 6.1 DCS-based multitank experimental system.

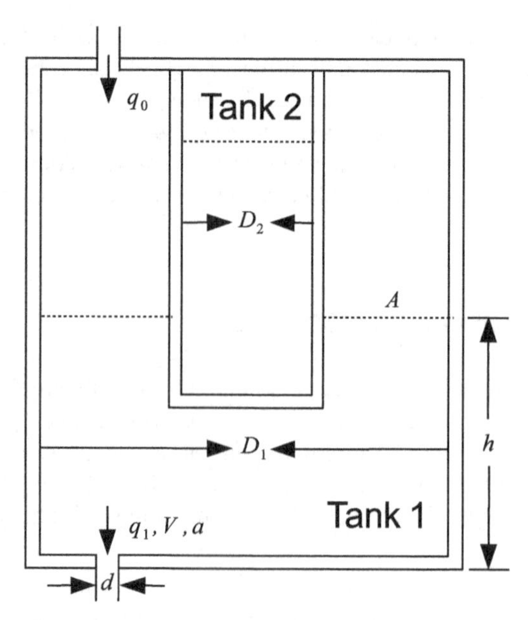

FIGURE 6.2 Equivalent diagram of single tank.

(FCS) is composed of a field control unit, input–output modules, and communication interfaces. A multitank process is regarded as controlled, including the water level process of the left single tank and the water flow process of the right twin tanks, respectively. In the two subprocesses, each input applies to the respective subprocess only as far as the decentralized control of the two subprocesses is considered [123]. In the water level process, a supersonic wave sensor is utilized to measure the water level of the single tank and inflow water is controlled to cause the water level to track to the reference input. For the water flow process, water flow is measured using a flow meter and the opening rate of an electric valve is regarded as the controlled variable. First, models of the water level process and the water flow process are set up as follows.

The equivalent diagram of the left single tank is shown in Figure 6.2. We consider that the water level changes from 30 to 72 cm because the water does not spill over from the tank and the supersonic wave sensor can work normally in this range.

The parameters' of the water level process are shown in Table 6.1. Then, sectional area A of tank 1 and area a of drain pipe are

$$A = \frac{D_1^2\pi}{4} - \frac{D_2^2\pi}{4} \qquad a = \frac{d^2\pi}{4}$$

Considering the mass balance of the water in tank 1, the following equation is obtained:

$$A\dot{h} = q_0 - aV \tag{6.1}$$

TABLE 6.1 Water Level Process Notations

Symbol	Definition	Unit
A	Tank cutting area	m^2
h	Water level in tank 1	m
a	Outflow entrance cutting area	m^2
P_a	Pressure in tank 1	hP_a
D_1	Diameter of tank 1	3.185×10^{-1} m
q_0	Inflow	m^3/min
D_2	Diameter of tank 2	1.143×10^{-1} m
q_1	Outflow	m^3/min
d	Diameter of outlet	1.2×10^{-2} m
ρ	Density of water	1.0×10^3 kg/m^3
V	Velocity of outflow	m/min
g	Gravitational acceleration	9.8 m/sec^2

Based on the Bernoulli theorem, the relationship between the water level and the flow is obtained as

$$\frac{q_0^2}{2gA^2} + h + \frac{P_a}{\rho g} = \frac{V^2}{2g} + \frac{P_a + \Delta P_a}{\rho g} \tag{6.2}$$

where $q0/A$ is the velocity of varying water levels and ΔP_a is the variation of the pressure. In the system, variation of the pressure P_a is very small such that ΔP_a can be assumed as zero. Then, the water level process is derived as

$$\dot{h} = \frac{1}{A} q_0 - \frac{a}{A} \sqrt{\frac{q_0^2}{A^2} + 2gh} \tag{6.3}$$

Consider that q_0 is much less than A in the water level process equipment such that $q_0^2/A^2 \approx 0$. For convenience of expression, we define that $u_{d1}(t) = q_0$ and $y_1(t) = h$. Then the water level process model is expressed in the form

$$\dot{y}_1(t) = \frac{1}{A} u_{d1}(t) - \frac{a}{A} \sqrt{2gy_1(t)} \tag{6.4}$$

where u_{d1} is control input and y_1 is process output.

According to the Bernoulli theorem, water flow through an electric valve can be described in the form [126]

$$Q = C_v f(x) \sqrt{\frac{\Delta P_v}{G_v}} \tag{6.5}$$

where the parameters are given in Table 6.2 and $f(x)$ is the flow characteristic, which is key to modeling the flow through the valve and depends on the kind of valve:

TABLE 6.2 Water Flow Subprocess Parameters

Symbol	Definition	Unit
Q	Volumetric flow rate	L/min
C_v	Flow coefficient of valve	L/min · psi
P_v	Pressure drop across valve	psi
G_v	Specific gravity of fluid	1.0
x	Valve opening rate	%

(1) $f(x) = x$ for linear valve control, (2) $f(x) = \sqrt{x}$ for quick-opening valve control, and (3) $f(x) = R_v^{x-1}$ for equal-percentage valve control, where R_v denotes the valve range (ratio of the maximum to minimum controllable flow rate).

In the designed system, an electric valve with equal-percentage flow characteristic is considered. Therefore, water flow through the valve is calculated as

$$Q = C_v R_v^{x-1} \sqrt{\frac{\Delta P_v}{G_v}} = \frac{C_v}{R_v} \sqrt{\frac{\Delta P_v}{G_v}} e^{(\ln R_v)x} \tag{6.6}$$

We define the input $u_{d2}(t) = x$ and the output $y_2(t) = Q$. The coefficients k and b are given as

$$k = \frac{C_v}{R_v} \sqrt{\frac{\Delta P_v}{G_v}} \qquad b = \ln R_v$$

Then, the water flow model can be reexpressed as

$$y_2(t) = k e^{b u_{d2}(t)} \tag{6.7}$$

Some basic definitions and notation for operator and right coprime factorization are first introduced.

Let U and Y be linear spaces over the field of real numbers and let U_s and Y_s be the normed linear subspaces, called the stable subspaces of U and Y, respectively. An operator is a mapping between the two spaces. For example, $Q : U \to Y$ is an operator mapping from U to Y. Assume that $\mathcal{D}(Q)$ and $\mathcal{R}(Q)$ are the domain and range of Q. Let Z be the family of real-valued measurable functions defined on $[0, \infty)$, which is a linear space. For each constant $T \in [0, \infty)$, let P_T be the projection operator mapping from Z to another linear space, Z_T, of measurable functions such that

$$f_T(t) := P_T(f)(t) = \begin{cases} f(t) & t \leq T \\ 0 & t > T \end{cases} \tag{6.8}$$

where $f_T(t) \in Z_T$ is called the truncation of $f(t)$. Let U^e and Y^e be two extended linear spaces associated respectively with two given Banach spaces U_B and Y_B of

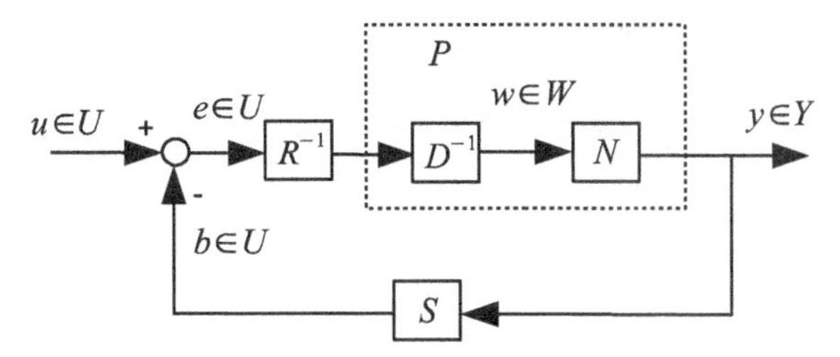

FIGURE 6.3 Nonlinear feedback control system.

measurable functions defined on the time domain $[0, \infty)$. Then, the operator $Q : U \to Y$ is said to be BIBO stable or, simply, stable if $Q(U_s) \subseteq Y_s$ and domain $\mathcal{D}(Q) \subseteq U^e$ and range $\mathcal{R}(Q) \subseteq Y^e$ [17]. In this chapter, nonlinear cases are considered and stable operators are always BIBO stable. It is worth mentioning that it is different operation between mapping operation and multiply operation. For example, a mapping relationship is expressed as $R_1(u_{d1})(t)$ when the time-varying variable $u_{d1}(t)$ is mapped by the operator R_1. When it is a multiply operation between the time-varying variable $u_{d1}(t)$ and the variable B_1, it is expressed as $B_1 u_{d1}(t)$.

A nonlinear feedback control system with a given process P is shown in Figure 6.3. The process operator P is said to have a right factorization composed of two operators N and D in the form $P = ND^{-1}$, where operators N and D^{-1} can be either linear or nonlinear. Suppose P is a nonlinear operator, for instance, both N and D are nonlinear operators in general. If, furthermore, the two operators N and D together satisfy the Bezout identity $SN + RD = L$ for some operators S and R, where L is a unimodular operator, then the right factorization is said to be coprime.

For the given process P, suppose that there are uncertainties ΔP, where ΔP is bounded. The right factorization of the uncertain process can be described as $P + \Delta P = (N + \Delta N)D^{-1}$. Right coprime factorization design of the uncertain process is shown in Figure 6.4. If the right coprime factorization of the uncertain process $P + \Delta P$ satisfies $SN + RD = L$ and $\|(S(N + \Delta N) - SN)L^{-1}\| < 1$, then the uncertain feedback control system is stable, where $\| \cdot \|$ is the Lipschitz operator norm and L the is unimodular operator.

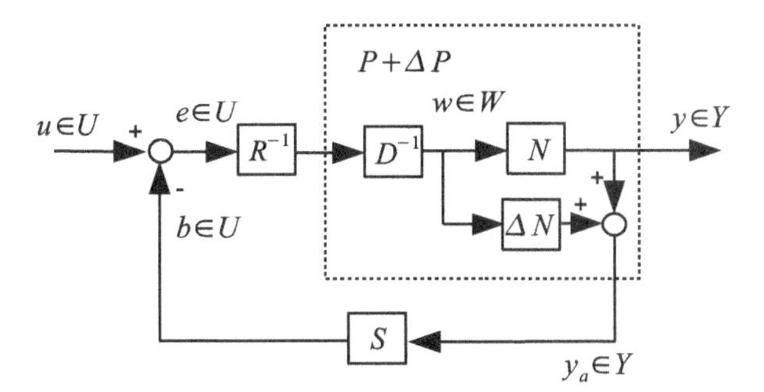

FIGURE 6.4 Nonlinear feedback control system with uncertainties.

The right coprime factorization design and robust stable condition discussed above are mostly investigated from a theoretical point of view. However, how to realize the right coprime factorization controllers and apply them to an industrial process using the DCS device has not been investigated from a large-scale industrial application point of view. In the DCS device, there are only some simple calculation blocks, such as the addition block, multiplication block, and division block. However, from the above water level and water flow processes, we can see that some nonlinear functions may exist in controllers using robust right coprime factorization, such as the logarithm function and differential function, which cannot be implemented in a straightforward way. How to properly realize these controllers in the DCS device to guarantee the robust stability of the nonlinear feedback control system and obtain satisfactory tracking performance is a significant issue. Therefore, in this chapter, operator-based robust right coprime factorization design and its realization are studied for the above multitank experimental process by using the DCS device.

6.3 ROBUST RIGHT COPRIME FACTORIZATION DESIGN AND CONTROLLER REALIZATION

Design of the multitank process using operator-based robust right coprime factorization is shown in Figure 6.5, where the multitank process consists of the water level process P_1 and the water flow process P_2 and ΔS_i is a parasitic term caused by the approximated realization, respectively. Denote the input space, output space, and quasi-state space as U, Y, W, respectively, and the signals of reference input, error, quasi-state, and process output as u, e, w, and y_a, respectively. The real process $\tilde{P}_i : U \rightarrow Y$ and nominal process P_i have right factorization $\tilde{P}_i = (N_i + \Delta N_i)D_i^{-1}$ and $P_i = N_i D_i^{-1}$ $(i = 1, 2)$, where N_i, ΔN_i, and D_i are stable and D_i^{-1} is invertible and the uncertainty of the process ΔN_i is unknown but the upper and lower bounds are known. Let S_i and R_i^{-1} represent the designed controllers. From the water level and water flow processes, we can see that some nonlinear functions may exist in the designed controllers. However, from an industrial viewpoint, only some simple calculation functions are included in the DCS device such that the nonlinear functions in the controllers cannot be implemented directly, such as the logarithm function and differential function. So, we must realize approximately the right coprime factorization controllers with nonlinear functions, which will lead to some parasitic terms. As a result, the parasitic terms and process uncertainties should be considered

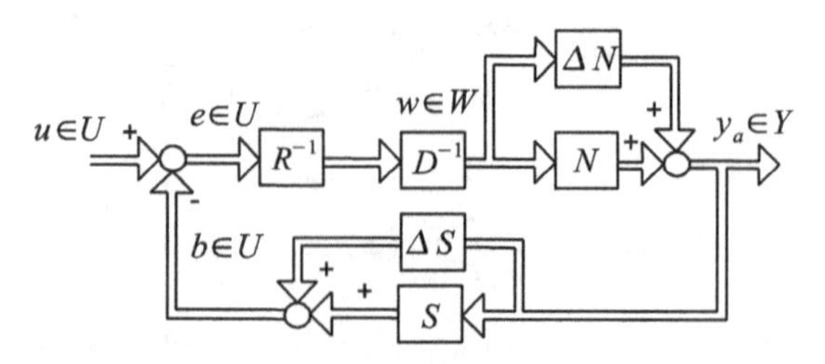

FIGURE 6.5 Robust right coprime factorization of the multitank process.

simultaneously. For simplicity, nonlinear functions can be placed in one controller such that another controller need not to be approximated. In this chapter, nonlinear functions are considered to be placed in controller S_i, which leads to parasitic term ΔS_i caused by approximated realization. In the real control system, because there are parasitic term ΔS_i and the uncertainty ΔN_i simultaneously, the robust stability condition might not be satisfied such that the robust stability of the system is not ensured. Therefore, a new robust stable condition is derived to guarantee the robust stability of the nonlinear feedback control system with the parasitic terms ΔS_i and the uncertainty ΔN_i.

Using robust right coprime factorization, the robust stability of the nonlinear feedback control system shown in Figure 6.5 can be guaranteed by the following theorem.

Theorem 6.1 The multitank process control system using operator-based robust right coprime factorization is shown in Figure 6.5. Assume that Bezout identities of the nominal process and the real process are $S_i N_i + R_i D_i = L_i \in \mathcal{U}(W, U)$ and $(S_i + \Delta S_i)(N_i + \Delta N_i) + R_i D_i = \tilde{L}_i$ $(i = 1, 2)$, respectively. Let $[(S_i + \Delta S_i)(N_i + \Delta N_i) - S_i N_i]L^{-1} \in \mathrm{Lip}(G)$, where G is a subset of U^e. If the condition

$$\|[(S_i + \Delta S_i)(N_i + \Delta N_i) - S_i N_i]L_i^{-1}\| < 1 \qquad i = 1, 2 \tag{6.9}$$

is satisfied, then the nonlinear feedback control system is robust stable, where L_i is the unimodular operator and $\|\cdot\|$ is the Lipschitz operator norm.

Proof In the system shown in Figure 6.5, S_i and R_i^{-1} denote designed controllers. Nonlinear functions are considered to be placed in the controller S_i such that the controller S_i is realized approximately, which leads parasitic term ΔS_i in the real system. Then, the feedback signal $b_i(t)$ is calculated as

$$\begin{aligned} b_i(t) &= (S_i + \Delta S_i)(N_i + \Delta N_i)(w_i)(t) \\ &= [S_i(N_i + \Delta N_i) + \Delta S_i(N_i + \Delta N_i)](w_i)(t) \qquad i = 1, 2 \end{aligned} \tag{6.10}$$

We define $[S_i(N_i + \Delta\tilde{N}_i)](w_i)(t) = [S_i(N_i + \Delta N_i) + \Delta S_i(N_i + \Delta N_i)](w_i)(t)$. Therefore, the parasitic term ΔS_i is equivalent to a part of the uncertainties $\Delta\tilde{N}_i$ shown in Figure 6.6, where b' and y'_a are the new feedback signal and process output,

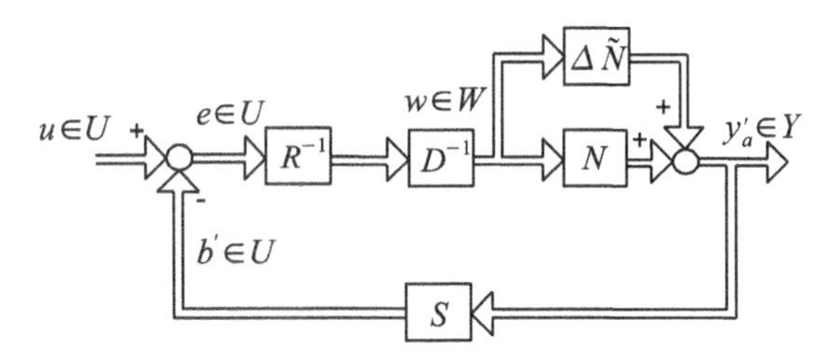

FIGURE 6.6 Equivalent diagram of Figure 6.5.

respectively. This indicates that the parasitic term ΔS_i is treated as the process uncertainty and $\Delta\tilde{N}_i$ is stable.

For any reference signal u_i, the control system has that

$$u_i(t) = [S_i(N_i + \Delta\tilde{N}_i)](w_i)(t) + R_i D_i(w_i)(t) \qquad i = 1, 2 \tag{6.11}$$

From $S_i N_i + R_i D_i = L_i$ and $(S_i + \Delta S_i)(N_i + \Delta N_i) + R_i D_i = \tilde{L}_i$, we have

$$\tilde{L}_i = L + [S_i(N_i + \Delta\tilde{N}_i) - S_i N_i] \qquad i = 1, 2 \tag{6.12}$$

Based on [2], it shows that $\tilde{L}_i \in \mathcal{U}(W, U)$, that is, the system shown in Figure 6.5 is well-posed. So, for any $u_i \in U_s$, it has that $w_i \in W_s$. Further, $y'_{ai} = (N_i + \Delta\tilde{N}_i)(w_i)$, $b'_i = S_i(y_{ai})'$ and $e_i = u_i - b'_i$. Thus the stability of S_i, N_i, $\Delta\tilde{N}_i$, R_i, and D_i implies that $y'_{ai} \in Y_s$, $b'_i \in U_s$, and $e_i \in U_s$. Then the system is stable. That is, under the condition (6.9), the robust stability of the overall nonlinear feedback control system can be guaranteed. ■

Note that in the above theorem all nonlinear functions are uncertainties and are assumed to be put into the controller S_i and the operator N_i in Figure 6.7. If the controller R_i also contains nonlinear functions and there exist the uncertainty ΔD_i, then $S_i N_i + R_i D_i = L_i$ and $\|[(S_i + \Delta S_i)(N_i + \Delta N_i) - S_i N_i + (R_i + \Delta R_i)(D_i + \Delta D_i) - R_i D_i]L_i^{-1}\| < 1$ are substituted for the robust stable condition. Furthermore, consider the process without uncertainties, namely, the model is exact, $\Delta N_i = 0$. Then, the robust right coprime factorization design of the exact process is shown in Figure 6.7. The Bezout identities of the nominal process and the real process are $S_i N_i + R_i D_i = L_i$ and $(S_i + \Delta S_i)N_i + R_i D_i = \tilde{L}_i$, respectively. Therefore, the robust stable condition is simplified as the form $\|\Delta S_i N_i L_i^{-1}\| < 1$ $(i = 1, 2)$. Further, according to a different controller definition, the condition (6.9) can be reselected. In this book, the discussion on this issue is omitted.

According to the design shown in Figure 6.5, right factorizations of the water level and water flow processes are given in the forms

$$y_1(t) = P_1(u_{d1})(t) = N_1 D_1^{-1}(u_{d1})(t)$$
$$y_2(t) = P_2(u_{d2})(t) = N_2 D_2^{-1}(u_{d2})(t)$$

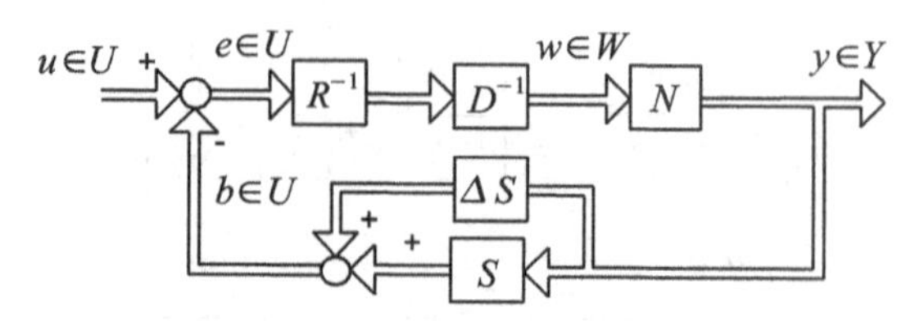

FIGURE 6.7 Robust right coprime factorization of exact process.

The right factorization operators of the nominal processes are calculated as

$$N_1(w_1)(t) = w_1(t) \qquad D_1(w_1)(t) = A\dot{w}_1(t) + a\sqrt{2gw_1(t)}$$

$$N_2(w_2)(t) = kw_2(t) \qquad D_2(w_2)(t) = \frac{1}{b}\ln w_2(t)$$

where operators N_1, N_2, D_1, and D_2 are stable under limited control input and D_1 and D_2 are invertible. According to the Bezout identities

$$S_iN_i + R_iD_i = L_i \quad \text{where } L_i = I_i \tag{6.13}$$

the right coprime factorization controllers are derived as

$$R_1(u_{d1})(t) = B_1u_{d1}(t) \qquad S_1(y_1)(t) = y_1(t) - B_1[A\dot{y}_1(t) + a\sqrt{2gy_1(t)}]$$

$$R_2(u_{d2})(t) = B_2u_{d2}(t) \qquad S_2(y_2)(t) = \frac{y_2(t)}{k} - \frac{B_2}{b}\ln\frac{y_2(t)}{k}$$

where B_1 and B_2 are designed controller parameters, $I_i(w_i)(t)$ is the identity operator, and operators R_1, S_1, R_2, and S_2 are stable under limited control input. It is worth mentioning that the initial state should also be considered, that is, $SN(w_0)(t_0) + RD(w_0)(t_0) = I(w_0)(t_0)$ should be satisfied. In this chapter, we select the initial time $t_0 = 0$ and $w_0 = w_0(t_0)$. In these controllers, there are some nonlinear functions, such as the differential function in controller S_1 and the logarithm function in controller S_2. In the chapter, right coprime factorization controllers are considered using the DCS device. In the DCS device, these functions cannot be realized directly. Therefore, they need to be realized approximately, which leads to the parasitic term ΔS_i in the real controller. The differential function and natural logarithm function are realized approximately using the inertia differential function and Taylor expansion, respectively. Robust stability is analyzed as follows.

In the right coprime factorization design of the water level process, R_1 can be implemented in a straightforward way but the differential function in controller S_1 cannot be realized directly. That is, the differential function must be realized approximately in practical control system design. Here, an inertia differential element using RC series current shown in Figure 6.8 is applied to realize the approximate

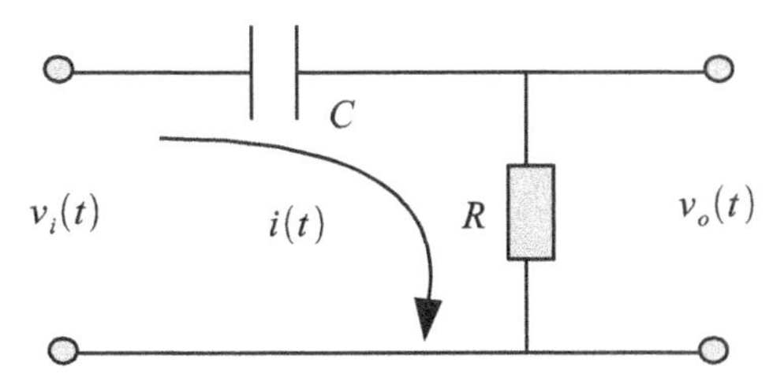

FIGURE 6.8 Circuit of approximate differential function.

differential function, where RC is defined as time constant T. Then, the approximate differential is expressed as

$$\frac{dv_i(t)}{dt}\,|_T = \frac{1}{T}v_o(t) = \frac{1}{T}v_i(t) - \frac{1}{T^2}e^{-t/T}\int v_i(\tau)e^{\tau/T}\,d\tau \tag{6.14}$$

where $v_i(t)$ and $v_o(t)$ denote the input and output of the differential function, respectively. So, the approximate differential of the water level is

$$\dot{y}_1(t) = \frac{1}{T}y_1(t) - \frac{1}{T^2}e^{-t/T}\int y_1(\tau)e^{\tau/T}\,d\tau \tag{6.15}$$

Then, the controller S_1 is realized as the following approximate form:

$$\tilde{S}_1(y_1)(t) = y_1(t) - B_1\left\{\frac{A}{T}y_1(t) - \frac{A}{T^2}e^{-t/T}\int y_1(\tau)e^{\tau/T}\,d\tau + a\sqrt{2gy_1(t)}\right\} \tag{6.16}$$

However, the approximated realization leads to the parasitic term ΔS_1 in the real controller. For the water level control system, the parasitic term ΔS_1 is calculated as

$$\Delta S_1(y_1)(t) = B_1 A\left\{\dot{y}_1(t) - \frac{1}{T}y_1(t) + \frac{1}{T^2}e^{-t/T}\int y_1(\tau)e^{\tau/T}\,d\tau\right\} \tag{6.17}$$

For the above parasitic term ΔS_1, under condition (6.9), the nonlinear feedback control system of the water level process is robust stable. Further, if the condition

$$\|\Delta S_1 N_1(w_1)\| = \left\|B_1 A\left(\dot{w}_1 - \frac{1}{T}w_1(t) + \frac{1}{T^2}e^{-t/T}\int w_1(\tau)e^{\tau/T}\,d\tau\right)\right\| < 1 \tag{6.18}$$

is satisfied, then the control system without water level process uncertainty is robust stable. In the right coprime factorization design of the water flow process, R_2 can be implemented in a straightforward manner but the natural logarithm function in controller S_2 cannot be realized directly. In this chapter, the approximate realization of the natural logarithm function is considered using the Taylor expansion. The Taylor expansion of the natural logarithm function $\ln y_2(t)$ in the neighborhood of $y_2(t) = 1$ is

$$\ln y_2(t) = [y_2(t) - 1] - \frac{[y_2(t) - 1]^2}{2} + E_n \tag{6.19}$$

where $y_2(t)$ belongs to the closed interval $(0, 2]$ and E_n is the remainder term

$$E_n = \frac{\ln^{(3)}(\xi)}{3!}(y_2(t) - 1)^3 \qquad \xi \in (0, 2] \tag{6.20}$$

It is noted that when $y_2(t)$ does not belong to the closed interval $(0, 2]$, we can transform it into this closed interval by dividing a constant. If we do not consider the remainder term, then the controller S_2 can be realized in the approximate form

$$\tilde{S}_2(y_2)(t) = \frac{y_2(t)}{k} - \frac{B_2}{2b}[4y_2(t) - y_2^2(t) - 3 - 2\ln k] \tag{6.21}$$

As in the water level process, the approximate realization leads to the parasitic term ΔS_2 in the real controller. For the water flow control system, the parasitic term is

$$\Delta S_2(y_2)(t) = \frac{B_2}{b}E_n \tag{6.22}$$

If the condition (6.9) is satisfied, then the nonlinear feedback control system of the water flow process with the parasitic term ΔS_2 is robust stable. Further, under the condition

$$\|\Delta S_2 N_2(w_2)\| = \left\| \frac{B_2 \ln^{(3)}\{kw_2\}}{6b}(kw_2 - 1)^3 \right\| < 1 \tag{6.23}$$

the robust stability of the water flow control system without the water flow process uncertainty is guaranteed.

Moreover, based on [7], the tracking control system shown in Figure 6.9 is designed to satisfy the following condition, where M denotes the tracking controller and the new reference signal is denoted by r. Between signals r and u, operator M is designed to realize tracking performance and space change from space Y to U:

$$(N + \Delta\tilde{N})\tilde{L}^{-1}M(r)(t) = I(r)(t) \tag{6.24}$$

For the water level and water flow processes, the approximate tracking controllers are given as

$$M_1(r_1)(t) = r_1(t) \qquad M_2(r_2)(t) = \frac{1}{k}r_2(t)$$

The main difference between this chapter and the results in [7] is that the parasitic terms caused by the approximate realization of the controllers and the process unknown uncertainties are considered simultaneously.

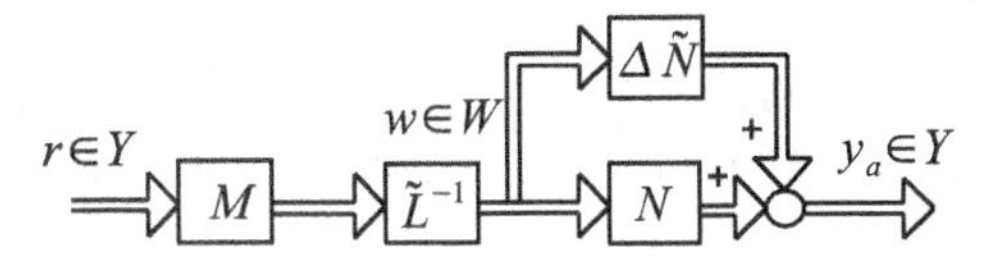

FIGURE 6.9 Tracking control system.

6.4 EXPERIMENTAL RESULTS

In this section, the multitank experimental system consisting of the water level and water flow processes shown in Figure 6.1 is employed to illustrate the effectiveness of the presented approach. In the system, Yokogawa's CENTUM CS 3000, a type of DCS device, is utilized. We consider that the water level changes from 30 to 72 cm because a supersonic wave sensor can normally work in this range and the water does not spill over from the tank. The reference input of the water flow is set in the neighborhood of 1 L/min and the Taylor expansion of the natural logarithm function is in the neighborhood of $y_2(t) = 1$.

Parameters of the water level process are given in Table 6.1 such that the tank 1 cutting area and the outflow entrance cutting area are calculated as $A = 6.94 \times 10^{-2}$ m^2 and $a = 1.13 \times 10^{-4}$ m^2, respectively. For the water flow process, the model parameters are identified using the experimental parameter identification approach. In this section, the least-squares identification rule is employed to obtain the model parameters. The input and output data are measured by manual operation. The initial values of k and b are assumed to be zero. The parameter identification result is derived, that is, $k = 0.225$ and $b = 0.034$.

In the CS3000, the control program is developed using a special environment based on a function block. Each function block can be monitored and the parameters of each function block can be adjusted online using the human interface station shown in Figure 6.10. The control and monitor programs are designed in an engineering station

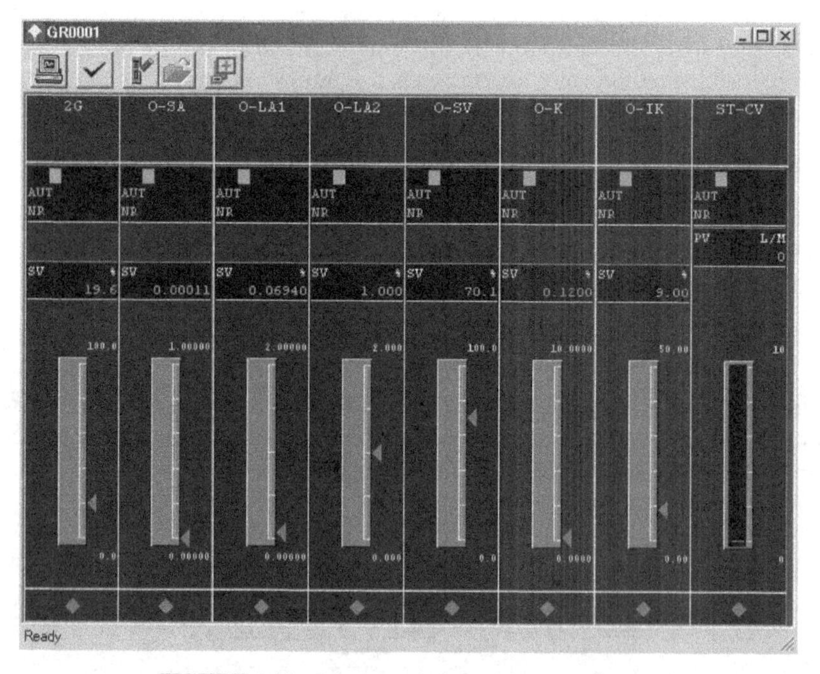

FIGURE 6.10 Monitor window of function blocks.

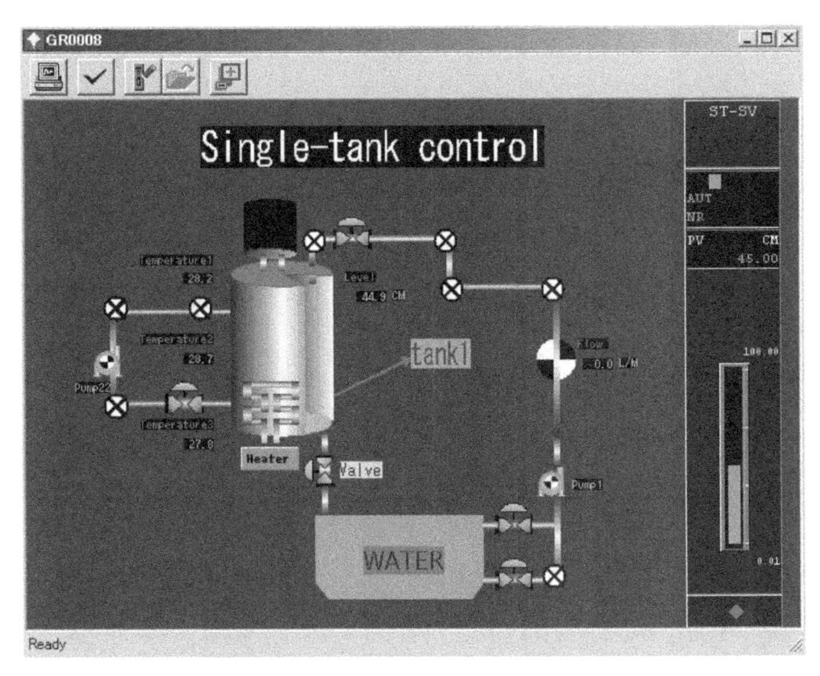

FIGURE 6.11 Monitor window of water level process.

and downloaded to the field control station and workstation by a V-net network which is a special communication network for the CS3000.

Figures 6.11 and 6.12 are the monitor windows of the water level and water flow processes, respectively. All of the variables related to the process can be displayed in the same monitor window. Also, intuitive graphics or animation can be designed to show changes in the process. Namely, it provides the functions of integrated data monitoring and processing. For example, the reference input, control input, and process output of each process are monitored in their own windows.

In this section, the reference inputs of the water level and the water flow are 45 cm and 1.0 L/min, respectively. The right coprime factorization controllers are designed according to the Bezout identity (6.13) and condition (6.9). The inertia differential element is used to realize the approximate differential function and the approximate realization of the natural logarithm function is considered using a Taylor expansion such that the real controllers $\tilde{S}_1$ and $\tilde{S}_2$ are shown in (6.16) and (6.21), respectively. The design parameters $B_1 = 0.13$ and $B_2 = 0.3$ are obtained to satisfy tracking performance. The control results of the water level and water flow processes are shown in Figures 6.13 and 6.14, where y_{ai}, u_{di}, and r_i are the real process output, the control input, and the reference input, respectively. In Figure 6.13, we can see that the water level can track to the desired level and the process output is able to provide good dynamic performance. This shows clearly the benefit of the control scheme. From Figure 6.14, accurate water flow control can be achieved. There is a good performance for the set-point response. That is, the process outputs can track to the

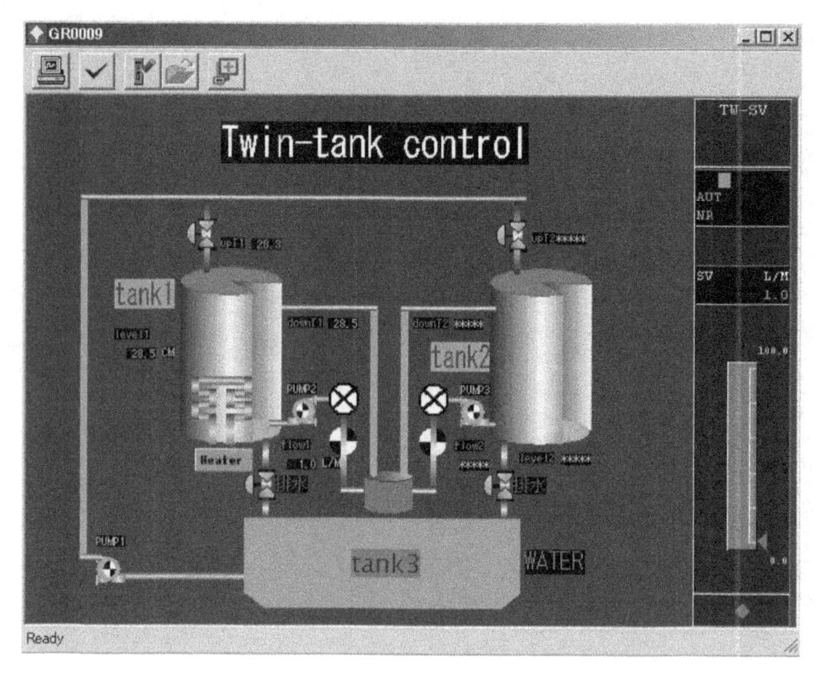

FIGURE 6.12 Monitor window of water flow process.

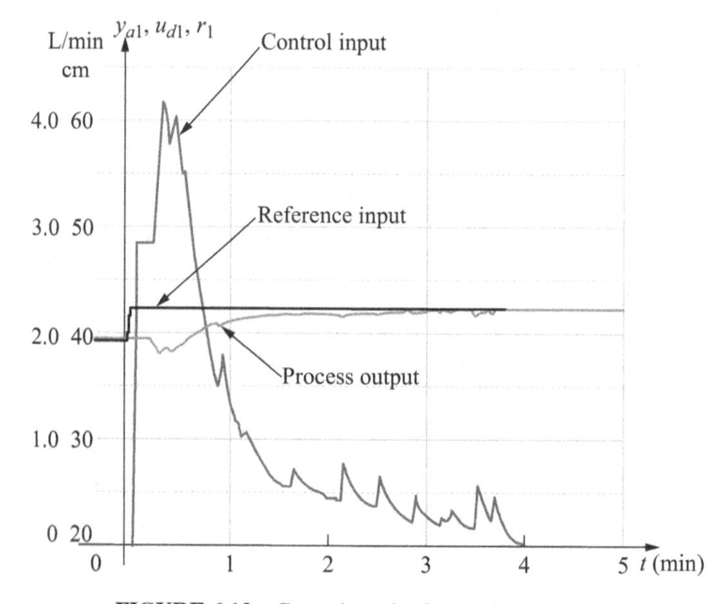

FIGURE 6.13 Control result of water level process.

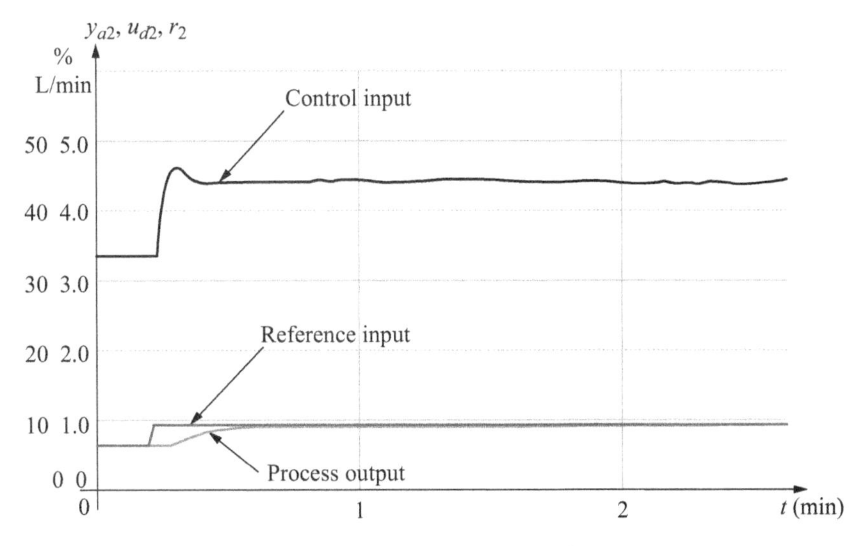

FIGURE 6.14 Control result of water flow process.

reference inputs. Also, it is clear that the controlled systems are robust stable under the process uncertainty and parasitic term caused by the approximated realization. That is, these results show that the robust right coprime factorization design and its realization of the multitank process control system are effective. Moreover, according to the experimental data, we also show that the robust stability condition (6.9) is satisfied, as shown in Figures 6.15 and 6.16, where $(S_i + \Delta S_i)(N_i + \Delta N_i)$ and $S_i N_i$ are obtained by using the real measured feedback signal and the calculated feedback

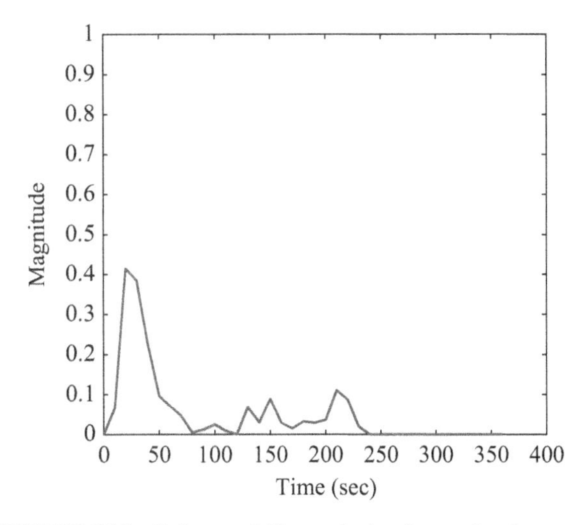

FIGURE 6.15 Robust stability analysis of water level process.

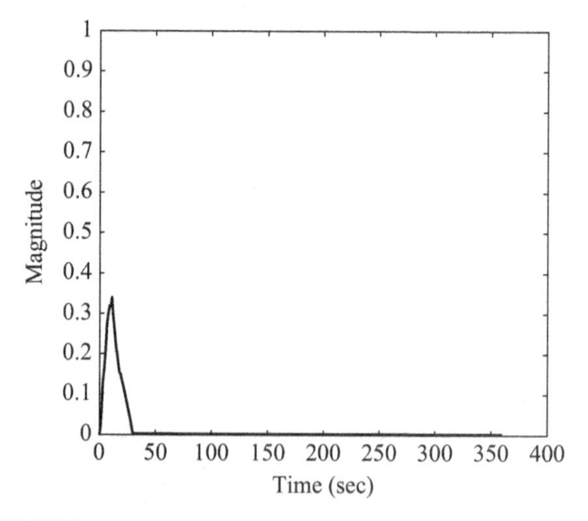

FIGURE 6.16 Robust stability analysis of water flow process.

signal based on the model. So, the results of the robust stability analysis are expressed by a time sequence. These results show that the effects of the parasitic term caused by the approximate realization can be eliminated gradually by the method.

6.5 SUMMARY

In this chapter, the multitank process experimental system using the DCS device is set up, including the water level and water flow processes. Theoretical modeling of the two processes is given according to the Bernoulli theorem. Then, the robust nonlinear design method using operator-based robust right coprime factorization is considered for the two processes. Also, from a large-scale industrial application viewpoint, the designed controller approximated realization is investigated using the CENTUM CS3000. For the parasitic terms caused by the approximated realization and the process uncertainties, a robust stable condition is derived to guarantee the robust stability of the nonlinear feedback control system. Finally, the experimental results show the effectiveness of the design scheme.

REFERENCES

1. R. J. P. de Figueiredo and G. Chen, *Nonlinear Feedback Control Systems: An Operator Theory Approach*, New York: Academic, 1993.
2. M. Deng, A. Inoue, and K. Ishikawa, "Operator-based nonlinear feedback control design using robust right coprime factorization," *IEEE Trans. Automatic Control*, vol. 51, no. 4, pp. 645–648, 2006.
3. G. Chen and Z. Han, "Robust right coprime factorization and robust stabilization of nonlinear feedback control systems," *IEEE Trans. Automatic Control*, vol. 43, no. 10, pp. 1505–1510, 1998.
4. M. Deng and A. Wang, "Robust nonlinear control design to an ionic polymer metal composite with hysteresis using operator based approach," *IET Control Theory Appl.*, vol. 6, no. 17, pp. 2667–2675, 2012.
5. N. Bu and M. Deng, "System design for nonlinear plants using operator-based robust right coprime factorization and isomorphism," *IEEE Trans. Automatic Control*, vol. 56, no. 4, pp. 952–957, 2011.
6. M. Deng, N. Bu, and A. Inoue, "Output tracking of nonlinear feedback systems with perturbation based on robust right coprime factorization," *Int. J. Innovative Comput. Inf. Control*, vol. 5, no. 10(B), pp. 3359–3366, 2009.
7. M. Deng and A. Inoue, "Networked nonlinear control for an aluminum plate thermal process with time-delays," *Int. J. Syst. Sci.*, vol. 39, no. 11, pp. 1075–1080, 2008.
8. M. Deng, S. Wen, and A. Inoue, "Operator-based robust nonlinear control for a Peltier actuated process," *Measur. Control*, vol. 44, no. 4, pp. 116–120, 2011.
9. R. Ortega, A. Loria, P. J. Nicklasson, and H. Sira-Ramirez, *Passivity-Based Control of Euler-Lagrange Systems*, London: Springer-Verlag, 1998.
10. D. Casagrande, A. Astolfi, and R. Ortega, "Asymptotic stabilization of passive systems without damping injection: A sampled integral technique," *Automatica*, vol. 47, no. 2, pp. 262–271, 2011.
11. R. Kristiansen, P. J. Nicklasson, and J. T. Gravdahl, "Spacecraft coordination control in 6DOF: Integrator backstepping vs passivity-based control," *Automatica*, vol. 44, no. 11, pp, 2896-2901, 2008.
12. I. Mizumoto, S. Ohdaira, and Z. Iwai, "Output feedback strict passivity of discrete-time nonlinear systems and adaptive control system design with a PFC," *Automatica*, vol. 46, no. 9, pp. 1503–1509, 2010.

Operator-Based Nonlinear Control Systems: Design and Applications, First Edition. Mingcong Deng.

13. R. Ortega, A. J. van der Schaft, F. Castanos, and A. Astolfi, "Control by interconnection and standard passivity-based control of Port-Hamiltonian systems," *IEEE Trans. Automatic Control*, vol. 53, no. 11, pp. 2527–2542, 2008.
14. R. Ortega, A. J. van der Schaft, I. Mareels, and B. Maschke, "Putting energy back in control," *IEEE Trans. Automatic Control*, vol. 21, no. 2, pp. 18–23, 2001.
15. C. A. Desoer and R. W. Liu, "Global parametrization of feedback systems," *Syst. Control Lett.*, vol. 1, no. 4, pp. 249–251, 1982.
16. J. Hammer, "Nonlinear systems: Stability and rationality," *Int. J. Control*, vol. 40, pp. 1–35, 1984.
17. A. B. D. Paice, J. B. Moore, and R. Horowitz, "Nonlinear feedback systems stability via coprime factorization analysis," *J. Math. Syst. Estimat. Control*, vol. 2, pp. 293–321, 1992.
18. E. D. Sontag, "Smooth stabilization implies coprime factorization," *IEEE Trans. Automatic Control*, vol. 34, pp. 435–443, 1989.
19. M. S. Verma, "Coprime fractional representations and stability of nonlinear feedback systems," *Int. J. Control*, vol. 48, no. 3, pp. 897–918, 1984.
20. M. Vidyasagar, *Control System Synthesis—A Factorization Approach*, Cambridge, MA: MIT Press, 1985.
21. M. Deng, A. Inoue, K. Ishikawa, and Y. Hirashima, "Tracking of perturbed nonlinear plants using robust right coprime factorization approach," in *Proceedings of the American Control Conference*, Boston, ISBN: 0-7803-8335-4, pp. 3666–3670, 2004.
22. G. Zames, "Functional analysis applied to nonlinear feedback systems," *IEEE Trans. Circuit Theory*, vol. CT-10, no. 3, pp. 392–404, 1963.
23. T. Hu and Z. Lin, *Control Systems with Actuator Saturation: Analysis and Design*, Boston, MA: Birkhauser, 2001.
24. M. M. Seron, G. C. Goodwin, and S. F. Graebe, "Control system design issues for unstable linear systems with saturated inputs," *IEE Proc. Control Theory Appl.*, vol. 142, pp. 335–344, 1995.
25. A. Teel, "Anti-windup for exponentitally unstable linear systems," *Int. J. Robust Nonlin. Control*, vol. 9, pp. 701–716, 1999.
26. S. Crawshaw and G. Vinnicombe, "Anti-windup for local stability of unstable linear plants," *Proc. ACC*, pp. 645–650, 2002.
27. P. J. Campo, M. Morari, and C. N. Nett, "Multivariable anti-windup and bumpless transfer: A general theory," *Proc. ACC*, pp. 1706–1711, 1989.
28. D. C. Youla, J. J. Bongiorno, and H. A. Jabr, "Modern Wiener-Hopf design of optimal contrfollers—Part 2: The multivariable case," *IEEE Trans. Automatic Control*, vol. 21, pp. 319–338, 1976.
29. M. Sebek and V. Kucera, "Polynomial approach to quadratic tracking in discrete linear systems," *IEEE Trans. Automatic Control*, vol. 27, pp. 1248–1250, 1982.
30. H. Demircioglu and P. J. Gawthrop, "Continuous-time generalized predictive control (CGPC)," *Automatica*, vol. 27, pp. 55–74, 1991.
31. M. Deng, A. Inoue, A. Yanou, and Y. Hirashima, "Continuous-time anti-windup generalized predictive control of non-minimum phase processes with input constraints," *Proc. IEEE CDC*, pp. 4457–4462, 2003.
32. Z. Han and G. Chen, "Dynamic right coprime factorization and observer design for nonlinear systems," *Latin Am. Appl. Res.*, vol. 32, pp. 327–336, 2002.

33. M. Cannon and B. Kouvaritakis, "Infinite horizon predictive control of constrained continuous-time linear systems," *Automatica*, vol. 36, pp. 943–955, 2000.
34. Z. Iwai, I. Mizumoto, and M. Deng, "A parallel feedforward compensator virtually realizing almost strictly positive real plant," *Proc. 33rd IEEE CDC*, vol. 3, pp. 2827–2832, 1994.
35. I. Mizumoto and Z. Iwai, "Simplified adaptive model output following control for plants with unmodelled dynamics," *Int. J. Control*, vol. 64, pp. 61–68, 1996.
36. H. Kaufman, I. Bar-Kana, and K. Sobel, *Direct Adaptive Control Algorithms? Theory and Applications*, New York: Springer, 1997.
37. Z. Iwai and I. Mizumoto, "Realization of simple adaptive control by using parallel feedforward compensator," *Int. J. Control*, vol. 59, pp. 1543–1565, 1994.
38. M. Deng, Z. Iwai, and I. Mizumoto, "Robust parallel compensator for output feedback stabilization of processes with uncertainties," *Proc. IFAC Low Cost Automation*, pp. TS1-7–TS1-12, 1998.
39. M. Deng, Z. Iwai, and I. Mizumoto, "Robust parallel compensator design for output feedback stabilization of plants with structured uncertainties," *Syst. Control Lett.*, vol. 36, pp. 193–198, 1999.
40. M. Deng, I. Mizumoto, Z. Iwai, and S. L. Shah, "Model output following control based on CGT approach for plants with time delays," *Int. J. Syst. Sci.*, vol. 30, no. 1, pp. 69–75, 1999.
41. R. Gessing, "Parallel compensator versus Smith predictor for control of the plants with delay," *Bull. Polish Acad. Sci. Tech. Sci.*, vol. 56, no. 4, pp. 339–345, 2008.
42. A. Osunleke, M. Deng, and A. Inoue, "A CAGPC controller design for systems with input windup and disturbances," *Int. J. Innovative Comput. Inform. Control*, vol. 5, no. 10(B), pp. 3517–3526, 2009.
43. A. Osunleke, M. Deng, and A. Yanou, "A design procedure for control of strictly proper non-minimum phase processes with input constraints and disturbance," *Int. J. Modelling, Identification Control*, vol. 13, no. 1/2, pp. 46–55, 2011.
44. A. Osunleke, M. Deng, A. Inoue, and A. Yanou, "Adaptive robust anti-windup generalized predictive control (RAGPC) of non-minimum phase systems with input constraints and disturbance," in *Proceedings of the IEEE AFRICON 2009*(CD-ROM), Nairobi, Kenya, 2009.
45. M. Deng, A. Inoue, and K. Edahiro, "Fault detection in a thermal process control system with input constraints using robust right coprime factorization approach," *Proc. IMechE, Part I: J. Syst. Control Eng.*, vol. 221, no. 6, pp. 819–831, 2007.
46. W. Zhang and S. Ge, "A global implicit function theorem without initial point and its applications to control of non-affine systems of high dimensions," *J. Math. Anal. Appl.*, vol. 313, pp. 251–261, 2006.
47. T. C. Yang, "Networked control system: A brief survey," *IEE Proc. Control Theory Appl.*, vol. 153, pp. 403–412, 2006.
48. S. H. Yang, "Internet-based control: The next generation of control systems," *Measur. Control*, vol. 38, no. 1, page 11, February, 2004.
49. C. A. Desoer and M. G. Kabuli, "Right factorizations of a class of time-varying nonlinear systems," *IEEE Trans. Automatic Control*, vol. 33, pp. 755–757, 1988.
50. G. Gnavi, "Factorization of reciprocal Hilbert port operators," *J. Math. Anal. Appl.*, vol. 114, no. 2, pp. 385–397, 1986.

51. A. D. B. Paice and J. B. Moore, "On the Youla-Kucera parameterization for nonlinear systems," *Syst. Control Lett.*, vol. 14, no. 2, pp. 121–129, 1990.
52. A. D. B. Paice and J. B. Moore, "Robust stabilisation of nonlinear plants via left coprime factorizations," *Syst. Control Lett.*, vol. 15, no. 2, pp. 125–135, 1990.
53. Z. Han and G. Chen, "Dynamic right coprime factorization for nonlinear systems," *Nonlin. Anal.*, vol. 30, no. 5, pp. 3113–3120, 1997.
54. M. Deng, A. Inoue, and S. Goto, "Operator based thermal control of an aluminum plate with a Peltier device," *Int. J. Innovative Comput. Inform. Control*, vol. 4, no. 12, pp. 3219–3229, 2008.
55. A. Isidori, *Nonlinear Control Systems*, 3rd ed., Berlin: Springer, 1995.
56. Z. Lin, "Almost disturbance decoupling with global asymptotic stability for nonlinear systems with disturbance-affected unstable zero dynamics," *Syst. Control Lett.*, vol. 33, pp. 163–169, 1998.
57. Y. Z. Tsypkin and B. T. Polyak, "High-gain robust control," *Eur. J. Control*, vol. 5, pp. 3–9, 1999.
58. S. Xu, J. Lam, and Y. Zou, Y., "New results on delay-dependent robust H_∞ control for systems with time-varying delays," *Automatica*, vol. 42, pp. 343–348, 2006.
59. H. L. Xing, C. C. Gao, G. Y. Tang, and D. Li, "Variable structure sliding mode control for a class of uncertain distributed parameter systems with time-varying delays," *Int. J. Control*, vol. 82, pp. 287–297, 2009.
60. X. Yan, C. Edwards, and S. Spurgeon, "Decentralised robust sliding mode control for a class of nonlinear interconnected systems by static output feedback," *Automatica*, vol. 40, pp. 613–620, 2004.
61. J. Ball and A. J. van der Schaft, "J-inner-outer factorization, J-spectral factorization, and robust control for nonlinear systems," *IEEE Trans. Automatic Control*, vol. 41, pp. 379–392, 1996.
62. P. M. Frank, "Fault diagnosis in dynamic systems using analytical and knowledge-based redundancy—A survey and some new results," *Automatica*, vol. 26, pp. 459–474, 1990.
63. J. Korbicz, J. M. Koscielny, Z. Kowalczuk, and W. Cholewa, *Fault Diagnosis, Models, Artificial Intelligence, Applications*, Springer-Verleg Berlin Heidelberg, 2004.
64. R. N. Clark, "Instrument fault detection," *IEEE Trans. Aerospace Electron. Syst.*, vol. AES-14, pp. 456–465, 1978.
65. H. Wang, Z. J. Huang, and S. Daley, "On the use of adaptive updating rules for actuator and sensor fault diagnosis," *Automatica*, vol. 33, pp. 217–225, 1997.
66. M. Deng, A. Inoue, and Y. Baba, "Operatar-based non-linear vibration control system design of a flexible arm with piezoelectric actuator," *Int. J. Adv. Mechatronic Syst.*, vol. 1, no. 1, pp. 71–76, 2008.
67. M. Nordin and P.-O. Gutman, "Controlling mechanical systems with backlash—A survey," *Automatica*, vol. 38, no. 10, pp. 1633–1649, 2002.
68. G. Tao and P. V. Kokotovic, "Continuous-time adaptive control of systems with unknown backlash," *IEEE Trans. Automatic Control*, vol. 40, no. 6, pp. 1083–1087, 1995.
69. R. R. Selmic and F. L. Lewis, "Backlash compensation in nonlinear systems using dynamic inversion by neural networks," *Asian J. Control*, vol. 2, no. 2, pp. 76–87, 2000.

70. J. Zhou, C. Zhang, and C. Wen, "Robust adaptive output control of uncertain nonlinear plants with unknown backlash nonlinearity," *IEEE Trans. Automatic Control*, vol. 52, no. 3, pp. 503–509, 2007.

71. C. Y. Su, Y. Stepanenko, J. Svoboda, and T. P. Leung, "Robust adaptive control of a class of nonlinear systems with unknown backlash-like hysteresis," *IEEE Trans. Automatic Control*, vol. 45, no. 12, pp. 2427–2432, 2000.

72. G. Tao and P. V. Kokotovic, "Adaptive control of systems with backlash," *Automatica*, vol. 29, no. 2, pp. 323–335, 1993.

73. A. Barreiro and A. Baños, "Input-output stability of systems with backlash," *Automatica*, vol. 42, no. 6, pp. 1017–1024, 2006.

74. K. Mine and T. Adachi, "An analysis of self-oscillation in a thermal process with unsymmetrical backlash and phase leading phenomena," *Control Eng.*, vol. 10, no. 7, pp. 344–351, 1966 (in Japanese).

75. A. Palla, C. J. Bockisch, O. Bergamin, and D. Straumann, "Dissociated hysteresis of static ocular counterroll in humans," *J. Neurophysiol.*, vol. 95, no. 4, pp. 2222–2232, 2006.

76. T. Tao, Y. Liang, and M. Taya, "Bio-inspired actuating system for swimming using shape memory alloy composites," *Int. J. Automat. Comput.*, vol. 3, no. 4, pp. 366–373, 2006.

77. Y. Zhang, J. He, J. Yang, S. Zhang, and K. Low, "A computational fluid dynamics (CFD) analysis of an undulatory mechanical fin driven by shape memory alloy," *Int. J. Automat. Comput.*, vol. 3, no. 4, pp. 374–381, 2006.

78. M. Brokate and J. Sprekels, *Hysteresis and Phase Transitions*, New York: Springer-Verlag, 1996.

79. C. Su, Q. Wang, X. Chen, and S. Rakheja, "Adaptive variable structure control of a class of nonlinear systems with unknown Prandtl-Ishlinskii hysteresisa," *IEEE Trans. Automatic Control*, vol. 50, no. 12, pp. 2069–2074, 2005.

80. X. Chen, "Robust control for the systems preceded by hysteresis," in *Proceedings of 25th IASTED International Conference of Modelling, Identification, and Control*, Lanzarote, Spain, pp. 173–178, 2006.

81. M. Deng, H. Yu, and Z. Iwai, "Simple robust adaptive control for structured uncertainty plants with unknown dead-zone," in *Proceedings of 40th IEEE Conference on Decision and Control*, Orlando, FL, pp. 1621–1626, 2001.

82. C. Jiang, M. Deng, and A. Inoue, "A novel modeling of nonlinear plants with hysteresis described by non-symmetric play operator," in *Proceedings of 7th World Congress on Intelligent Control and Automation*, Chongqing, China, pp. 2221–2224, 2008.

83. M. Janaideh, S. Rakheja, and C. Su, "Modeling rate-dependent symmetric and asymmetric hysteresis loops of smart actuators," *Int. J. Adv. Mechatronic Syst.*, vol. 1, no. 1, pp. 32–43, 2008.

84. C. Han and S. Wen, "Control system design of mine hoists by using robust right coprime factorisation," *Int. J. Adv. Mechatronic Syst.*, vol. 1, no. 3, pp. 214–222, 2009.

85. M. Deng, S. Bi, and S. Wen, "Operator based actuator fault detection system design of a thermal process," *Int. J. Computer Appl. Technol.*, vol. 45, no. 2/3, pp. 148–155, 2012.

86. N. Viswanadham and R. Srichander, "Fault detection using unknown-input observers," *Control Theory Adv. Technol.*, vol. 3, no. 2, pp. 91–101, 1987.

87. G. Duan and A. Wu, "Robust fault detection in linear systems based on full-order state observers," paper presented at the 23rd Chinese Control Conference, Wuxi, pp. 993–997, 2004.

88. F. Caccavalc and L. Villani, "An adaptive observer for fault diagnosis in nonlinear discrete-time systems," paper presented at the 2004 American Control Conference, Boston, pp. 2463–2468, 2004.

89. R. J. Patton, P. M. Frank, and R. N. Clark, *Fault Diagnosis in Dynamic Systems, Theory and Applications*, Englewood Cliffs, NJ: Prentice-Hall, 1989.

90. A. Inoue, M. Deng, and S. Yoshinaga, "Fault detection by using an adaptive observer," paper presented at the 2005 International Conference on Control, Automation and Systems, Korea, pp. 710–713, 2005.

91. C. Jiang, M. Deng, and A. Inoue, "A novel modeling of nonlinear plants with hysteresis described by non-symmetric play operator," in *Proceedings of 7th World Congress on Intelligent Control and Automation*, Chongqing, China, pp. 2221–2224, 2008.

92. M. Deng, C. Jiang, A. Inoue, and C. Su, "Operator based robust control for nonlinear systems with Prandtl-Ishlinskii hysteresis," *Int. J. Syst. Sci.*, vol. 42, no. 4, pp. 643–652, 2011.

93. P. Krejci and K. Kuhnen, "Inverse control of systems with hysteresis and creep," *Proc. Inst. Elect. Eng. Control Theory Appl.*, vol. 148, pp. 185–192, 2001.

94. M. Hashimoto, M. Takeda, H. Sagawa, I. Chiba, and K. Sato, "Shape memory alloy and robotic actuators," *J. Robotic Syst.*, vol. 2, pp. 3–25, 1985.

95. J. Stevens and G. Buckner, "Actuation and control strategies for miniature robotic surgical systems," *J. Dynamic Syst. Measurement Control*, vol. 127, pp. 537–549, 2005.

96. D. Grant and V. Hayward, "Variable structure control of shape memory alloy actuators," *IEEE Control Syst. Mag.*, vol. 17, no. 3, pp. 80–88, 1997.

97. L. C. Brinson, "One-dimensional constitutive behavior of shape memory alloys: Thermomechanical derivation with non-constant material functions and redefined martensite internal variable," *ASME J. Intelligent Material Syst. Struct.*, vol. 4, no. 2, pp. 229–242, 1993.

98. D. Madill and D. Wang, "Modeling and L_2-stability of a shape memory alloy position control system," *IEEE Trans. Control Syst. Technol.*, vol. 6, no. 4, pp. 473–481, 1998.

99. S. Choi, Y. Han, J. Kim, and C. Cheong, "Force tracking control of a flexible gripper featuring shape memory alloy actuators," *Mechatronics*, vol. 11, pp. 677–690, 2001.

100. S. Majima, K. Kodama, and T. Hasegawa, "Modeling of shape memory alloy actuator and tracking control system with the model," *IEEE Trans. Control Syst. Technol.*, vol. 9, no. 1, pp. 54–59, 2001.

101. M. Moallem and V. A. Tabrizi, "Tracking control of an antagonistic shape memory alloy actuator pair," *IEEE Trans. Control Syst. Technol.*, vol. 17, no. 1, pp. 184–190, 2009.

102. A. Baz, K. Imam, and J. McCoy, "Active vibration control of flexible beams using shape memory actuators," *J. Sound Vibration*, vol. 140, no. 3, pp. 437–456, 1990.

103. K. Yuse and Y. Kikushima, "Development and experimental consideration of SMA/CFRP actuator for vibration control," *J. Sensors Actuators*, vol. 122, pp. 99–107, 2005.

104. S. Saito, M. Deng, M. Minami, C. Jiang, and A. Yanou, "Operator-based vibration control system design of a flexible arm using an SMA actuator with hysteresis," *SICE J. Control Measurement Syst. Integration*, vol. 5, no. 2, pp. 115–126, 2012.

105. X. Chen, T. Hisayama, and C. Su, "Pseudo-inverse-based adaptive control for uncertain discrete time systems preceded by hysteresis," *Automatica*, vol. 45, pp. 469–476, 2009.

106. Q. Wang and C. Su, "Robust adaptive control of a class of nonlinear systems including actuator hysteresis with Prandtl-Ishlinskii presentations," *Automatica*, vol. 42, no. 5, pp. 859–867, 2006.

107. S. Saito, M. Deng, A. Inoue, and C. Jiang, "Vibration control of a flexible arm experimental system with hysteresis of piezoelectric actuator," *Int. J. Innovative Comput. Information Control*, vol. 6, no. 7, pp. 2965–2975, 2010.

108. M. Shahinpoor and K. Kim, "Ionic polymer-metal composites: I. Fundamentals," *Smart Mater. Struct.*, vol. 10, no. 4, pp. 819–833, 2001.

109. M. Shahinpoor, "Ionic polymer-conductor composites as biomimetic sensors, robotic actuators and artificial muscles—A review," *Electrochim. Acta*, vol. 48, nos. 14–16, pp. 2343–2353, 2003.

110. M. Shahinpoor and K. Kim, "Ionic polymer-metal composites: III. Modeling and simulation as biomimetic sensors, actuators, transducers, and artificial muscles," *Smart Mater. Struct.*, vol. 13, no. 6, pp. 1362–1388, 2004.

111. Z. Chen, D. R. Hedgepeth, and X. Tan, "A nonlinear control-oriented model for ionic polymer-metal composite actuators," *Smart Mater. Struct.*, vol. 18, no. 5, pp. 1–9, 2009.

112. S. Oh and H. Kim, "A study on the control of an IPMC actuator using an adaptive fuzzy algorithm," *KSME Int. J.*, vol. 18, no. 1, pp. 1–11, 2004.

113. N. Bhat and W. Kim, "Precision position control of ionic polymer metal composite," *Proc. Conf. 2004 American Control*, pp. 740–745, 2004.

114. N. Bhat and W. Kim, "Precision force and position control of ionic polymer metal composite," *J. Syst. Control Eng.*, vol. 218, no. 6, pp. 421–432, 2004.

115. C. Jiang, M. Deng, and A. Inoue, "Robust stability of nonlinear plants with a non-symmetric Prandtl-Ishlinskii hysteresis model," *Int. J. Automat. Comput.*, vol. 7, no. 2, pp. 213–218, 2010.

116. J. Seo, R. Venugopal, and J. Kenne, "Feedback linearization based control of a rotational hydraulic drive," *Control Eng. Pract.*, vol. 15, pp. 1495–1507, 2007.

117. B. Labibi, H. Marquez, and T. Chen, "Decentralized robust output feedback control for control affine nonlinear interconnected systems," *J. Process Control*, vol. 19, pp. 865–878, 2009.

118. H. Li, M. Y. Chow, and Z. Sun, "Optimal stabilizing gain selection for networked control systems with time delays and packet losses," *IEEE Trans. Control Syst. Technol.*, vol. 17, pp. 1154–1162, 2009.

119. M. P. C. Parte, M. Cirre, E. F. Camacho, and M. Berenguel, "Application of predictive sliding mode controllers to a solar plant," *IEEE Trans. Control Syst. Technol.*, vol. 16, pp. 819–825, 2008.

120. M. M. Polycarpou and P. Ioannou, "A robust adaptive nonlinear control design," *Automatica*, vol. 32, pp. 423–427, 1996.

121. S. Wen, M. Deng, and A. Inoue, "Application of robust right coprime factorization approach to a distributed process control system," in *Proc. IEEE International Conference on Automation and Logistics*, Shengyang, pp. 504–508, 2009.

122. M. Deng, S. Bi, and A. Inoue, "Robust nonlinear control and tracking design for multi-input multi-output nonlinear perturbed plants," *IET Control Theory Appl.*, vol. 3, pp. 1237–1248, 2009.
123. Y. Zhou and Y. Wu, "Output feedback control for MIMO nonlinear systems using factorization," *Nonlin. Anal.*, vol. 68, pp. 1362–1374, 2008.
124. Fisher, *Control Valve Handbook*, 4th ed., Fisher Controls International, USA, 2005.
125. M. Deng, C. Jiang, and A. Inoue, "Operator-based robust control for nonlinear plants with uncertain non-symmetric backlash," *Asian J. Control*, vol. 13, pp. 317–327, 2011.
126. S. Bi and M. Deng, "Operator based robust control design for nonlinear plants with perturbation," *Int. J. Control*, vol. 84, pp. 815–821, 2011.

INDEX

Operator-Based Nonlinear Control Systems: Design and Applications, First Edition. Mingcong Deng.

IEEE PRESS SERIES ON SYSTEMS SCIENCE AND ENGINEERING

The focus of this series is to introduce the advances in theory and applications of systems science and engineering to industrial practitioners, researchers, and students. This series seeks to foster system-of-systems multidisciplinary theory and tools to satisfy the needs of the industrial and academic areas to model, analyze, design, optimize and operate increasingly complex man-made systems ranging from control systems, computer systems, discrete event systems, information systems, networked systems, production systems, robotic systems, service systems, and transportation systems to Internet, sensor networks, smart grid, social network, sustainable infrastructure, and systems biology.

Reinforcement and Systemic Machine Learning for Decision Making
Parag Kulkarni

Remote Sensing and Actuation Using Unmanned Vehicles
Haiyang Chao, YangQuan Chen

Hybrid Control and Motion Planning of Dynamical Legged Locomotion
Nasser Sadati, Guy A. Dumont, Kaveh Akbari Hamed, and William A. Gruver

Design of Business and Scientific Workflows: A Web Service-Oriented Approach
MengChu Zhou and Wei Tan

Operator-based Nonlinear Control Systems: Design and Applications
Mingcong Deng